Functional Polymers for Nanomedicine

RSC Polymer Chemistry Series

Series Editors:
Professor Ben Zhong Tang (Editor-in-Chief), The Hong Kong University of Science and Technology, Hong Kong, China
Professor Alaa S. Abd-El-Aziz, University of Prince Edward Island, Canada
Professor Stephen L. Craig, Duke University, USA
Professor Jianhua Dong, National Natural Science Foundation of China, China
Professor Toshio Masuda, Fukui University of Technology, Japan
Professor Christoph Weder, University of Fribourg, Switzerland

Titles in the Series:
1: Renewable Resources for Functional Polymers and Biomaterials
2: Molecular Design and Applications of Photofunctional Polymers and Materials
3: Functional Polymers for Nanomedicine

How to obtain future titles on publication:
A standing order plan is available for this series. A standing order will bring delivery of each new volume immediately on publication.

For further information please contact:
Book Sales Department, Royal Society of Chemistry, Thomas Graham House, Science Park, Milton Road, Cambridge, CB4 0WF, UK
Telephone: +44 (0)1223 420066, Fax: +44 (0)1223 420247,
Email: booksales@rsc.org
Visit our website at www.rsc.org/books

Functional Polymers for Nanomedicine

Edited by

Youqing Shen
Zhejiang University, Hangzhou, China
Email: shenyq@zju.edu.cn

RSC Publishing

RSC Polymer Chemistry Series No. 3

ISBN: 978-1-84973-620-6
ISSN: 2044-0790

A catalogue record for this book is available from the British Library

Published by The Royal Society of Chemistry,
Thomas Graham House, Science Park, Milton Road,
Cambridge CB4 0WF, UK

Registered Charity Number 207890

For further information see our web site at www.rsc.org

Printed in the United Kingdom by Henry Ling Limited, Dorchester, DT1 1HD, UK

Preface

It has been already more than a decade since nanotechnology was introduced to the field of controlled drug delivery. The nanotechnology push by various funding agencies throughout the world has resulted in significant advances in the drug delivery field, as evidenced by the explosive increase in the number of publications on the subject during the last several years. The term "nanotechnology" has permeated into every aspect of drug delivery research. The thrust of nanotechnology, at least in the drug delivery field, has been the use of nanosized drug delivery systems, which are collectively called "nanoparticles" or "nanovehicles". The goal of utilizing nanovehicles is to develop drug delivery systems that can enhance the efficacy of various drugs by making them more bioavailable through targeted delivery, and thus reducing side effects. The question we have to ask now is whether the nanovehicles have fulfilled their intended function of producing more clinically useful drug delivery systems. The answer is unfortunately "not quite".

Controlled drug delivery began in the early 1950s, and now we are literally entering the third generation (3G) of the discipline. The first generation (1G) during 1950-1980 has been most fruitful in producing clinically useful sustained release formulations, which are mainly in the areas of oral and transdermal formulations. The second generation (2G) drug delivery systems have not been as successful, however. This is mainly due to the difficulty in overcoming the biological barriers when drug delivery systems, such as pulmonary and injectable systems, are introduced into the body. One of the expectations of the nanovehicles is to overcome the biological barriers. The nanovehicle revolution, which began near the end of the 2G, is currently continuing into the 3G systems. It is time to review the progress made by the nanotechnology-based drug delivery systems in the last decade to prepare for the next decade or two.

Developing better drug delivery systems for the future requires understanding the reasons for successful formulations and the causes for the lack of expected outcomes. The book "Functional Polymers for Nanomedicine", edited by Professor Youqing Shen, is designed for this purpose. In retrospect, it appears that the field of controlled drug delivery has given blind trust to the nanotechnology-based formulations without any critical assessments. The book tries to rectify this by examining the current bottlenecks in the development of

RSC Polymer Chemistry Series No. 3
Functional Polymers for Nanomedicine
Edited by Youqing Shen
© The Royal Society of Chemistry 2013
Published by the Royal Society of Chemistry, www.rsc.org

various nanovehicles. One of the most important goals of nanomedicine is to achieve targeted drug delivery, which is essential in treatment of tumors as well as gene therapy. The goal is to deliver a drug to the target site in large enough quantity lethal to cancer cells with minimized side effects. To do this, we need to be able to control drug release kinetics, and this in turn requires good formulation stability in the blood followed by desired tissue distribution. The nanovehicles also need to deliver a wide range of drugs including small, hydrophobic drugs (both poorly soluble and water soluble drugs) to large hydrophilic drugs (peptides, proteins, siRNA, and DNA). The book describes unique advantages and limitations of a variety of drug delivery vehicles, such as liposomes, polymer micelles, dendrimers, polymersomes, polymer-drug conjugates, polymer-drug complexes, and core-shell-corona micelles. The book also deals with the current misunderstanding of targeted drug delivery and provides paradigm-changing new ideas.

One of the reasons we study history is to find out what happened and not to repeat the same mistakes, thereby making the future better. This book is about learning from the failures, as much as the progresses made in the last ten years and finding ways to design better drug delivery systems in the next ten years. Naturally, this book is expected to be a valuable guide for the drug delivery scientists, especially those who are relatively new in the area.

Kinam Park
Purdue University

Contents

Chapter 1 Targeted Drug Delivery in Oncology: Current Paradigm and Challenges   **1**
Darren Lars Stirland and You Han Bae

1.1	Targeted Drug Delivery	1
	1.1.1 Origins of Targeted Drug Delivery	2
	1.1.2 Progress in Targeted Drug Delivery	2
1.2	Current Paradigm	3
1.3	Challenges to Current Paradigm	5
	1.3.1 Challenges Present in the Carrier	6
	1.3.2 Challenges Present in the Target	9
1.4	Revolution	14
	References	15

Chapter 2 Targeted Nanomedicines: Challenges and Opportunities   **20**
Xinpeng Ma, Gang Huang, Yiguang Wang and Jinming Gao

2.1	Introduction	20
2.2	Passive Targeting by Stealth Nanomedicines	21
	2.2.1 Nanomedicine Clearance by the Reticuloendothelial System	22
	2.2.2 Tumor Penetration by Nanomedicines	22
2.3	Active Targeting by Surface-Functionalized Nanomedicines	23
	2.3.1 Cancer Specificity of Active Targeting Nanomedicines	24
	2.3.2 Increased Clearance of Active Targeting Nanomedicines	26

RSC Polymer Chemistry Series No. 3
Functional Polymers for Nanomedicine
Edited by Youqing Shen
© The Royal Society of Chemistry 2013
Published by the Royal Society of Chemistry, www.rsc.org

        2.3.3   Tumor Accumulation: Passive *vs.* Active
                Targeting Nanomedicines   26
    2.4   Conclusion and Future Perspectives   27
    Acknowledgements   28
    References   28

Chapter 3   Rational Design of Translational Nanocarriers   32
            Qihang Sun, Maciej Radosz and Youqing Shen

    3.1   The Three Key Elements for Translational Nanomedicine   32
    3.2   The *2R2S* Capability of Nanocarriers   35
        3.2.1   *2R*: Drug Retention in Circulation *versus*
                Intracellular Release   35
        3.2.2   *2S*: Stealthy in Circulation and Tumor
                Penetration *versus* Sticky to Tumor Cells   44
    3.3   The Material Excipientability and Production Process
          Scale-Up Ability   51
    3.4   Challenges of Rational Design for Translational
          Nanomedicine   52
    3.5   Conclusion   53
    References   53

Chapter 4   Functional Polymers for Gene Delivery   63
            Xuan Zeng, Ren-Xi Zhuo and Xian-Zheng Zhang

    4.1   Introduction   63
    4.2   Polyethylenimine-Based Gene Vectors   65
        4.2.1   Low-Toxicity Polyethylenimine   66
        4.2.2   Cell-Targeted Polyethylenimine   73
        4.2.3   Other Polyethylenimine Derivatives   76
    4.3   Chitosan-Based Gene Vectors   77
        4.3.1   PEI-Modified Chitosans   78
        4.3.2   Cell-Targeted Chitosans   79
        4.3.3   Other Chitosan Derivatives   79
    4.4   Dendrimer-Based Gene Vectors   81
        4.4.1   Polyamidoamine Dendrimers   82
        4.4.2   Polypropylenimine Dendrimers   85
        4.4.3   Poly(L-lysine) Dendrimers   86
    4.5   Polypeptide Gene Vectors   87
        4.5.1   Normal Peptide-Based Vectors   87
        4.5.2   Cell-Penetrating Peptides   89
        4.5.3   Nuclear Localization Signal   91
        4.5.4   Asp-Based Peptides   92

    4.6   Other Gene Vectors   92
           4.6.1   Lipid-Based Vectors   92
           4.6.2   Polyallylamine   94
           4.6.3   Linear Poly(amidoamine)s   95
           4.6.4   Multi-layer Complexes   96
           4.6.5   Polycarbonates   98
           4.6.6   Nanoparticles   99
           4.6.7   Other Types   101
    4.7   Future Trends   106
           4.7.1   Stem Cell Transfection   106
           4.7.2   Combinatorial Vectors   108
           4.7.3   Virus Mimic Vectors   110
           4.7.4   Therapeutic Genes   111
    4.8   Conclusion   112
    Acknowledgements   112
    References   112

Chapter 5   Functional Hyperbranched Polymers for Drug and Gene Delivery   121
    Yue Jin and Xinyuan Zhu

    5.1   Introduction   121
    5.2   Preparation of Functional HBPs   122
           5.2.1   Preparation of HBPs   122
           5.2.2   Functionalization of HBPs   123
    5.3   Functionality of Delivery   126
           5.3.1   Responsiveness   126
           5.3.2   Targeting   130
           5.3.3   Imaging   135
           5.3.4   Biodegradability and Biocompatibility   136
           5.3.5   Multifunctionality   136
    5.4   Applications in Drug and Gene Delivery   137
           5.4.1   Application as Drug Carriers   137
           5.4.2   Application as Gene Vectors   139
    5.5   Summary   139
    Acknowledgements   140
    References   140

Chapter 6   Functional Polymersomes for Controlled Drug Delivery   144
    Fenghua Meng, Ru Cheng, Chao Deng and Zhiyuan Zhong

    6.1   Introduction   144
    6.2   Stimuli-Responsive Polymersomes   147
    6.3   Chimaeric Polymersomes   150
    6.4   Biomimetic Polymersomes   151

6.5   Tumor-Targeting Polymersomes   153
6.6   Conclusion and Perspectives   154
Acknowledgements   154
References   154

Chapter 7   Polymeric Micelle-Based Nanomedicine for siRNA Delivery   158
Xi-Qiu Liu, Xian-Zhu Yang and Jun Wang

7.1   Introduction   158
7.2   Barriers to the Efficacy of siRNA Therapeutics   160
7.3   Polymeric Micelles for siRNA Delivery   163
       7.3.1   Polymeric Micelles Based on Amphiphilic
               Polymers for siRNA Delivery   164
       7.3.2   Smart Responsive Micelles for siRNA Delivery 169
7.4   Co-delivery of siRNA and Drugs Based on Polymeric
       Micelles   178
7.5   Future Perspectives   181
7.6   Conclusion   183
References   184

**Chapter 8   Polysaccharide/Polynucleotide Complexes for Cell-Specific
DNA Delivery   190**
Shinichi Mochizuki and Kazuo Sakurai

8.1   Introduction   190
8.2   Characterization of the SPG/DNA Complex   192
       8.2.1   Preparation of the SPG/DNA Complex   192
       8.2.2   Solution Properties and Characterization   192
       8.2.3   Thermal Stability of the Complexes   194
8.3   Application of the Complex to ODN Delivery   196
       8.3.1   Uptake of the Complex by Macrophages   196
       8.3.2   IL-12 Secretion Due to Administration of Cpg-
               ODN/SPG Complexes   196
       8.3.3   LPS-Induced TNF-α Suppression by the AS-
               ODN/SPG Complex *in vitro* and *in vivo*   199
       8.3.4   A New Therapy for Inflammatory Bowel
               Disease Using Antisense Macrophage-
               Migration Inhibitory Factor   199
8.4   Conclusion   203
References   204

Chapter 9   Design of Complex Micelles for Drug Delivery   207
Rujiang Ma and Linqi Shi

9.1   Introduction   207
9.2   Core–Shell–Corona Micelles for Drug Delivery   208
9.3   Complex Micelles with Surface Channels for Drug
      Delivery   216
9.4   Polyion Complex Micelles for Drug Delivery   221
References   225

Chapter 10   Zwitterionic Polymers for Targeted Drug Delivery   227
Weifeng Lin, Zheng Wang and ShengFu Chen

10.1   Introduction   227
10.2   Principles Toward Protein-Resistant Zwitterionic
       Polymers   228
10.3   Phosphorylcholine-Based Polymers for Drug Delivery   233
10.4   CBMA-Based Polymers for Drug Delivery   239
10.5   Conclusion and Perspectives   241
References   241

Chapter 11   Polymer-Based Prodrugs for Cancer Chemotherapy   245
Qihang Sun, Jinqiang Wang, Maciej Radosz and Youqing Shen

11.1   Introduction   245
11.2   Design of Polymer-Based Prodrugs   246
       11.2.1   Linkers   246
       11.2.2   Modifiers   248
       11.2.3   Drawbacks of Current Polymer-Based
                Prodrugs   249
11.3   New Strategies for Polymer Prodrugs   250
       11.3.1   Self-Assembling Prodrugs   250
       11.3.2   Prodrug Micelles   253
       11.3.3   Drug Polymers   254
11.4   Future Challenges   255
References   257

**Chapter 12   Nonviral Vector Recombinant Mesenchymal Stem Cells: A
              Promising Targeted-Delivery Vehicle in Cancer Gene Therapy   261**
Yu-Lan Hu, Ying-Hua Fu, Yasuhiko Tabata and Jian-Qing Gao

12.1   Introduction   261
12.2   Gene Recombination of MSCs   264
       12.2.1   Viral Vectors   265

        12.2.2 Nonviral Vectors   265
        12.2.3 Three-Dimensional and Reverse Transfection
                Systems   266
    12.3 MSCs as a Promising Targeted-Delivery Vehicle in
          Cancer Gene Therapy   267
        12.3.1 Rationale for Using MSCs as a Vehicle for
                Gene Delivery   267
        12.3.2 Targeting of MSCs to Tumor Cells   268
        12.3.3 MSCs as Tumor Target Vehicles for Gene
                Delivery   271
    12.4 Future Perspectives   273
    Acknowledgements   274
    References   275

**Chapter 13   Near-Critical Micellization for Nanomedicine: Enhanced
              Drug Loading, Reduced Burst Release   281**
              Jade Green, Maciej Radosz and Youqing Shen

    13.1 Introduction   281
    13.2 Early Feasibility Studies on Model Systems   282
    13.3 Extension to PEG-*b*-PCL   287
    13.4 Optimizing the NCM Solvent   290
    13.5 Loading PEG-*b*-PCL with a Cancer Drug   293
    13.6 NCM: A Remedy for Burst Release?   295
    13.7 Conclusion and Future Research Questions   299
    Acknowledgements   301
    References   301

Subject Index   302

CHAPTER 1

Targeted Drug Delivery in Oncology: Current Paradigm and Challenges

DARREN LARS STIRLAND[a] AND YOU HAN BAE*[b]

[a] Department of Bioengineering, College of Engineering, University of Utah, Salt Lake City, UT 84108, USA; [b] Department of Pharmaceutics and Pharmaceutical Chemistry, College of Pharmacy, University of Utah, Salt Lake City, UT 84108, USA
*E-mail: you.bae@utah.edu

1.1 Targeted Drug Delivery

Targeted drug delivery seeks to improve the therapeutic index, that is, lower the toxicity but increase the efficacy of a drug. Some designs try to minimize side effects to allow a higher dose and increased therapeutic effects. Other designs focus on increasing efficacy to require less drug that could cause side effects. The methods of targeted drug delivery often control both when and where the drug is effective. If the drug could be presented only to the disease in the body, then there would be no side effects and efficacy would be improved with high concentrations in the target area. In this chapter, we focus on the topic of targeted drug delivery in cancer therapy. Research has been ongoing for years to accumulate a significant amount of knowledge into the field and fill the literature with the term "targeted drug delivery." Some have very aggressive claims of succeeding at targeted drug delivery and others, perhaps more accurately, state a goal of improving targeted drug delivery. While clear

RSC Polymer Chemistry Series No. 3
Functional Polymers for Nanomedicine
Edited by Youqing Shen
© The Royal Society of Chemistry 2013
Published by the Royal Society of Chemistry, www.rsc.org

delineations of what targeted drug delivery is and is not would help, there are challenges with the current state of targeted drug delivery that extend beyond defining the term. These challenges are present in both the carrier and the target. Carrier technology has improved greatly, yet still has trouble delivering drug to the target. The target, cancer in the clinical setting, is still resisting treatment. The challenges associated with this problem need to be addressed in order to move forward. The systems may be targeted by design, but they are not hitting the target fully and exclusively.

1.1.1   Origins of Targeted Drug Delivery

The current paradigm of targeted drug delivery is linked to its origins, which are tied to chemotherapy and immunology. Paul Ehrlich was a pioneer in chemotherapy and is known for the metaphor of a magic bullet. The vision of the magic bullet was inspired from his ability to selectively stain bacteria cultures.[1] He reasoned a toxic molecule could be tied to the stains to selectively kill only that target. Targeted drug delivery has been guided by Ehrlich's vision, particularly in the field of cancer therapy.[2] In the context of cancer and chemotherapy, a magic bullet carrying an anticancer drug is administered to the patient to provide exclusive delivery to the cancer. In contemplating the metaphor of a magic bullet, perhaps the body's immune system fits best. It has both mobility and specificity. Immune cells can follow chemical gradients and have specificity with antibodies and cell receptors. Ehrlich himself stated that cancer would be more prevalent if not for the immune system.[3]

1.1.2   Progress in Targeted Drug Delivery

Advances in immunology and the advent of monoclonal antibodies have become an important part of pursuing the vision of a magic bullet. With immunostaining, the targeting application seems flawless and provides motivation for application in cancer therapy. In targeted drug delivery there is a targeting aspect and a therapeutic aspect. Sometimes these monoclonal antibodies can provide a therapeutic effect on their own. However, monoclonal antibodies and other targeting designs are usually incorporated in a variety of therapeutic carriers, such as microemulsions, inorganic nanoparticles, viruses, and polymers. Liposomes are composed of lipids which assemble into vesicles with a bilayer capable of carrying drug molecules. Gold and iron oxide nanoparticles are among the more popular inorganic molecules used. Viral carriers, made by modifying existing viruses or by using certain aspects from them, have also been used in targeted drug delivery of chemotherapeutics.[4,5] All of these have dimensions on the scale of nanometers and can be described as nanoparticles. While colloidal chemistry and even targeted drug delivery have a long history, nanotechnology and its application in nanomedicine are quickly becoming popular topics. Polymer therapeutics is another major trend in research that seeks after the properties of a magic bullet. Helmut Ringsdorf suggested a standard model that could be used to improve targeted drug delivery by focusing on different components of the polymer to give abilities for imaging, targeting, and drug loading.[6] The Ringsdorf model has been an inspiration for many polymer designs in drug delivery for cancer therapy.[7] It has led to a trend of polymer therapeutics, with researchers devising various ways to give properties to polymers. The versatility in chemistry and molecular architecture is one of the advantages of polymers in targeted drug delivery. With polymer therapeutics and the newly emerging field of nanomedicine, the possibilities are only limited by one's imagination.

Comparative PubMed searches show an exponential increase in the number of articles related to polymer therapeutics and nanomedicine.[8] Companies are willing to invest large amounts for research and development of these products because of their potential to return large profits. Review articles list various targeted drug delivery approaches that are in clinical trials, which include polymer–drug conjugates, monoclonal therapeutics, some that specialize in multidrug resistance, and those classified as nanoparticle-based therapeutics.[9–12]

1.2 Current Paradigm

The current paradigm associated with targeted drug delivery shapes the design of the drug carriers. Currently, there are certain properties thought to maximize drug delivery to the tumor. First, a stable carrier for the drug can help reduce side effects and increase the therapeutic effect. A stable carrier will mean that the drug is protected from the body and that normal (nontargeted) tissues are protected from the drug. Second, the carrier will accumulate in the tumor *via* the enhanced permeability and retention (EPR) effect. Third, the carrier can target the tumor based on both environmental and cellular components.

In order to arrive at the tumor, the drug needs a stable carrier. The majority of anticancer drugs are hydrophobic and do not dissolve in aqueous solutions. The first requirement of a stable carrier is simply the ability to carry the drug. As the carriers transport the drug, they need to form a stable barrier between the drug and the body. Protecting the body from the drug requires that the drug not interact with nontargeted cells, tissues, or organs. Protecting the drug from the body provides long circulation in the bloodstream. Increasing the blood circulation half-life of the carrier can increase chances of interactions with the target. To achieve long blood circulation, it needs to avoid interaction with the reticuloendothelial system (RES), also known as the mononuclear phagocyte system (MPS). Cell uptake by the MPS will decrease the efficacy of the treatment. There should also be protection against blood-borne proteins that would lead to inactivation, destabilization, or opsonization. The carrier should also be designed to limit accumulation in the kidney, liver, spleen, and other non-targeted organs. Administration of the chemotherapeutic agent

Taxol[®] is a good example of the need for good carriers. Taxol[®] consists of paclitaxel solubilized in Cremephor EL[®], which is a highly toxic mixture of castor oil and ethanol. This excipient can cause hypersensitivity reactions in patients undergoing chemotherapy and can be a treatment limiting problem.[13] Furthermore, the active pharmaceutical ingredient, paclitaxel, is known to cause dose limiting neurotoxicity at high doses.[14] If stability and long circulation are achieved, it will be able to travel through the body and eventually reach the site of the tumor. Here, carrier technology can encourage treatment preferentially to the tumor by using environmental characteristics or cellular components, such as surface proteins, as targets.

The carrier can target environmental characteristics of the tumor. Some of the environmental characteristics provide a means for passive accumulation *via* the EPR effect. The EPR effect is possible because of two hallmarks of cancer: unchecked growth and continued angiogenesis.[15] Continuous growth of the tumor leads to a chaotic tumor environment, with cells in hypoxic regions producing angiogenic factors which stimulate the production of new blood vessels. These new blood vessels are poorly formed and have gaps or fenestrations in the endothelium that allow passage of macromolecules into the tumor from the blood.[16] Increased mass transport from the blood vessels is beneficial for a tumor that is starved of nutrients. Therapies can take advantage of this hyperpermeability that allows nanoparticles to accumulate in the tumor where this leakiness occurs. Furthermore, retention in the tumor is aided by the lack of functional lymphatics that would normally drain the tissue.[17]

Once in the tumor environment, molecular targeting is a method used to provide treatment preferentially to tumors. Some cancer cell types overexpress certain surface markers. These markers are present in other cells, but can be much more abundant in some tumors. Overexpression of folate receptor and human epidermal growth factor receptor 2 (HER2) have been shown to be a predictor of poor prognosis of breast cancers.[18,19] These and other over-expressed proteins have become a target implemented into the designs of drug carrier technologies.[20–22] Alternatively, some carriers use a membrane penetration mechanism, such as the TAT peptide, to increase cell internaliza-tion in a nonspecific way.[23] While techniques not relying on protein expression can increase efficacy, methods must be used to block activity outside the tumor environment. To achieve this, the carrier's actions can designed to be triggered by environmental characteristics of the tumor. Some drug carriers use pH-sensitive groups to provide a triggered action in the acidic environment.[24] The low pH is a result of metabolism in hypoxic conditions that exist in the core of most solid tumors.[25] As not all tumors possess a markedly low pH, it has been shown that a glucose infusion in nondiabetic patients can lower the pH in tumors.[26] It also might be possible to use the high concentration of matrix metalloproteinases (MMPs) in the extracellular matrix (ECM) of tumors for triggering an active form of the carrier.[27]

Thus, the current paradigm is to have the carrier stably transport the drug while circulating through the bloodstream so that it can accumulate in the tumor, where a targeting mechanism will provide selective treatment. This paradigm has a plethora of opportunities for polymer therapeutics and nanomedicine. Designs can be custom tailored to fit certain aspects of the paradigm. What often happens is a therapeutic approach will focus on one specific aspect of the paradigm to increase efficacy. Unfortunately, this can result in shortcomings in other areas. Overall, there is still a low rate of clinical success for targeted drug delivery.

1.3 Challenges to Current Paradigm

Methods for targeted drug delivery have trouble showing that they are safe, effective, and better than what is currently used for treatment. Currently, ligand–receptor targeting drug carriers have challenges getting approved. For example, Mylotarg® is an antibody–drug conjugate that was approved for clinical use to treat leukemia. However, it has since been withdrawn due to poor results from a post-approval clinical trial, and continued postmarketing surveillance reveals lack of improved efficacy and unacceptable side effects. The antibody has no trouble binding to its target antigen of CD33, but lack of antigen exclusivity to the cancer and accumulation in the liver doomed this product.[28,29] To make successful targeted drug delivery products, the assumptions underlying these magic bullets will have to be questioned.

For solid tumors, an additional barrier seems to be intratumoral distribution. Once the carrier reaches the site of the tumor, it might be tempting to assume it will then invariably reach and enter the cancer cells. However, there are various factors that prevent the nanoparticle from penetrating into the tumor core.[30] Figure 1.1 shows how accumulation in the tumor is possible but also how penetration is limited. Tumors grown in window chambers show heterogeneous permeation and the accumulation of nanoparticles adjacent to the blood vessel.[31] Other studies also show that the permeability varies and that liposomes do not diffuse far from blood vessels.[32] Distribution beyond the vicinity of blood vessels is a challenge even for viruses and free drug molecules.[33,34]

Challenges to successful targeted drug delivery come from carrier dependent factors and tumor dependent factors. Current targeted drug delivery formulations are not magic bullets. Biodistribution and the EPR effect depend on carrier characteristics and these carriers are not always well characterized or stable or safe. Also, if good results with a drug carrier are seen in a mouse model, the success does not translate into the clinical model. The tumor resists treatment from the environmental to the cellular level. Aspects of the paradigm need to be examined in order to see improvements. There have been critical examinations of targeted drug delivery for quite some time now,[35] and certain challenges are beginning to be understood.

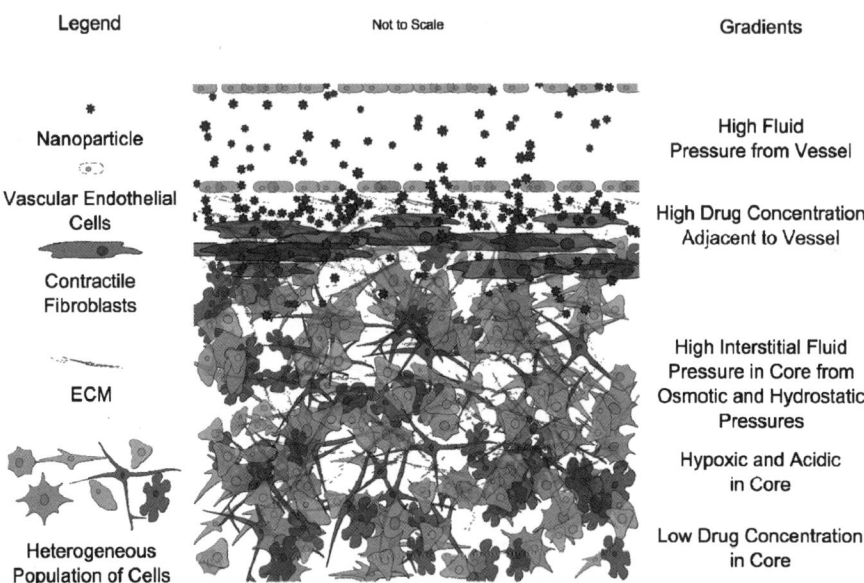

Legend   Not to Scale   Gradients

Nanoparticle

Vascular Endothelial
Cells

Contractile
Fibroblasts

ECM

Heterogeneous
Population of Cells

High Fluid
Pressure from Vessel

High Drug Concentration
Adjacent to Vessel

High Interstitial Fluid
Pressure in Core from
Osmotic and Hydrostatic
Pressures

Hypoxic and Acidic
in Core

Low Drug Concentration
in Core

Figure 1.1   How accumulation in the tumor is possible. There is evidence for
fenestrations in tumor vasculature that allow some macromolecules to
extravasate. However, the extent of extravasation and intratumoral
distribution of macromolecules is debatable. Nanoparticle distribution is
affected by the vasculature, the ECM, interstitial fluid pressure, and the
presence of cells.

1.3.1 Challenges Present in the Carrier

In Ehrlich's day a "magic bullet" seemed an appropriate analogy for the vision
he had. Present-day scientists sometimes interpret this analogy to something
more akin to a homing missile. However, current targeting drug delivery
designs are not target-seeking as they cannot sense the target from a distance
and adjust their trajectory to home in on it. Furthermore, current designs are
not capable of self-propulsion. Instead, they are opportunistic in that they rely
on being carried to the target by convective blood flow and diffusion through
tissues. There are also challenges associated with the characteristics of the
carrier that can have an effect on its stability.

1.3.1.1 Challenges in Stability and Tumor Accumulation

As mentioned earlier, a carrier needs to be stable and this will have an effect on
safety and efficacy. For example, while a strongly positively charged molecule
aids in entering the cell, high doses of cationic carriers can be toxic.[36]
Essentially, the biocompatibility of the carrier is determined by the chemistry
of the monomer and degradation products.[37] Viruses also serve as effective
carriers but have risks—one of which being an immune response against the

viral components. Stability upon dilution in the blood may be challenged for some microemulsion carriers such as liposomes or micelles.[38] Other potential problems that all carriers face when introduced into the blood include destabilization from high salt concentrations, adsorption of proteins, interactions with lipids, opsonization, and phagocytosis from cells. When the drug interacts with normal tissue because of the carrier's instability or is unable to reach targeted tissue because it lacks long circulation, the targeted drug delivery has failed. A common technique to increase circulation time is to add extremely hydrophilic poly(ethylene glycol) (PEG) to oppose adsorption or molecular interactions.[39,40] PEG interacts with water to make it thermodynamically favorable for the PEG chains to extend and limit adsorption of proteins.[41] However, long circulation is only relative and there can still be some uptake by the MPS and other organs as well as recognition from antibodies.[42] This is evident when PEG only increases the blood circulation half-life of liposomes in blood by just hours.[43]

Carrier size is another property that is often tuned to maximize EPR and blood circulation half-life. Carrier sizes are usually sought to be large enough to avoid renal clearance and small enough to escape out of fenestrations in the tumor vasculature.[44,45] Only a fraction of the drug carrier's population will reach the tumor as it is a matter of probability on whether the drug carrier will happen to be on the right path to hit the narrow opening in the vascular structure or if it will have to circulate through the body and risk uptake by other means. Uptake can occur in other organs that have vascular fenestrations, which vary in size by organ type.[46] Any particles too large to pass through fenestrations can be engulfed by resident macrophages. Polymer carriers will also possess a size distribution, as it is extremely difficult to avoid polydispersity with synthesized polymers and assembled carriers. Furthermore, many carriers can become more polydisperse during storage or while circulating in the blood. Following extravasation into the tumor, size also determines how different nanoparticles will penetrate into the core. Generally, a smaller molecule can more easily extravasate and diffuse to penetrate into the tumor; however, it can also diffuse away more quickly. On the other hand, a larger particle will be retained more readily in the tumor environment, but has a slower rate of extravasation and likely will not penetrate into the tumor core. Figure 1.2 explains the concept how carrier localization is dependent more on probability than targeting mechanisms.

The probability of any given nanoparticle reaching the tumor, extravasating, entering a cancer cell, and ultimately causing an effect within that cell is currently unknown. Imaging and tracking techniques have become increasingly popular to study this issue, leading to the growth of theragnostics where diagnosis is combined with therapy. Fluorescence resonance energy transfer may be particularly useful as it will to determine whether microemulsions or other complexes have dissociated.[47] Furthermore, Nomoto *et al.* have developed a new imaging method to allow *in vivo* real time observation of how PEGylation limits aggregation.[48] Tracking the dissociation or aggregation

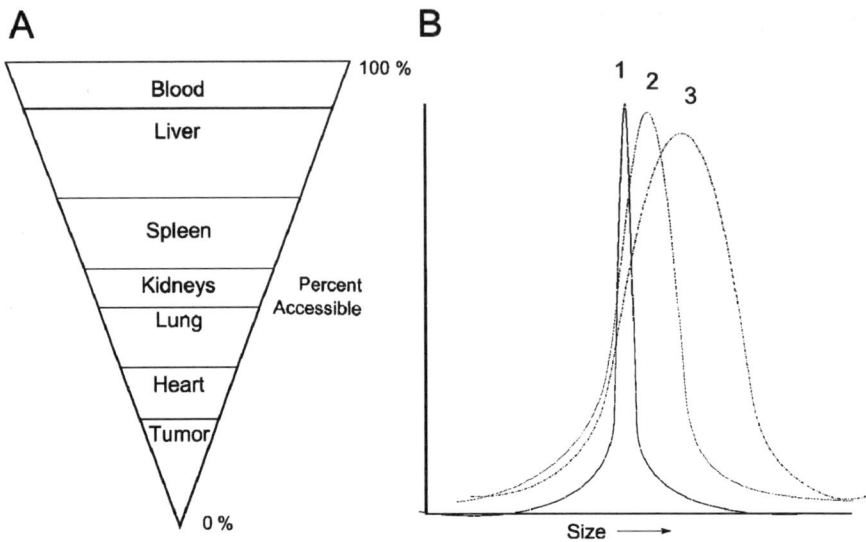

Figure 1.2 Illustration of the concept that probability plays a bigger role than targeting components. (A) The pyramid represents the accessibility to the different organs based on tissue size and ability to uptake different sized particles. (B) The Gaussian curve represents how the polydispersity of the carrier will result in a range of sizes. The carrier can start out with a low polydispersity index (1), but aggregate to become larger and more polydisperse (2 and 3).

of these drug carriers is important, as it directly affects the polydispersity and biodistribution of these particles.

1.3.1.2 Challenges in Targeting Mechanisms

Long circulation is also desired to benefit affinity-based targeting. Targeting moieties that bind to overexpressed proteins are used to give the drug specificity to the cancer to create the precision of the magic bullet. However, some studies have shown that carriers with targeting ligands are more limited in tumor penetration.[49] It is likely that these ligands experience more binding events that inhibit diffusion. On the other hand, more specific targeting with monoclonal antibodies for HER2 has been shown to increase tumor internalization.[50] Unfortunately, it is sometimes difficult, yet very important, to find cell surface proteins that are exclusive to the cancer. As discussed earlier, antigen exclusivity was part of the problem that Mylotarg® faced. Furthermore, it was assumed that targeting HER2 could direct drug delivery to cardiac tissues that express HER2 as part of repair.[51–53] Eligibility for HER2 treatment involves checking the levels of overexpression of these surface proteins and assigning a score. However, for the highest score of 3+ the FDA needs only 10% of tumor cells to stain positively for HER2, while an updated scoring systems defines it as a uniform staining of more than 30%.[54]

Often, proof of concept can be shown with *in vitro* studies and the drug carrier properties can be adjusted and perfected to show good results in a murine model. Unfortunately, success is not only dependent on the drug carrier.

1.3.2 Challenges Present in the Target

It may be possible that some fairly effective magic bullets have been made. While some may have failed because the wrong target was selected by design, it is also possible that failure occurred because of target mutation and resistance. A good portion of the targeting is passive and relies on the EPR effect. While EPR has a lot of evidence in animal models, it could be studied more extensively in human models as it does not always have a reliable effect.[55] Figure 1.3 illustrates how therapeutics do not accumulate exclusively to nor distribute evenly throughout the tumor. Studies in human patients evaluating effectiveness of targeted delivery with scintigraphy show the difficulties of avoiding accumulation in normal tissues.[56]

The unreliable EPR effect in clinical tumors shows the disparity between human and animal models. Animal models help cancer research but there must be some improvements in the models used to see significant improvements in

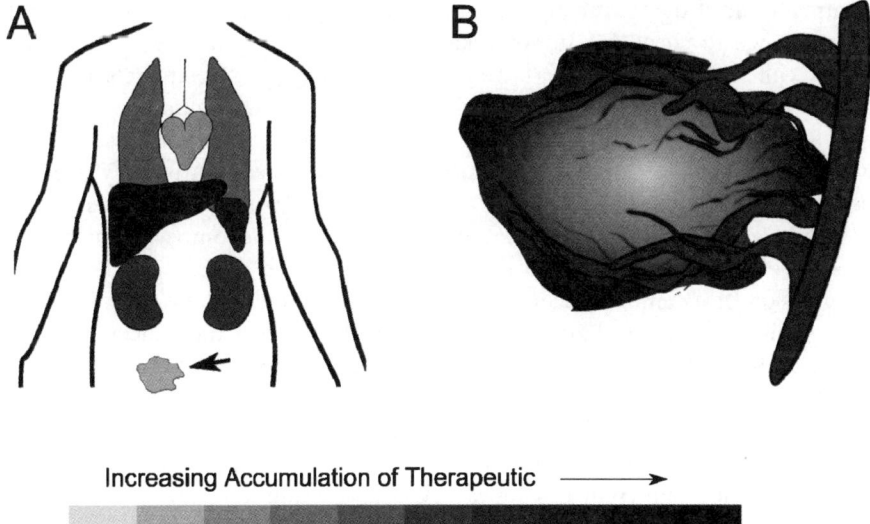

Increasing Accumulation of Therapeutic ⟶

Figure 1.3 Representation of how typical clinical results show lack of preferential targeting. (A) Targeted therapies accumulate in other organs and the total accumulated amount is greater than that found in the tumors. (B) The accumulation in the tumor is limited to the vascularized periphery and does not penetrate to the core of the tumor.

the clinic. Homer Pearce (once head of cancer research and clinical investigation at Eli Lilly and Company) said, "If you look at the millions and millions and millions of mice that have been cured, and you compare that to the relative success, or lack thereof, that we've achieved in the treatment of metastatic disease clinically, you realize that there just has to be something wrong with those models".[57] In the clinic, the normal procedure for humans is surgery and chemotherapy. If the patient cannot receive surgery or if normal chemotherapy fails, then nanomedicine is used as a last resort. At this point the tumor is considered very aggressive and resistant; however, the same cannot be said of the tumors that are cured in many mice models.

Some of the disparity is present in the methods used to grow the tumors in animal models. The rate and size of tumor growth in animal models is not representative of clinical cancers. Clinical tumors have many years to undergo genetic mutations, slow growth, and evasion of the immune system.[58] While mouse models can be controlled to develop cancers in a random nature and on a more representative timescale, for convenience and efficiency, many mouse models form tumors approximately 10 days following inoculation.[59,60] Still, mice have a higher rate of metabolism that will result in faster tumor growth and requires a more frequent dosing regimen as a countermeasure. Additionally, the ratio of tumor to body weight is also much larger in mouse models compared to human patients. Having a large tumor mass may contribute to lower pH and an increased EPR effect for mouse models. Also, study animals are sacrificed post-treatment to ascertain the extent of the treatment and if the treatment is positive. When the animal is sacrificed, there is no chance to check for tumor relapse. The environment where the tumor grows will influence results as well. Unfortunately, mouse models are not always orthotropic. Breast cancer cells injected to grow a tumor in the hind quarters of a mouse will not accurately portray what a tumor in breast tissue will do. Furthermore, since most murine tumors are xenografts, they need to be placed in immunodeficient mice. Tumors grown in immunodeficient mice are less aggressive than tumors that are grown in immunocompetent mice. The process of immunoediting explains how the immune system can drive the progression of a tumor to a more resistant and aggressive state. Studies have shown that tumor formation is much more likely when the cancer cell line comes from a tumor grown in an immunocompetent mouse.[61]

While the immune system provides a great example of targeted cancer therapy, the cancer is still able to escape and continue growing.[62,63] Immune evasion has become an emerging hallmark of cancer and is made possible by the heterogeneity inherent in cancer.[64] Understanding how cancer resists the immune system may lead to improved therapeutics in the future.

1.3.2.1   Cancer is Heterogeneous

There is both intertumoral and intratumoral heterogeneity, with differences in the cancer type, environment, individual cells, and signaling. Ovarian cancer is classified into different disease types by the type of cell it originates from.[65] In neuroepithelial tissues there are more than 100 entities that have been discovered.[66] This cancer heterogeneity has been imaged in tissue specimens using antibody staining combined with quantum dots.[67] Heterogeneity in the tumor environment also explains how EPR is not equally present in all tumors.

Variability in the environment is apparent in the tumor vasculature (Figure 1.4). Beyond shape, architecture, and density, tumors show differences in vascular permeability based on the type of tumor, location in the body, location in the tumor, hormone concentrations, and state of growth or regression of the tumor.[68–70] Unfortunately, heterogeneity in tumor vasculature makes it difficult to know specific size cutoffs for carrier sizes targeting the EPR effect. Still, it may be possible to try to increase permeability in the tumor with hyperthermia or with iRGD peptides, the latter having been classified as tumor penetrating peptides.[71,72] It also may help to artificially increase the vascular pressure to slightly improve tumor extravasation.[73,74]

Once the particle extravasates, it is confronted with a wall of dense packing in the tumor environment. Part of this environment is the ECM that is found in a state of flux, where the ECM of normal tissue is being broken down to make room for cancer cells that are secreting proteins for new ECM to support the growing tumor.[75] The support comes in the form of survival signals, inhibition of the immune system, limited diffusion of drug particles, and enabling cancer invasion and metastasis.[76–78] This new ECM is analogous to the biofilm that bacterial colonies will form for protection and continued growth. The ECM has a higher stiffness and viscosity due to more fibronectin, collagen I, and crosslinking agents which limit diffusion and also have an effect on the interstitial fluid pressure (IFP).[79–81]

While absolute values will vary patient to patient, the IFP is higher towards the center of the tumor environment.[82,83] The pressure results from both hydraulic and osmotic factors. There is a constant hydraulic pressure from the arterioles and capillaries that drives fluid and solutes into the tumor environment. Solutes flow into the interstitium from the blood vessels and accumulate into the dense packing of matrix and cells to create an osmotic pressure. Lack of functional lymphatics prevents an outlet and equalization of these pressures. Hence, the high IFP presents an opposition for macromolecules that could otherwise be transported by bulk flow and penetrate into the tumor core.[84] Interstitial pressure also seems to be affected by contractile stromal cells in the tumor environment.[85–87]

There is a diverse collection of cells, cancerous and noncancerous, in the tumor environment. This diversity in cell types plays a role in how cancers resist treatment. For example, immune cells such as myeloid-derived suppressor cells and CD25+ T cells can be present in tumors and suppress the immune response against cancer cells.[88,89] While there are different types of cells in the tumor environment, there are also differences among the cancer cells. Among the cancer cell population, there is a subdivision of cells that are cancer stem-like cells.[90] The properties of stem-like cancer cells make them a

A

B

Animal	Tumor line	Permeability ($\times 10^{-7}$ cm/s)
Fisher rat	R3230AC	1.7 ± 0.6
C3H mouse	MCaIV	2.9 ± 1.5
SCID mouse	MCaIV	1.9 ± 0.5
SCID mouse	U87	3.8 ± 1.2
SCID mouse	HGL21	0.11 ± 0.05

C

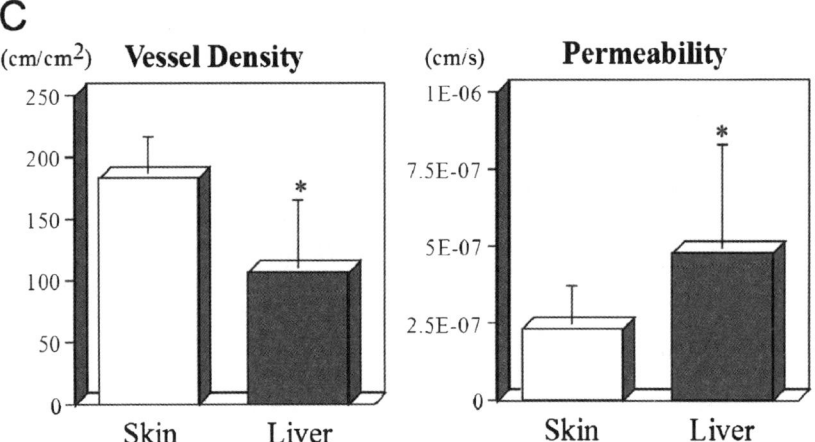

Figure 1.4 Heterogeneity in vascular permeability. Tumor heterogeneity affects the vessel density and permeability that EPR is dependent on. Among others, extravasation varies with (a) location inside tumor, (b) cell type, and (c) location of the tumor in the body. (A) Image of extravasated liposomes showing permeability varies with location inside the tumor (Reproduced from Yuan *et al.*[32] with permission from the American Association for Cancer Research). (B) Vascular permeability also varies with different cell lines or hosts (Adapted from Yuan *et al.*[69] with permission from the American Association for Cancer Research). (c) The same tumor line (LS174T) was grown in two different sites and resulted in differences in vessel density and permeability; *$P < 0.05$ when compared against values for skin group (Reproduced from Fukumura *et al.*[70] with permission from Elsevier).

unique challenge in cancer therapy. They contribute to pro-survival signaling pathways, are able to differentiate into other cells, and are more tumorigenic when injected into a new host—which may imply metastatic potential.[91–93] Cancer stem-like cells seem always to be present and in equilibrium with other cancer cells.[94] While the source of these cells is debated (they may come from a mutated stem cell or a normal cancer cell mutates to become stem cell-like), the fact remains that they have characteristics that make them capable of resistance and renewal.[95,96] Efforts to categorize and target cancer stem cells have been made by seeking after specific surface markers.[97,98] However, it could be beneficial to use a "magic shotgun" that is selective toward more than just one target to effect a heterogeneous population of cancer cells.[99]

The heterogeneity in the environment and cell population results from the molecular biology of cancer. The differences in cellular phenotype and states of growth and quiescence are influenced by stochastic variations of protein concentrations and interactions that lead to dynamics in molecular signaling, some of which originate from environmental cues.[100] Furthermore, genomic and epigenetic instability is characteristic of cancer cells.[101] These dynamic and heterogeneous characteristics play a role in how cancer is able to resist treatments.[102]

1.3.2.2 Multidrug Resistance

Tumor heterogeneity at the cellular level has a direct effect on multidrug resistance. Mutations cause diversity in cell characteristics and that diversity provides an opportunity to select for survival. Essentially, when a heterogeneous mixture of cells is treated with a drug, some will be able to resist the treatment. Furthermore, with treatment in solid tumors, cells at the periphery of the tumor often die off, leading to a smaller, yet more resistant, tumor mass. The apparent increase in resistance in the core could be due to conditioning from limited drug exposure, adaptations to survive in the already harsh environment, or the cells being more resistant inherently—as is the case with cancer stem-like cells.

Resistance comes from factors that range from the tumor environment down to cellular changes and mechanisms involving DNA. As stated before, only low amounts of drug are able to enter into the core of the tumor. Cancer cells are more easily able to adopt drug resistance to a low dose. This is apparent even in cultured cell lines, as drug resistant lines are made by adding low doses of the drug to the growth medium.[103,104] Furthermore, the core of the tumor is hypoxic and acidic and the cells have already adapted to survive to these harsh conditions. Some of these adaptations and cellular mechanisms for resistance are: detoxification *via* active efflux pumps, exocytosis, sequestration in acidic vesicles, DNA repair, compensation in chemical pathways, triggering of specific oncogenes, and even quiescence—as it is easy for a cancer cell to resist the effects of doxorubicin or paclitaxel when it is not actively dividing.[105]

As seen in Figure 1.5, cancer cells that are able to survive will be prepared to resist future treatment, repopulate the tumor, and metastasize. If the cancer relapses, more aggressive cell types have been selected and the prognosis for the patient is poor.

The source of a cancerous tumor could be mutated stem cells or somatic cells that have mutated to become stem-like. Initial treatment can lead to a reduction of the cancer. Complete elimination is avoided with heterogeneous expression of targeted antigen, or by incomplete distribution, or by innate cellular resistance. Despite continued treatment, resistant cells will survive and

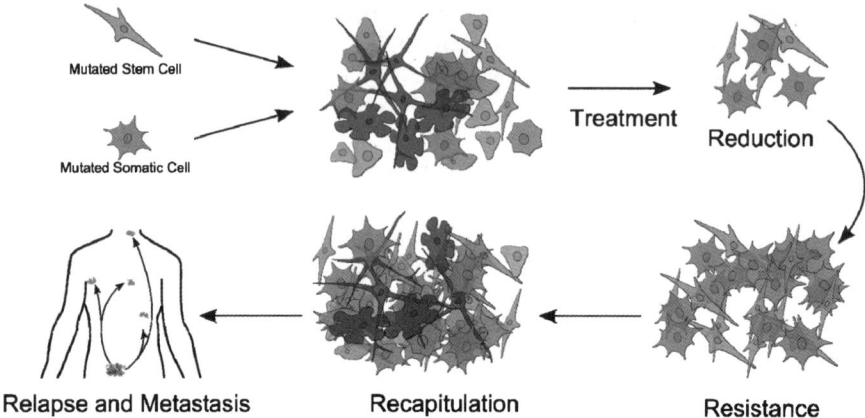

Relapse and Metastasis   Recapitulation   Resistance

Figure 1.5   Illustration of how anticancer therapy can lead to reduction, resistance, recapitulation, and relapse of the cancer. It is analogous to immunosurveillance's elimination, equilibrium, and escape.[106]

multiply to repopulate the tumor. Recapitulation of the heterogeneous population occurs by phenotypic expansion. Finally, a resistant and aggressive cancer will be capable of relapse and metastasis.

1.4   Revolution

In 1962, Thomas Kuhn wrote a book about paradigms and scientific revolution.[107] Right now we are in a state of collecting knowledge, with only slow improvements in targeted drug delivery occurring. One can hope that a shift of paradigm can be a catalyst for a massive leap forward in targeted drug delivery. That may mean abandoning, or at least not using too broadly, the terminology of targeted drug delivery. A more accurate description for some targeted designs may be "affinity therapy", which was used over 30 years ago.[108] When about 5% or less of the injected dose gets to the targeted tumor in the clinical setting, it is clear that existing dogma needs to be questioned.

It may be possible to provide something close to a magic bullet since monoclonal antibody formulations of Zevalin® and BEXXAR® have reached the market after showing effects in clinical trials. However, it must be recognized that these therapies are directed towards cancer cells circulating in the blood stream. This environment is discrete compared to cancer cells in solid tumors, as the therapeutic begins with 100% bioavailability in the blood and flow encourages contact. Worth noting is the approval of Trastuzumab® that has been approved for HER2-positive breast cancer. However, access is still unique as it is approved for treating metastatic cancer or as an adjuvant therapy. Thus, changing the focus of the targeting might be considered. It may be better to focus on changing the tumor environment or implementing immunomodulation to induce a state where the body can combat the cancer with its own resources.

Current anticancer treatments are used because they balances the risk-to-benefit ratio. A minimally effective treatment is allowed because it is better than the alternative of death. Along with this perspective, chemotherapy could be compared to lobotomies. When a cure for cancer is achieved, chemotherapy will look similarly barbaric in retrospect.

The desired effect of targeted drug delivery is to increase the therapeutic index by delivering the drug to the tumor and not to other organs. There are clever technologies addressing specific problems. Sometimes, changes can be made to environments that are less ideal for drug delivery. The ability to track and provide in-depth characterization, such as better assays for compatibility and stability in whole blood, are important for all of these drug carriers. By combining technologies into multifunctional approaches there is more hope of solving a complex problem like cancer. However, in order to be successful in the market, this versatility needs to be balanced with simplicity and reproducibility.

References

1. P. Ehrlich, *Studies in Immunity*, Plenum Press, New York, 1906.
2. K. Strebhardt and A. Ullrich, *Nat. Rev. Cancer*, 2008, **8**, 473–480.
3. P. Ehrlich, *Ned. Tijdschr. Geneeskd.*, 1909, **5**, 273–279.
4. Y. Ren, S. M. Wong and L. Y. Lim, *Pharm. Res.*, 2010, **27**, 2509–2513.
5. S. Boeckle and E. Wagner, *AAPS J.*, 2006, **8**, E731–E742.
6. H. Ringsdorf, *J. Polym. Sci. Polym. Symp.*, 1975, **51**, 135–153.
7. A. Nori and J. Kopecek, *Adv. Drug Delivery Rev.*, 2005, **57**, 609–636.
8. M. J. Vicent, H. Ringsdorf and R. Duncan, *Adv. Drug Delivery Rev.*, 2009, **61**, 1117–1120.
9. R. Duncan and M. J. Vicent, *Adv. Drug Delivery Rev.*, 2010, **62**, 272–282.
10. R. Duncan and R. Gaspar, *Mol. Pharmaceutics*, 2011, **8**, 2101–2141.
11. M. Liscovitch and Y. Lavie, *IDrugs*, 2002, **5**, 349–355.
12. P. P. Adiseshaiah, J. B. Hall and S. E. McNeil, *Wiley Interdiscip. Rev.: Nanomed. Nanobiotechnol.*, 2009, **2**, 99–112.
13. R. B. Weiss, R. C. Donehower, P. H. Wiernik, T. Ohnuma, R. J. Gralla, D. L. Trump, J. R. Baker, D. A. Van Echo, D. D. Von Hoff and B. Leyland-Jones, *J. Clin. Oncol.*, 1990, **8**, 1263–1268.
14. A. du Bois, M. Schlaich, H.-J. Lück, A. Mollenkopf, U. Wechsel, M. Rauchholz, T. Bauknecht and H.-G. Meerpohl, *Supportive Care Cancer*, 1999, **7**, 354–361.
15. D. Hanahan and R. A. Weinberg, *Cell*, 2000, **100**, 57–70.
16. H. Maeda, J. Wu, T. Sawa, Y. Matsumura and K. Hori, *J. Controlled Release*, 2000, **65**, 271–284.
17. A. J. Leu, D. A. Berk, A. Lymboussaki, K. Alitalo and R. K. Jain, *Cancer Res.*, 2000, **60**, 4324–4327.

18. L. C. Hartmann, G. L. Keeney, W. L. Lingle, T. J. H. Christianson, B. Varghese, D. Hillman, A. L. Oberg and P. S. Low, *Int. J. Cancer*, 2007, **121**, 938–942.
19. D. J. Slamon, G. M. Clark, S. G. Wong, W. J. Levin, A. Ullrich and W. L. McGuire, *Science*, 1987, **235**, 177–182.
20. K. R. Kalli, A. L. Oberg, G. L. Keeney, T. J. H. Christianson, P. S. Low, K. L. Knutson and L. C. Hartmann, *Gynecol. Oncol.*, 2008, **108**, 619–626.
21. M. N. Koopaei, R. Dinarvand, M. Amini, H. Rabbani, S. Emami, S. N. Ostad and F. Atyabi, *Int. J. Nanomed.*, 2011, **6**, 1903–1912.
22. J. S. Ross and G. S. Gray, *Clin. Leadersh. Manag. Rev.*, 2003, **17**, 333–340.
23. M. Silhol, M. Tyagi, M. Giacca, B. Lebleu and E. Vivès, *Eur. J. Biochem.*, 2002, **269**, 494–501.
24. E. S. Lee, Z. Gao, D. Kim, K. Park, I. C. Kwon, Y. H. Bae, E. Seong, I. Chan and Y. Han, *J. Controlled Release*, 2008, **129**, 228–236.
25. A. I. Minchinton and I. F. Tannock, *Nat. Rev. Cancer*, 2006, **6**, 583–592.
26. A. J. Thistlethwaite, G. A. Alexander, D. J. Moylan and D. B. Leeper, *Int. J. Radiat. Oncol., Biol., Phys.*, 1987, **13**, 603–610.
27. C. Wong, T. Stylianopoulos, J. Cui, J. Martin, V. P. Chauhan, W. Jiang, Z. Popović, R. K. Jain, M. G. Bawendi and D. Fukumura, *Proc. Natl. Acad. Sci. U. S. A.*, 2011, **108**, 2426–2431.
28. R. A. Larson, E. L. Sievers, E. A. Stadtmauer, B. Löwenberg, E. H. Estey, H. Dombret, M. Theobald, D. Voliotis, J. M. Bennett, M. Richie, L. H. Leopold, M. S. Berger, M. L. Sherman, M. R. Loken, J. J. M. van Dongen, I. D. Bernstein and F. R. Appelbaum, *Cancer*, 2005, **104**, 1442–1452.
29. P. F. Bross, J. Beitz, G. Chen, X. H. Chen, E. Duffy, L. Kieffer, S. Roy, R. Sridhara, A. Rahman, G. Williams and R. Pazdur, *Clin. Cancer Res.*, 2001, **7**, 1490–1496.
30. H. Holback and Y. Yeo, *Pharm. Res.*, 2011, **28**, 1819–1830.
31. E. S. Lee, Z. Gao and Y. H. Bae, *J. Controlled Release*, 2008, **132**, 164–170.
32. F. Yuan, M. Leunig, S. K. Huang, D. A. Berk, D. Papahadjopoulos and R. K. Jain, *Cancer Res.*, 1994, **54**, 3352–3356.
33. E. Smith, J. Breznik and B. D. Lichty, *Human Gene Ther.*, 2011, **1060**, 1053–1060.
34. A. J. Primeau, A. Rendon, D. Hedley, L. Lilge and I. F. Tannock, *Clin. Cancer Res.*, 2005, **11**, 8782–8788.
35. D. R. Welch, *Cancer Treat. Rev.*, 1987, **14**, 351–358.
36. S. Dokka, D. Toledo, X. Shi, V. Castranova and Y. Rojanasakul, *Pharm. Res.*, 2000, **17**, 521–525.
37. K. E. Uhrich, S. M. Cannizzaro, R. S. Langer and K. M. Shakesheff, *Chem. Rev.*, 1999, **99**, 3181–3198.
38. Y. H. Bae and H. Yin, *J. Controlled Release*, 2008, **131**, 2–4.

39. R. Gref, A. Domb, P. Quellec, T. Blunk, R. H. Müller, J. M. Verbavatz and R. Langer, *Adv. Drug Delivery Rev.*, 1995, **16**, 215–233.

40. G. Pasut and F. M. Veronese, *Adv. Drug Delivery Rev.*, 2009, **61**, 1177–1188.

41. M. Morra, in *Water in Biomaterials Surface Science*, ed. M. Morra, Wiley, New York, 2001, pp. 307–332.

42. T. Ishida, X. Wang, T. Shimizu, K. Nawata and H. Kiwada, *J. Controlled Release*, 2007, **122**, 349–355.

43. A. L. Klibanov, K. Maruyama, V. P. Torchilin and L. Huang, *FEBS Lett.*, 1990, **268**, 235–237.

44. H. S. Choi, W. Liu, P. Misra, E. Tanaka, J. P. Zimmer, B. Itty Ipe, M. G. Bawendi and J. V. Frangioni, *Nat. Biotechnol.*, 2007, **25**, 1165–1170.

45. F. Alexis, E. Pridgen, L. K. Molnar and O. C. Farokhzad, *Mol. Pharmaceutics*, 2008, **5**, 505–515.

46. M. K. Danquah, X. A. Zhang and R. I. Mahato, *Adv. Drug Delivery Rev.*, 2011, **63**, 623–639.

47. I. L. Medintz, A. R. Clapp, H. Mattoussi, E. R. Goldman, B. Fisher and J. M. Mauro, *Nat. Mater.*, 2003, **2**, 630–638.

48. T. Nomoto, Y. Matsumoto, K. Miyata, M. Oba, S. Fukushima, N. Nishiyama, T. Yamasoba and K. Kataoka, *J. Controlled Release*, 2011, **151**, 104–109.

49. H. Lee, H. Fonge, B. Hoang, R. M. Reilly and C. Allen, *Mol. Pharmaceutics*, 2010, **7**, 1195–1208.

50. D. B. Kirpotin, D. C. Drummond, Y. Shao, M. R. Shalaby, K. Hong, U. B. Nielsen, J. D. Marks, C. C. Benz and J. W. Park, *Cancer Res.*, 2006, **66**, 6732–6740.

51. K. F. Lee, H. Simon, H. Chen, B. Bates, M. C. Hung and C. Hauser, *Nature*, 1995, **378**, 394–398.

52. A. Negro, *Recent Prog. Hormone Res.*, 2004, **59**, 1–12.

53. E. A. Perez and R. Rodeheffer, *J. Clin. Oncol.*, 2004, **22**, 322–329.

54. M. Brunelli, E. Manfrin, G. Martignoni, S. Bersani, A. Remo, D. Reghellin, M. Chilosi and F. Bonetti, *Am. J. Clin. Pathol.*, 2008, **129**, 907–911.

55. M. N. Wente, J. Kleeff, M. W. Büchler, J. Wanders, P. Cheverton, S. Langman and H. Friess, *Invest. New Drugs*, 2005, **23**, 339–347.

56. K. J. Harrington, S. Mohammadtaghi, P. S. Uster, D. Glass, A. M. Peters, R. G. Vile and J. S. W. Stewart, *Clin. Cancer Res.*, 2001, **7**, 243–254.

57. C. Leaf, *Fortune*, 2004, **149**, 6.

58. S. Yachida, S. Jones, I. Bozic, T. Antal, R. Leary, B. Fu, M. Kamiyama, R. H. Hruban, J. R. Eshleman, M. A. Nowak, V. E. Velculescu, K. W. Kinzler, B. Vogelstein and C. A. Iacobuzio-Donahue, *Nature*, 2010, **467**, 1114–1117.

59. C. K. Osborne, K. Hobbs and G. M. Clark, *Cancer Res.*, 1985, **45**, 584–590.

60. J. Jonkers and A. Berns, *Nat. Rev. Cancer*, 2002, **2**, 251–265.

61. V. Shankaran, H. Ikeda, A. T. Bruce, J. M. White, P. E. Swanson, L. J. Old and R. D. Schreiber, *Nature*, 2001, **410**, 1107–1111.

62. G. P. Dunn, A. T. Bruce, H. Ikeda, L. J. Old and R. D. Schreiber, *Nat. Immunol.*, 2002, **3**, 991–998.

63. R. D. Schreiber, L. J. Old and M. J. Smyth, *Science*, 2011, **331**, 1565–1570.

64. D. Hanahan and R. A. Weinberg, *Cell*, 2011, **144**, 646–674.

65. M. Köbel, S. E. Kalloger, N. Boyd, S. McKinney, E. Mehl, C. Palmer, S. Leung, N. J. Bowen, D. N. Ionescu, A. Rajput, L. M. Prentice, D. Miller, J. Santos, K. Swenerton, C. B. Gilks and D. Huntsman, *PLoS Med.*, 2008, **5**, e232.

66. D. Louis, H. Ohgaki, O. Wiestler, W. Cavenee, P. Burger, A. Jouvet, B. Scheithauer and P. Kleihues, *Acta Neuropathol.*, 2007, **114**, 97–109.

67. J. Liu, S. K. Lau, V. A. Varma, R. A. Moffitt, M. Caldwell, T. Liu, A. N. Young, J. A. Petros, A. O. Osunkoya, T. Krogstad, B. Leyland-Jones, M. D. Wang and S. Nie, *ACS Nano*, 2010, **4**, 2755–2765.

68. S. K. Hobbs, W. L. Monsky, F. Yuan, W. G. Roberts, L. Griffith, V. P. Torchilin and R. K. Jain, *Proc. Natl. Acad. Sci. U. S. A.*, 1998, **95**, 4607–4612.

69. F. Yuan, H. A. Salehi, Y. Boucher, U. S. Vasthare, R. F. Tuma and R. K. Jain, *Cancer Res.*, 1994, **54**, 4564–4568.

70. D. Fukumura, F. Yuan, W. L. Monsky, Y. Chen and R. K. Jain, *Am. J. Pathol.*, 1997, **151**, 679–688.

71. K. N. Sugahara, T. Teesalu, P. P. Karmali, V. R. Kotamraju, L. Agemy, D. R. Greenwald and E. Ruoslahti, *Science*, 2010, **328**, 1031–1035.

72. G. Kong, R. D. Braun and M. W. Dewhirst, *Cancer Res.*, 2000, **60**, 4440–4445.

73. H. Maeda, Y. Noguchi, K. Sato and T. Akaike, *Cancer Sci.*, 1994, **85**, 331–334.

74. C. Li, Y. Miyamoto, Y. Kojima and H. Maeda, *Br. J. Cancer*, 1993, **67**, 975–980.

75. Q. Yu and I. Stamenkovic, *Genes Dev.*, 2000, **14**, 163–176.

76. G. A. McQuibban, *Science*, 2000, **289**, 1202–1206.

77. F. G. Giancotti, *Science*, 1999, **285**, 1028–1033.

78. L. A. Liotta and C. N. Rao, *Ann. N. Y. Acad. Sci.*, 1985, **460**, 333–344.

79. D. Barkan, J. E. Green and A. F. Chambers, *Eur. J. Cancer*, 2010, **46**, 1181–1188.

80. K. R. Levental, H. Yu, L. Kass, J. N. Lakins, M. Egeblad, J. T. Erler, S. F. T. Fong, K. Csiszar, A. Giaccia, W. Weninger, M. Yamauchi, D. L. Gasser and V. M. Weaver, *Cell*, 2009, **139**, 891–906.

81. P. A. Netti, D. A. Berk, M. A. Swartz, A. J. Grodzinsky and R. K. Jain, *Cancer Res.*, 2000, **60**, 2497–2503.

82. J. R. Less, M. C. Posner, Y. Boucher, D. Borochovitz, N. Wolmark and R. K. Jain, *Cancer Res.*, 1992, **52**, 6371–6374.

83. R. Gutmann, M. Leunig, J. Feyh, A. E. Goetz, K. Messmer, E. Kastenbauer and R. K. Jain, *Cancer Res.*, 1992, **52**, 1993–1995.
84. R. K. Jain and L. T. Baxter, *Cancer Res.*, 1988, **48**, 7022–7032.
85. R. A. Clark, J. M. Folkvord, C. E. Hart, M. J. Murray and J. M. McPherson, *J. Clin. Invest.*, 1989, **84**, 1036–1040.
86. R. Montesano and L. Orci, *Proc. Natl. Acad. Sci. U. S. A.*, 1988, **85**, 4894–4897.
87. G. Gabbiani, *J. Pathol.*, 2003, **200**, 500–503.
88. D. I. Gabrilovich and S. Nagaraj, *Nat. Rev. Immunol.*, 2009, **9**, 162–174.
89. S. Onizuka, I. Tawara, J. Shimizu, S. Sakaguchi, T. Fujita and E. Nakayama, *Cancer Res.*, 1999, **59**, 3128–3133.
90. M. Al-Hajj, M. S. Wicha, A. Benito-Hernandez, S. J. Morrison and M. F. Clarke, *Proc. Natl. Acad. Sci. U. S. A.*, 2003, **100**, 3983–3988.
91. S. Bao, Q. Wu, R. E. McLendon, Y. Hao, Q. Shi, A. B. Hjelmeland, M. W. Dewhirst, D. D. Bigner and J. N. Rich, *Nature*, 2006, **444**, 756–760.
92. P. C. Hermann, S. L. Huber, T. Herrler, A. Aicher, J. W. Ellwart, M. Guba, C. J. Bruns and C. Heeschen, *Cell Stem Cell*, 2007, **1**, 313–323.
93. J. Zhou, J. Wulfkuhle, H. Zhang, P. Gu, Y. Yang, J. Deng, J. B. Margolick, L. A. Liotta, E. Petricoin and Y. Zhang, *Proc. Natl. Acad. Sci. U. S. A.*, 2007, **104**, 16158–16163.
94. D. Iliopoulos, H. A. Hirsch, G. Wang and K. Struhl, *Proc. Natl. Acad. Sci. U. S. A.*, 2011, **108**, 1397–1402.
95. B.-B. S. Zhou, H. Zhang, M. Damelin, K. G. Geles, J. C. Grindley and P. B. Dirks, *Nat. Rev. Drug Discovery*, 2009, **8**, 806–823.
96. H. Clevers, *Nat. Med.*, 2011, **17**, 313–319.
97. D. Bonnet and J. E. Dick, *Nat. Med.*, 1997, **3**, 730–737.
98. L. Jin, K. J. Hope, Q. Zhai, F. Smadja-Joffe and J. E. Dick, *Nat. Med.*, 2006, **12**, 1167–1174.
99. B. L. Roth, D. J. Sheffler and W. K. Kroeze, *Nat. Rev. Drug Discovery*, 2004, **3**, 353–359.
100. H. Kitano, *Nat. Rev. Cancer*, 2004, **4**, 227–235.
101. P. A. Jones and S. B. Baylin, *Nat. Rev. Gene*, 2002, **3**, 415–428.
102. T. Tian, S. Olson, J. M. Whitacre and A. Harding, *Integr. Biol.*, 2011, **3**, 17–30.
103. E. S. Lee, K. Na and Y. H. Bae, *J. Controlled Release*, 2005, **103**, 405–418.
104. T. Minko, P. Kopečková and J. Kopeček, *J. Controlled Release*, 1999, **59**, 133–148.
105. M. M. Gottesman, *Annu. Rev. Med.*, 2002, **53**, 615–627.
106. G. P. Dunn, L. J. Old and R. D. Schreiber, *Immunity*, 2004, **21**, 137–148.
107. T. S. Kuhn, *The Structure of Scientific Revolutions*, University of Chicago Press, Chicago, 1996.
108. M. Wilchek, *Makromol. Chem.*, 1979, **2**, 207–214.

CHAPTER 2

Targeted Nanomedicines: Challenges and Opportunities

XINPENG MA, GANG HUANG, YIGUANG WANG AND JINMING GAO*

Department of Pharmacology, Harold C. Simmons Comprehensive Cancer Center, University of Texas Southwestern Medical Center at Dallas, Dallas, TX 75390, USA
*E-mail: jinming.gao@utsouthwestern.edu

2.1 Introduction

Cancer is one of the leading public health risks in the world. In the United States alone, one in four deaths is caused by this disease. A total of over 1.6 million new cancer cases and 0.58 million deaths are expected in 2012.[1] Current cancer treatments include surgical resection, radiation therapy, and chemotherapy.[2] In the past 50 years, small molecular anticancer drugs have been the mainstay treatment for many cancers at the advanced stage, but have shown almost no reduction in the death rate. These chemotherapeutic drugs lack specific targeting to the cancer cells and, consequently, they cause severe side-effects in patients by killing healthy cells. Therefore, it would be desirable to develop therapeutic systems that can directly target and kill cancerous cells selectively over the normal cells. More than a decade into the era of "targeted therapy," only a few small molecular drugs are truly amenable to this approach. Imatinib for chronic myeloid leukemia, erlotinib for epidermal growth factor receptor mutated non-small-cell lung cancer (NSCLC), and Trastuzumab® for human epidermal growth factor receptor-2-positive breast

RSC Polymer Chemistry Series No. 3
Functional Polymers for Nanomedicine
Edited by Youqing Shen
© The Royal Society of Chemistry 2013
Published by the Royal Society of Chemistry, www.rsc.org

cancer account for only 5% of cancer treatment in the U.S. annually.[3] Furthermore, these targeted approaches suffer from acquired resistance after prolonged treatment.

In the past decades, nanomedicines, nanoscopic therapeutics consisting of active pharmaceutical ingredients and delivery carriers, have shown great potential to provide paradigm-shifting solutions to improve the outcome of cancer diagnosis and therapy.[4-8] Compared to traditional small molecular drugs, nanomedicine (10–200 nm in diameter) has the advantages of unique pharmacological properties such as prolonged blood circulation time and reduced systemic toxicity,[9] high payloads of anticancer drugs and diagnostic agents, efficient cell uptake, and passive/active targeting to the tumor sites.

It is now widely accepted that nanomedicine can take advantage of the leaky tumor vasculature also known as the enhanced permeability and retention (EPR) effect to accumulate in the tumor tissues.[10,11] Compared to normal tissues, the endothelium lining of blood vessel walls is not well organized in tumors with numerous pores (200–1,200 nm) and high permeability to nanoparticles.[12,13] In addition, tumor tissues lack lymphatic drainage, which allows for the retention of nanoparticles.[14] In contrast, low molecular weight anticancer drugs are not able to be retained in tumors because of their ability to return to the circulation by diffusion. In multidrug-resistant (MDR) cancer cells that overexpress certain membrane-embedded drug efflux pumps such as P-glycoproteins (Pgp), the antitumor efficacy of nanomedicines is shown to be beneficial over free drugs.[15] Nanomedicine can prevent the encapsulated drugs from being recognized by efflux pumps.[16,17] Several drug-loaded nanomedicines to treat multidrug-resistant tumors are now in clinical trials.[18,19]

Despite the rapid progress, many technical and pharmacological challenges still exist in the successful implementation of nanomedicines in clinics. Finding the optimal physicochemical parameters that simultaneously confer molecular targeting, immune evasion, and controlled drug release is the most important challenge for successful clinical translation. The complex nanoparticle properties, including size, shape, and surface properties (*i.e.*, targeting ligand type and density), pharmaceutical properties such as drug loading and release kinetics, and *in vivo* physiological barriers to nanoparticle trafficking, have great impact on the safety and efficacy of nanomedicines for solid tumor treatment. This chapter will present current advances and discuss potential problems in the development of targeted nanomedicines.

2.2 Passive Targeting by Stealth Nanomedicines

Passive targeting (or physical targeting) describes that nanomedicines with prolonged blood circulation times allow for selective tumor accumulation and retention through the EPR effect. Although passive targeting approaches achieved great success (*e.g.*, clinical usage of stealth liposomes such as Doxil®), many questions and concerns have recently been raised. First, the EPR effect is a highly heterogeneous phenomenon, which varies from inside the same tumor and between different tumors, as well as from patient to patient.[20,21] Second, some tumor types (*e.g.*, pancreatic tumor) have low vascularity, which may limit the nanomedicine access to the tumor parenchyma. Finally, clearance by the reticuloendothelial system (RES) and the nanomedicine stability can also dramatically affect the delivery efficiency of the nanomedicine to the tumor site.

2.2.1 Nanomedicine Clearance by the Reticuloendothelial System

Opsonization is the process in which opsonin proteins bind to the surface of a foreign organism or particle so that they are more visible to phagocytic cells. Depending on their surface properties, such as surface charge and hydrophobicity, nanomedicines can be opsonized within minutes upon exposure to blood.[22] Subsequently, macrophages in the liver, spleen, and bone marrow rapidly take up opsonized nanomedicines.[23] As a result, the nanomedicines are removed from the bloodstream,[24] which will limit the passive targeting efficiency to tumors.

To prevent phagocytic clearance by the RES, the most commonly used strategy is to introduce poly(ethylene glycol) (PEG) or other neutral hydrophilic polymers (*e.g.*, polyvinylpyrrolidone, PVP) onto the surface of nanomedicines.[25,26] The resulting stealth nanomedicines contain a hydrated barrier that hinders the adsorption of opsonin proteins.[27] PEGylation effectively reduces the rate of RES uptake and increases the circulation half-life of various types of nanomedicines.[28,29] The most representative example is the PEGylated Doxil®, which prolongs the blood circulation time of doxorubicin to 45 hours in humans. Our labs pioneered the development of β-lapachone (β-lap) nanomedicines by loading the drug into poly(ethylene glycol)-*co*-poly(D,L-lactic acid) (PEG-PLA, MW 10,000 Da) polymeric micelles.[9] The pharmacokinetics of β-lap micelles were examined in athymic mice bearing A549 NSCLC xenografts. The blood concentration of β-lap micelles was prolonged with an elimination half-life ($t_{1/2,\beta}$) of 28 h, much longer than β-lap•HP-β-CD complexes ($t_{1/2,\beta} = 24$ min). As a result, β-lap micelles showed higher tumor accumulation and significantly improved antitumor efficacy than β-lap•HP-β-CD complexes.

2.2.2 Tumor Penetration by Nanomedicines

Systemic delivery of nanomedicines to tumors is a three-step process. Firstly, nanoparticles transport to different regions of the tumors *via* blood vessels. They must then cross the vessel wall and, finally, penetrate through the interstitial space to reach the cancer cells.[30] Delivery of therapeutic agents differs between tumor and normal tissues. The abnormal organization and structure of the tumor vasculature lead to tortuous and leaky vessels and heterogeneous vascular perfusion throughout the tumor.[31,32] In addition, hyperpermeability of the abnormal vasculature and lack of functional lymphatics lead to elevated levels of interstitial fluid pressure (IFP).[33] This interstitial hypertension, in turn, hinders drug delivery by abolishing the fluid pressure gradients that lead to rapid convective transport across the vessel wall and into interstitial space, while the complex extracellular matrix hinders diffusion. Typically, poor tumor penetration presents as a major barrier for effective delivery of nanomedicines into solid tumors.

The nanoparticle size, shape, and surface property all have significant impact on the tumor penetration of nanoparticles. Rational design of nanoparticle properties can improve tumor penetration by overcoming these barriers. Chilkoti and colleagues demonstrated the molecular weight (particle size) dependence on tumor penetration using macromolecular dextran as a model carrier system in a dorsal skin fold window chamber.[34] They found that increasing the molecular weight of dextran significantly reduced its vascular permeability, and dextrans of 3.3–10 kDa penetrated deeply (>35 μm) and were distributed homogeneously in the tumor tissues; however, dextrans of higher molecular weight (40–70 kDa) were observed only 15 μm from the vessel wall. Kataoka *et al.* demonstrated that only the 30 nm micelles could penetrate poorly permeable pancreatic tumors to achieve an antitumor effect, whereas 50–100 nm micelles showed much less tumor response.[35] In addition, they also demonstrated that the penetration and efficacy of larger micelles could be enhanced by using a transforming growth factor-β inhibitor to increase the permeability of the tumors.

To increase nanoparticle penetration into tumor tissue, Dai *et al.* developed a multistage nanoparticle delivery system in which 100-nm nanoparticles "shrink" to 10-nm nanoparticles after they extravasate from leaky regions of the tumor vasculature and are exposed to the tumor microenvironment.[36] The "shrunken" nanoparticles can more easily diffuse throughout the tumor interstitial space with a dense collagen matrix. Jain *et al.* recently reported that normalization of tumor vasculature improves the delivery of nanomedicines in a size-dependent manner.[37] They found that repairing the abnormal vessels in mammary tumors, by blocking vascular endothelial growth factor receptor-2, improves the delivery and anticancer efficacy of smaller nanoparticles (12 nm) while hindering the delivery of larger nanoparticles (125 nm).

Besides particle size, nanoparticle shape can also significantly influence the tumor penetration of nanoparticles. Jain *et al.* reported a shape-dependent tumor penetration of nanoparticles.[38] They found that nanorods penetrate tumors more rapidly than nanospheres due to improved transport through pores. More fundamental studies are warranted to systematically investigate the synergy in simultaneous control of nanoparticle size, shape, and surface properties, and their influence on systemic and local tumor pharmacokinetics.

2.3 Active Targeting by Surface-Functionalized Nanomedicines

Compared with normal cells, tumor cells undergo rapid proliferation and uncontrolled cell growth. Many cell surface receptors that respond to growth signals (*e.g.*, epidermal growth factor, EGF) or nutrient uptake for DNA or protein syntheses (*e.g.*, folic acid) are overexpressed. These receptors potentially provide useful cancer-specific molecular targets for targeted drug delivery. A variety of ligands such as monoclonal antibodies, peptides, or small molecules have been developed, with many of them used as therapeutics directly (*e.g.*, Erbitux®, a monoclonal antibody that targets the EGF receptor). The availability of these targeting ligands prompted an intensive research effort to develop nanoparticles conjugated with the ligands for tumor-targeted delivery. Active targeting nanomedicines have the potential to directly recognize cancer-specific receptors on the tumor cell surface, followed by receptor-mediated endocytosis to deliver drug payloads inside cancer cells. As a result, active targeting nanomedicines hold considerable promise to increase tumor selectivity with a potential decrease of adverse side effects in healthy tissues.[39–41]

Despite the therapeutic promise, active targeting nanomedicines must overcome numerous obstacles and physiological barriers before they can achieve the optimal targeting efficacy. First, before nanomedicines reach the cancer cells, they need to accumulate inside tumor tissues. In this case, passive targeting *via* the EPR effect may play a predominant role. Limitations such as fast RES clearance from blood, limited nanoparticle penetration, nanoparticle instability in blood, and heterogeneous tumor vasculature, as described in the passive targeting section (Section 2.2), all have great impact on the active targeting efficiency.[42,43] Second, introduction of targeting ligands can increase the clearance rates and affect other pharmacokinetic profiles of nanoparticles. The added complexity in surface functionalization may also be a limiting factor that increases the cost in commercial scale-up.[44,45] Recently, several research papers have reported that active targeting primarily increases intracellular uptake of nanomedicines and does not increase tumor accumulation as advocated by "magic bullets", whereas other reports indicate that tumor localization is dependent on the active targeting.[46–48] In the ensuring sections, we will review some of the above issues in more detail and discuss their importance in the design of active targeting nanomedicines.

2.3.1 Cancer Specificity of Active Targeting Nanomedicines

The basic principle of active targeting nanomedicines is the molecular recognition of cancer-specific receptors by ligand-encoded nanoparticles to increase the delivery of anticancer drugs. Ideally, the receptors should be uniquely expressed on the surface of cancer cells. However, in reality, receptor expression profiles are never black-and-white between cancer cells *vs.* normal cells, as most of the receptors are also expressed on the normal cell surface at a lower level compared to cancer cells. As such, even for antibodies, which provide the broadest opportunities in the diversity of targets and high specificity of interactions, their encoded nanoparticles still show moderate binding to normal cells and can cause nonspecific toxicity and side effects.[49]

For example, although some of upregulated targeting receptors such as folate and transferrin receptors correlate with metabolic rates and are overexpressed in various tumors, they are also expressed in fast-growing healthy cells such as some fibroblasts and epithelial cells.[50,51] This can lead to nonspecific targeting and increase toxicity in healthy tissues. More specifically, folate receptor-targeted liposomes have been extensively studied in cancer therapy.[52] While *in vitro* studies showed encouraging results, demonstrating that drugs delivered *via* folate receptor-targeted liposomes were rapidly internalized in cancer cells, the *in vivo* performance of these agents did not show much improved therapeutic efficacy in terms of prolonged survival.[53] In another example, PK2 is a polyHPMA-Gly-Phe-Leu-Gly–doxorubicin conjugate that contains the sugar galactosamine and which was used to target the asialoglycoprotein receptor (ASGPR) expressed on hepatomas. However, since ASGPR is also expressed on healthy hepatocytes, the targeted nanomedicines also nonspecifically accumulated in normal liver cells.[54]

In addition to receptor expression levels, the density of the receptors and the strength of the noncovalent interactions between targeting ligands and receptors are also important in achieving specific cell targeting. Park *et al.* found that a receptor density of 10^5 ERBB2 receptors per cell was necessary for an improved therapeutic effect of anti-ERBB2-targeted liposomal doxorubicin over nontargeted liposomal doxorubicin in a metastatic breast cancer model.[55] The targeting ligands on the nanomedicine surface need to engage in molecular interactions with receptors at a strength of 50–300 piconewtons (pN) and a density of $\sim 500 \text{ cm}^{-2}$.[19,56,57] Once attached on the cell surface, the target molecules are rapidly saturated, and their turnover time is typically about 20 minutes. To enhance binding avidity, multivalent binding provides an effective strategy to increase nanoparticle targeting to the cancer cells. The cooperative binding from multivalent interactions is much stronger than monovalent interactions. For example, dendrimer nanocarriers conjugated with an average of five folate molecules showed a 2,500- to 170,000-fold enhancement in dissociation constants (K_d) over free folate when attaching to folate-binding proteins immobilized on a surface.[19] Despite the increased binding avidity, it should be cautioned that polyvalent interactions from ligands on the nanoparticle surface may affect the receptor recycling efficiency during endocytosis. In some cases, nanoparticle binding will change receptors from recycling to the cell surface to lysosomal degradation, which will limit their availability for further cargo internalization.[47] For solid tumors, though higher binding affinity can increase targeting efficacy, this can also reduce penetration of nanocarriers due to a "binding-site barrier", where the nanocarrier binds to its target so strongly that penetration into the tissue is hampered.[58]

In several cancer-targeting nanomedicine formulations, the linkers between targeting ligands and drug carriers employ peptidase cleavable or acid labile bonds.[59] Nonspecific cleavage of the linkers may lead to the loss of the targeting moiety and thus influence the performance of the active targeting nanomedicines. For example, thiol–maleimide chemistry is a commonly used linker to attach targeting ligands on the surface of nanoparticles.[19] However, Junutula and co-workers engineered cysteines into a therapeutic HER2/neu antibody at three sites differing in solvent accessibility. They found that highly solvent-accessible sites rapidly lost conjugated thiol-reactive linkers in plasma owing to maleimide exchange with reactive thiols in albumin, free cysteine, or glutathione.[60]

2.3.2   Increased Clearance of Active Targeting Nanomedicines

Several recent studies have shown that the presence of targeting ligands such as folate, peptide, and antibodies on the surface of nanomedicines can decrease their blood circulation times.[53,61,62] This is not unexpected, since a prolonged circulation time of nanomedicines is a result of the shielding hydrophilic polymer (*e.g.*, PEG). When targeting ligands are conjugated on the distal end of the PEG chain, they sometimes result in recognition by the RES, with accelerated clearance by the liver.[63] Our previous study found that cRGD-encoded and cRGD-free superparamagnetic polymeric micelle (SPPM) formulations had comparable α-phase plasma half-lives ($t_{1/2,\alpha}$) at 0.34 ± 0.09 and 0.40 ± 0.34 h, respectively. However, cRGD-free SPPM had a significantly slower clearance in the β-phase, as represented by longer $t_{1/2,\beta}$ (9.2 ± 0.8 h) than cRGD-encoded SPPM (3.9 ± 0.8 h). We attribute this variation to the different functionalization of peptides (*i.e.*, cRGD *vs.* Cys) on the SPPM surface.[64] McNeeley *et al.* reported that adding just 0.15% folate to PEGylated liposomes significantly reduced the $t_{1/2}$ value from 18 to 6.7 h.[62] Nie and co-workers also reported that three types of peptide-coated gold nanoparticles had significantly shorter $t_{1/2}$ than nontargeted nanoparticles.[65] Clearly, the density of targeting ligands needs to be optimized to provide prolonged blood circulation times for tumor accumulation while allowing for cancer-specific receptor targeting to increase uptake in cancer cells.

2.3.3   Tumor Accumulation: Passive *vs.* Active Targeting Nanomedicines

There are still considerable debates on the relative contributions of active and passive targeting to the targeting efficiency and antitumor efficacy of nanoparticles to the tumor tissues and cancer cells.[66] Results from several research groups suggested that cancer-specific nanomedicines help improve the delivery efficiency of anticancer drugs, with increased dose accumulation at tumor sites compared to control nanoparticles lacking targeting ligands.[62,67] In contrast, several other groups have shown that the use of tumor-targeting ligands did not increase the total accumulation of nanoparticles in solid tumors, although it did increase receptor-mediated internalization in cancer cells.[42,48,54] In other words, active targeting may play a larger role in facilitating increased cellular uptake of nanomaterials than in increasing tumor localization.[68] Choi *et al.* found that the accumulation in tumors and other organs of PEGylated gold nanoparticles decorated with various amounts of human transferrin (TF) were independent of TF in mice bearing s.c. Neuro2A tumors at 24 h after i.v. tail-vein injections.[42] Park and co-workers also found that antibody-directed targeting did not increase the tumor localization of immunoliposomes, as both targeted and nontargeted liposomes achieved similarly high levels (7–8% injected dose/g tumor tissue) of tumor tissue accumulation in HER2-overexpressing breast cancer xenografts (BT-474).[50]

These debates point to the necessity to fundamentally understand the spatiotemporal patterns of active targeting nanomedicines and their dependence on the pathophysiology of the tumor microenvironment. Our recent studies revealed that cell targeting efficiency and tumor accumulation of cRGD-encoded superparamagnetic polymeric micelles (SPPM) to $\alpha_v\beta_3$-expressing angiogenic tumor vasculature in A549 tumor-bearing mice are highly dependent on time. Time-resolved MRI data clearly demonstrate that increased tumor accumulation of cRGD-encoded SPPM was observed at the vascular "hot spots" in tumors in the first 30 min after SPPM injection.[69] This is further corroborated by tissue distribution studies using ^{3}H-labeled SPPM, where a two-fold increase of nanoparticle accumulation was observed for cRGD-encoded SPPM over the nontargeted control at 60 min post-injection.[64] However, the accumulation of the cRGD-encoded SPPM did not show any significant differences compared to the cRGD-free SPPM in A549 tumor xenograft after 24 h (unpublished results). Presumably, we hypothesize that the longer circulation times of cRGD-free SPPM compensated for the tumor accumulation at longer times (*e.g.*, 24 h) *via* the passive targeting effect despite the short-term advantage of the active targeting nanoparticles. Similar compromises may also exist in folate and other targeted nanoparticles.[62,65] The temporal dependence of cell targeting and/or tumor accumulation of passive *vs.* activate targeting nanomedicines indicates much more careful investigation is necessary to evaluate their therapeutic contributions in cancer chemotherapy.

2.4   Conclusion and Future Perspectives

Although small molecular drugs are the mainstay of chemotherapy for cancer treatment, they suffer many limitations such as nonspecific tissue distribution, fast clearance from blood, and high acute toxicity. Nanomedicines are emerging as a powerful platform with integrated optimization of drug pharmacokinetics and pharmacodynamics for cancer therapy. Passive tumor targeting through the EPR effect has substantially improved therapeutic efficacy and increased the amount of drugs at tumor sites. Moreover, nanomedicines that can actively target cell-surface receptors with high affinity and cancer specificity have the potential to be most efficacious as the next generation of personalized medicines for cancer therapy. However, molecular and phenotypic heterogeneity is a significant challenge since not all cells within a given tumor will respond to a single therapeutic agent. Moreover, solid tumors are known to have a highly heterogeneous population of different cell types. Despite tremendous advances in tumor targeting strategies, many scientific, technological, and clinical challenges remain that will require a highly interdisciplinary and collaborative approach to overcome. Further studies are critically important to understand the systemic delivery of targeted nanomedicines in the context of complex tumor pathophysiology. Although we are still far from Paul Ehrlich's prediction of a "magic bullet", with advances in cancer biology and explosive developments in materials science and imaging technology, it is feasible that we can break through the current threshold and enter a new era of personalized nanomedicines for combined cancer diagnosis and therapy.

Acknowledgements

This work is supported by the National Institutes of Health (RO1 EB013149, RO1CA129011), the Cancer Prevention Research Institute of Texas (RP120897, RP120094), and the Texas FUSION fund.

References

1. R. Siegel, D. Naishadham and A. Jemal, *Ca–Cancer J. Clin.*, 2012, **62**, 10–29.
2. V. T. DeVita and S. A. Rosenberg, *New Engl. J. Med.*, 2012, **366**, 2207–2214.
3. A. Jemal, R. Siegel, E. Ward, Y. Hao, J. Xu and M. J. Thun, *Ca–Cancer J. Clin*, 2009, **59**, 225–249.
4. D. Peer, J. M. Karp, S. Hong, O. C. Farokhzad, R. Margalit and R. Langer, *Nat. Nanotechnol.*, 2007, **2**, 751–760.
5. B. Sumer and J. Gao, *Nanomedicine (London, U. K.)*, 2008, **3**, 137–140.
6. M. E. Davis, Z. Chen and D. M. Shin, *Nat. Rev. Drug Discovery*, 2008, **7**, 771–782.
7. D. Sutton, N. Nasongkla, E. Blanco and J. Gao, *Pharm. Res.*, 2007, **24**, 1029–1046.
8. E. Blanco, C. W. Kessinger, B. D. Sumer and J. Gao, *Exp. Biol. Med.*, 2009, **234**, 123–131.
9. E. Blanco, E. A. Bey, C. Khemtong, S.-G. Yang, J. Setti-Guthi, H. Chen, C. W. Kessinger, K. A. Carnevale, W. G. Bornmann, D. A. Boothman and J. Gao, *Cancer Res.*, 2010, **70**, 3896–3904.
10. Y. Matsumura and H. Maeda, *Cancer Res.*, 1986, **46**, 6387–6392.
11. V. Torchilin, *Adv. Drug Delivery Rev.*, 2011, **63**, 131–135.
12. F. Yuan, M. Leunig, S. K. Huang, D. A. Berk, D. Papahadjopoulos and R. K. Jain, *Cancer Res.*, 1994, **54**, 3352–3356.

13. S. K. Hobbs, W. L. Monsky, F. Yuan, W. G. Roberts, L. Griffith, V. P. Torchilin and R. K. Jain, *Proc. Natl. Acad. Sci. U. S. A.*, 1998, **95**, 4607–4612.
14. L. T. Baxter and R. K. Jain, *Microvasc. Res.*, 1990, **40**, 246–263.
15. A. R. Thierry, A. Dritschilo and A. Rahman, *Biochem. Biophys. Res. Commun.*, 1992, **187**, 1098–1105.
16. G. Sahay, D. Y. Alakhova and A. V. Kabanov, *J. Controlled Release*, 2010, **145**, 182–195.
17. S. Kunjachan, A. Błauż, D. Möckel, B. Theek, F. Kiessling, T. Etrych, K. Ulbrich, L. v. Bloois, G. Storm, G. Bartosz, B. Rychlik and T. Lammers, *Eur. J. Pharm. Sci.*, 2012, **45**, 421–428.
18. Y. Shen, E. Jin, B. Zhang, C. Murphy, M. Sui, J. Zhao, J. Wang, J. Tang, M. Fan, K. E. Van and W. Murdoch, *J. Am. Chem. Soc.*, 2010, **132**, 4259–4265.
19. S. Hong, P. R. Leroueil, I. J. Majoros, B. G. Orr, J. R. Baker, Jr. and M. M. Banaszak Holl, *Chem. Biol.*, 2007, **14**, 107–115.
20. Y. H. Bae and K. Park, *J. Controlled Release*, 2011, **153**, 198–205.
21. T. Lammers, F. Kiessling, E. Hennink Wim and G. Storm, *J. Controlled Release*, 2012, **161**, 175–187.
22. G. Gregoriadis, *J. Drug Targeting*, 2008, **16**, 520–524.
23. K. Winkler and L. T. Skovgård, *Clin. Physiol.*, 1984, **4**, 135–146.
24. T. Cedervall, I. Lynch, S. Lindman, T. Berggård, E. Thulin, H. Nilsson, K. A. Dawson and S. Linse, *Proc. Natl. Acad. Sci. U. S. A.*, 2007, **104**, 2050–2055.
25. Z. Amoozgar, J. Park, Q. Lin and Y. Yeo, *Mol. Pharmaceutics*, 2012, **9**, 1262–1270.
26. J. L. Dalsin and P. B. Messersmith, *Mater. Today*, 2005, **8**, 38–46.
27. M. Yallapu, S. Foy, T. Jain and V. Labhasetwar, *Pharm. Res.*, 2010, **27**, 2283–2295.
28. K. Chaudhari, M. Ukawala, A. Manjappa, A. Kumar, P. Mundada, A. Mishra, R. Mathur, J. Mönkkönen and R. Murthy, *Pharm. Res.*, 2012, **29**, 53–68.
29. M. L. Immordino, F. Dosio and L. Cattel, *Int. J. Nanomed.*, 2006, **1**, 297–315.
30. R. K. Jain and T. Stylianopoulos, *Nat. Rev. Clin. Oncol.*, 2010, **7**, 653–664.
31. R. K. Jain, *Sci. Am.*, 1994, **271**, 58–65.
32. R. K. Jain, *Science*, 2005, **307**, 58–62.
33. Y. Boucher, L. T. Baxter and R. K. Jain, *Cancer Res.*, 1990, **50**, 4478–4484.
34. M. R. Dreher, W. Liu, C. R. Michelich, M. W. Dewhirst, F. Yuan and A. Chilkoti, *J. Natl. Cancer Inst.*, 2006, **98**, 335–344.
35. H. Cabral, Y. Matsumoto, K. Mizuno, Q. Chen, M. Murakami, M. Kimura, Y. Terada, M. R. Kano, K. Miyazono, M. Uesaka, N. Nishiyama and K. Kataoka, *Nat. Nanotechnol.*, 2011, **6**, 815–823.

36. C. Wong, T. Stylianopoulos, J. Cui, J. Martin, V. P. Chauhan, W. Jiang, Z. Popovic, R. K. Jain, M. G. Bawendi and D. Fukumura, *Proc. Natl. Acad. Sci. U. S. A.*, 2011, **108**, 2426–2431.
37. V. P. Chauhan, T. Stylianopoulos, J. D. Martin, Z. Popovic, O. Chen, W. S. Kamoun, M. G. Bawendi, D. Fukumura and R. K. Jain, *Nat. Nanotechnol.*, 2012, **7**, 383–388.
38. V. P. Chauhan, Z. Popovic, O. Chen, J. Cui, D. Fukumura, M. G. Bawendi and R. K. Jain, *Angew. Chem. Int. Ed.*, 2011, **50**, 11417–11420.
39. V. Torchilin, *AAPS J.*, 2007, **9**, E128–E147.
40. A. Ray, N. Larson, D. B. Pike, M. Grüner, S. Naik, H. Bauer, A. Malugin, K. Greish and H. Ghandehari, *Mol. Pharmaceutics*, 2011, **8**, 1090–1099.
41. L. Basile, R. Pignatello and C. Passirani, *Curr. Drug Delivery*, 2012, **9**, 255–268.
42. C. H. J. Choi, C. A. Alabi, P. Webster and M. E. Davis, *Proc. Natl. Acad. Sci. U. S. A.*, 2010, **107**, 1235–1240.
43. T. M. Allen, *Nat. Rev. Cancer*, 2002, **2**, 750–763.
44. S. T. Stern, J. B. Hall, L. L. Yu, L. J. Wood, G. F. Paciotti, L. Tamarkin, S. E. Long and S. E. McNeil, *J. Controlled Release*, 2010, **146**, 164–174.
45. P. Debbage and G. C. Thurner, *Pharmaceuticals*, 2010, **3**, 3371–3416.
46. S. Bhattacharyya, R. Bhattacharya, S. Curley, M. A. McNiven and P. Mukherjee, *Proc. Natl. Acad. Sci. U. S. A.*, 2010, **107**, 14541–14546.
47. A. J. Schraa, R. J. Kok, A. D. Berendsen, H. E. Moorlag, E. J. Bos, D. K. F. Meijer, L. F. M. H. de Leij and G. Molema, *J. Controlled Release*, 2002, **83**, 241–251.
48. D. W. Bartlett, H. Su, I. J. Hildebrandt, W. A. Weber and M. E. Davis, *Proc. Natl. Acad. Sci. U. S. A.*, 2007, **104**, 15549–15554.
49. A. W. Tolcher, S. Sugarman, K. A. Gelmon, R. Cohen, M. Saleh, C. Isaacs, L. Young, D. Healey, N. Onetto and W. Slichenmyer, *J. Clin. Oncol.*, 1999, **17**, 478.
50. D. B. Kirpotin, D. C. Drummond, Y. Shao, M. R. Shalaby, K. Hong, U. B. Nielsen, J. D. Marks, C. C. Benz and J. W. Park, *Cancer Res.*, 2006, **66**, 6732–6740.
51. S. D. Weitman, R. H. Lark, L. R. Coney, D. W. Fort, V. Frasca, V. R. Zurawski and B. A. Kamen, *Cancer Res.*, 1992, **52**, 3396–3401.
52. A. R. Hilgenbrink and P. S. Low, *J. Pharm. Sci.*, 2005, **94**, 2135–2146.
53. A. Gabizon, H. Shmeeda, A. T. Horowitz and S. Zalipsky, *Adv. Drug Delivery Rev.*, 2004, **56**, 1177–1192.
54. L. W. Seymour, D. R. Ferry, D. Anderson, S. Hesslewood, P. J. Julyan, R. Poyner, J. Doran, A. M. Young, S. Burtles, D. J. Kerr, *et al J. Clin. Oncol.*, 2002, **20**, 1668–1676.
55. J. W. Park, K. Hong, D. B. Kirpotin, G. Colbern, R. Shalaby, J. Baselga, Y. Shao, U. B. Nielsen, J. D. Marks, D. Moore, D. Papahadjopoulos and C. C. Benz, *Clin. Cancer Res.*, 2002, **8**, 1172–1181.

56. W. Hanley, O. McCarty, S. Jadhav, Y. Tseng, D. Wirtz and K. Konstantopoulos, *J. Biol. Chem.*, 2003, **278**, 10556–10561.
57. O. A. Saleh and L. L. Sohn, *Proc. Natl. Acad. Sci. U. S. A.*, 2003, **100**, 820–824.
58. G. P. Adams, R. Schier, A. M. McCall, H. H. Simmons, E. M. Horak, R. K. Alpaugh, J. D. Marks and L. M. Weiner, *Cancer Res.*, 2001, **61**, 4750–4755.
59. L. Nobs, F. Buchegger, R. Gurny and E. Allémann, *J. Pharm. Sci.*, 2004, **93**, 1980–1992.
60. B.-Q. Shen, K. Xu, L. Liu, H. Raab, S. Bhakta, M. Kenrick, K. L. Parsons-Reponte, J. Tien, S.-F. Yu, E. Mai, D. Li, J. Tibbitts, J. Baudys, O. M. Saad, S. J. Scales, P. J. McDonald, P. E. Hass, C. Eigenbrot, T. Nguyen, W. A. Solis, R. N. Fuji, K. M. Flagella, D. Patel, S. D. Spencer, L. A. Khawli, A. Ebens, W. L. Wong, R. Vandlen, S. Kaur, M. X. Sliwkowski, R. H. Scheller, P. Polakis and J. R. Junutula, *Nat. Biotechnol.*, 2012, **30**, 184–189.
61. J. Wu, Q. Liu and R. J. Lee, *Int. J. Pharm.*, 2006, **316**, 148–153.
62. K. M. McNeeley, A. Annapragada and R. V. Bellamkonda, *Nanotechnology*, 2007, **18**, 385101.
63. P. M. Peiris and E. Karathanasis, *Oncotarget*, 2011, **2**, 430–432.
64. C. Khemtong, C. W. Kessinger, J. Ren, E. A. Bey, S.-G. Yang, J. S. Guthi, D. A. Boothman, A. D. Sherry and J. Gao, *Cancer Res.*, 2009, **69**, 1651–1658.
65. X. Huang, X. Peng, Y. Wang, Y. Wang, D. M. Shin, M. A. El-Sayed and S. Nie, *ACS Nano*, 2010, **4**, 5887–5896.
66. K. F. Pirollo and E. H. Chang, *Trends Biotechnol.*, 2008, **26**, 552–558.
67. N. Nasongkla, E. Bey, J. Ren, H. Ai, C. Khemtong, J. S. Guthi, S.-F. Chin, A. D. Sherry, D. A. Boothman and J. Gao, *Nano Lett.*, 2006, **6**, 2427–2430.
68. P. P. Adiseshaiah, J. B. Hall and S. E. McNeil, *Wiley Interdiscip. Rev.: Nanomed. Nanobiotechnol.*, 2010, **2**, 99–112.
69. C. W. Kessinger, O. Togao, C. Khemtong, G. Huang, M. Takahashi and J. Gao, *Theranostics*, 2011, **1**, 263–273.

CHAPTER 3

Rational Design of Translational Nanocarriers

QIHANG SUN[b], MACIEJ RADOSZ[b] AND
YOUQING SHEN*[a]

[a] Center for Bionanoengineering and State Key Laboratory of Chemical
Engineering, Department of Chemical and Biological Engineering, Zhejiang
University, Hangzhou 310027, P. R. China; [b] Department of Chemical and
Petroleum Engineering, Soft Materials Laboratory, University of Wyoming,
Laramie, WY 82071, USA
*E-mail: shenyq@zju.edu.cn

3.1 The Three Key Elements for Translational Nanomedicine

Nanometer-sized drug carriers, including polymer–drug conjugates, dendrimers, liposomes, polymer micelles, and nanoparticles, have been extensively investigated in drug delivery for cancer chemotherapy.[1,2] Cancer drug delivery is a process using nanocarriers with appropriate sizes (usually between several nanometers and 200 nm) and stealth properties to preferentially carry drugs to tumor tissues *via* the enhanced permeability and retention (EPR) effect.[2] However, despite improved pharmacokinetic properties and reduced adverse effects,[1,3] currently cancer drug delivery has only achieved modest therapeutic benefits.[3,4] Thus, the design of nanocarriers with more efficient drug delivery and thus higher therapeutic efficacy is still a pressing need.

The cancer drug delivery process can be divided into three stages, shown in Figure 3.1. Initially, the drug-loaded nanocarrier circulates in the blood compartments, including the liver and the spleen. When passing through

RSC Polymer Chemistry Series No. 3
Functional Polymers for Nanomedicine
Edited by Youqing Shen
© The Royal Society of Chemistry 2013
Published by the Royal Society of Chemistry, www.rsc.org

tumor blood vessels, the carrier may fall into the pores in the blood vessel wall and extravasate into the tumor tissue (EPR effect) (Figure 3.1A).[5,6] Next, it may further penetrate through the tumor tissue, which is nontrivial because of the high cell density and high interstitial fluid pressure (IFP) (Figure 3.1B).[7] Upon sticking to the surrounding cancer-cell membrane (Figure 3.1C), the carrier is expected to enter the cells *via* one or several possible pathways, and finally traverse the crowded intracellular structures and viscous cytosol to the targeted subcellular sites and release the carried drug cargo.

Thus, to achieve efficient drug delivery from the i.v. injection site to the target in the tumor cells, the nanocarrier must simultaneously meet two pairs of challenges (Figure 3.1): (a) the nanocarrier must retain the drug very tightly, ideally without any release, during the transport in the blood compartments and the tumor tissue, but must be able to efficiently release the drug once reaching the intracellular target to exert its pharmaceutical action; (b) the nanocarrier must be "slippery" or "stealthy" while in the blood compartments and in the tumor tissue until it reaches the targeted tumor cells. The stealth in the blood compartments enables it to effectively evade the reticuloendothelial system (RES) screening, particularly the capture by liver and spleen for a long blood circulation time. As the blood circulation time of the nanocarrier increases, so does its opportunity to pass the hyperpermeable tumor blood vessel and extravasation into the tumor tissue. After extravasating into the tumor, the nanocarrier must remain "stealthy" to penetrate deep into the center region to deliver the drug. This region lacks vascular perfusion

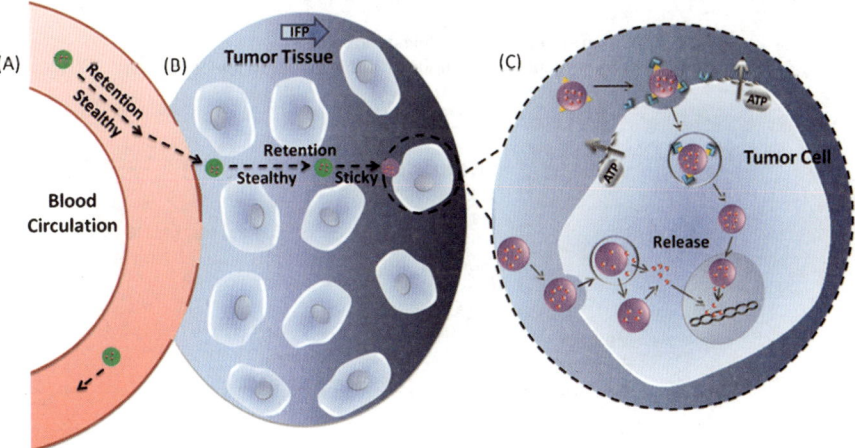

Figure 3.1   Cancer drug delivery process: (A) transport in the circulation, (B) transport through the tumor tissue, and (C) transport in the tumor cell. The nanocarrier must meet two pairs of challenges — For the drug: the nanocarrier must retain the drug very tightly during the transport in the blood compartments and the tumor tissue but efficiently release the drug once reaching the intracellular target; For the surface: the nanocarrier must be "very stealthy" during in the blood compartments for a long blood circulation time and remain "stealthy" in penetrating the tumor tissues but must become "sticky" or "cell binding" once interacting with tumor cells for efficient cellular uptake.

but harbors the most aggressive and resistant cells. On reaching the targeted cells the nanocarrier must become "sticky" or "cell binding" to interact with the tumor cell for efficient cellular uptake. A nanocarrier capable of simultaneously satisfying such opposite *2R2S* requirements at the right time and the right place, that is, "drug *R*etention in blood circulation *vs. R*elease in tumor cells (*2R*)" and "surface *S*tealthy in blood circulation and tumor tissues *vs. S*ticky to tumor cells (*2S*)", will deliver the drug specifically to the tumor cells, giving rise to high therapeutic efficacy and few side effects.

While the *2R2S* capability of a nanocarrier may render the resulting nanomedicine efficacious and potentially safe for clinical translation, two other elements, namely the feasibility of the nanocarrier materials to be proved for use as excipients (referred to as material excipientability) and the ability to establish scaled-up production processes for good manufacturing practice (GMP) for the nanocarrier and its formulation with the drug (nanomedicine) (referred to as process scale-up ability) are also indispensible for the nanomedicine to be truly translational from the benchtop to the bedside (Figure 3.2).[8] Most of our current research is focused on using new material design and chemistry to improve the *2R2S* capability; however, research aimed at translational applications should comprehensively consider the other two elements at an early stage.

Herein, we briefly review the approaches addressing nanocarriers' *2R2S* capability and summarize the factors affecting material excipientability and process scale-up ability, aimed at promoting the developments of truly translational nanomedicine for cancer drug delivery.

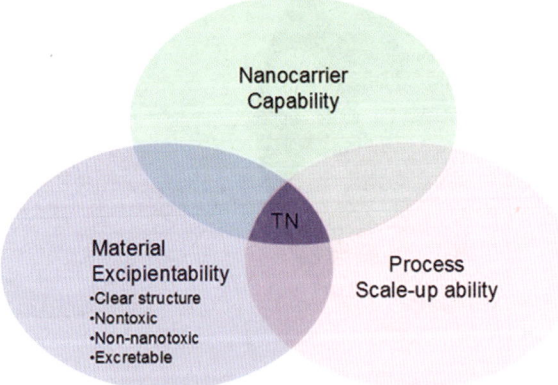

Drug: Retention in circulation *vs.* Release in cell (*2R*)
Surface: Stealthy in circulation & tumor tissues *vs.* Sticky to tumor cell (*2S*)

TN = Translational Nanomedicine

Figure 3.2   The three elements for translational nanomedicine: the nanocarrier should have the *2R2S* capability and its material should be suitable for excipient use (referred to as material excipientability); the production of the nanocarrier and its formulation with drug (nanomedicine) should be able to scale up for good manufacture process (GMP) (scale-up ability). Reprinted with permission from ref. 8. Copyright 2012 Elsevier.

3.2   The *2R2S* Capability of Nanocarriers

3.2.1   *2R*: Drug Retention in Circulation *versus* Intracellular Release

3.2.1.1   *Approaches to Minimize Premature Release from a Stable Carrier*

Figure 3.3 illustrates two examples, an ideal one for the case when the carrier retains the drug during transport in the blood compartments and the tumor tissue, but releases it in the tumor cells, and another for a typical case of undesirable burst release when the carrier releases its drug cargo prematurely while still circulating in the blood. Such a burst release is generally observed for polymer particles[9,10] and liposomes.[11,12] As a result, the drug is dumped in the blood compartments, which causes not only local or systemic toxicity, but also lowers the drug availability to the tumor and thereby the therapeutic efficacy.

Although the exact mechanism of burst release is still not fully understood, it is likely that drug-diffusion resistance can help explain and control it. A study on a model zero-order device indicated that the rate and extent of burst release from an otherwise stable carrier were affected by drug solubility and drug diffusion in an aqueous medium and by the drug loading content.[13] Such findings inspired more recent approaches to prevent burst release aimed at enhancing drug loading, inhibiting drug diffusion from the carrier, or both.

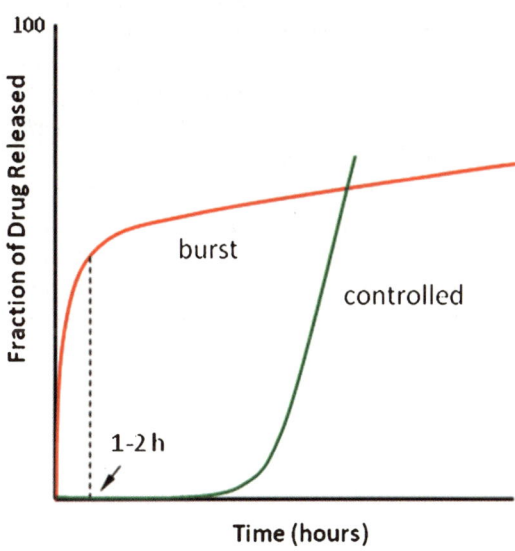

Figure 3.3   Sketch of ideal controlled release *vs.* premature burst release. Adapted with permission from ref. 8. Copyright 2012 Elsevier.

(1)   Using new chemical processes to fabricate structured nanoparticles.

Polymeric micelles encapsulate drugs mostly *via* physical trapping based on hydrophobic interactions. They are generally fabricated by coprecipitation of the hydrophobic drugs with the hydrophobic blocks of amphiphilic copolymers by dialysis or the solvent-evaporation method,[14] assuming that the drugs and the hydrophobic blocks precipitate simultaneously and thus the drugs are completely embedded in the hydrophobic micelle core. However, in many cases this is not a very realistic assumption, as either the drugs can precipitate first or the core can form first, which prevents proper drug encapsulation in the core. For example, when the core forms first, most drug molecules may precipitate around the core, which are prone to burst release upon dispersion in an aqueous solution.[15]

Building on this finding, we proposed that coating the core with an additional hydrophobic layer would impose an extra diffusion barrier and thereby minimize burst drug release. Using a stepwise pH-controlled process, three-layer onion-structured nanoparticles (3LNPs) were synthesized that consisted of a poly(ε-caprolactone) (PCL) core, a pH-responsive poly[2-(*N*,*N*-diethylamino)ethyl methacrylate] (PDEA) middle layer, and a polyethylene glycol (PEG) outer coronal layer.[16] Compared to the conventional core-corona micelles, such 3LNPs were found to exhibit a significantly lower burst release of camptothecin (CPT) at physiological pH due to the effective barrier of the hydrophobic PDEA barrier.

The conventional method for preparing polymeric micelles through liquid solvent evaporation or dialysis offers little control of micellization *versus* drug precipitation. However, this can be accomplished with a near-critical fluid micellization (NCM) method to prepare drug-loaded polymeric micelles.[17] The solvating power of a near-critical fluid solvent is easily tunable with pressure. Thus, more selective and flexible micellization can be controlled by adjusting the pressure alone. At high pressures, drugs and polymers were molecularly homogenous in a near-critical solvent, whereas at moderate pressures micellization/drug encapsulation occurred (Figure 3.4A). With this process, PEG-PCL micelles, formed in a near-critical dimethyl ether/trifluoromethane, could be loaded with paclitaxel (PTX) as high as 12 wt% (Figure 3.4B). More recently, we prepared three-layered micelles formed by a stepwise NCM process that exhibited little, if any, burst release despite the high drug loading content (Figure 3.4C, D).[18] The biggest advantage of this NCM is that it uses the conventional Food and Drug Administration (FDA) approved materials to obtain high drug loading micelles with minimized burst or even burst-free. Such products are also free of contamination from organic solvents.

(2)   Drug conjugation

The second approach to eliminate burst release is by conjugating drugs to the carriers *via* covalent bonds. Because the drug must be released once at the target, the covalent bonds or linkers must be cleavable in the tumor-cell environment. For instance, doxorubicin (DOX) was conjugated to a poly(L-aspartic acid) [P(Asp)] block in the block copolymer PEG-*b*-P(Asp) through amide[19,20] or hydrazone linkers.[21,22] Drugs can also be conjugated to the ends of hydrophobic blocks.[23,24] The resulting micelles formed from PEG-*block*-poly(L-amino acid)

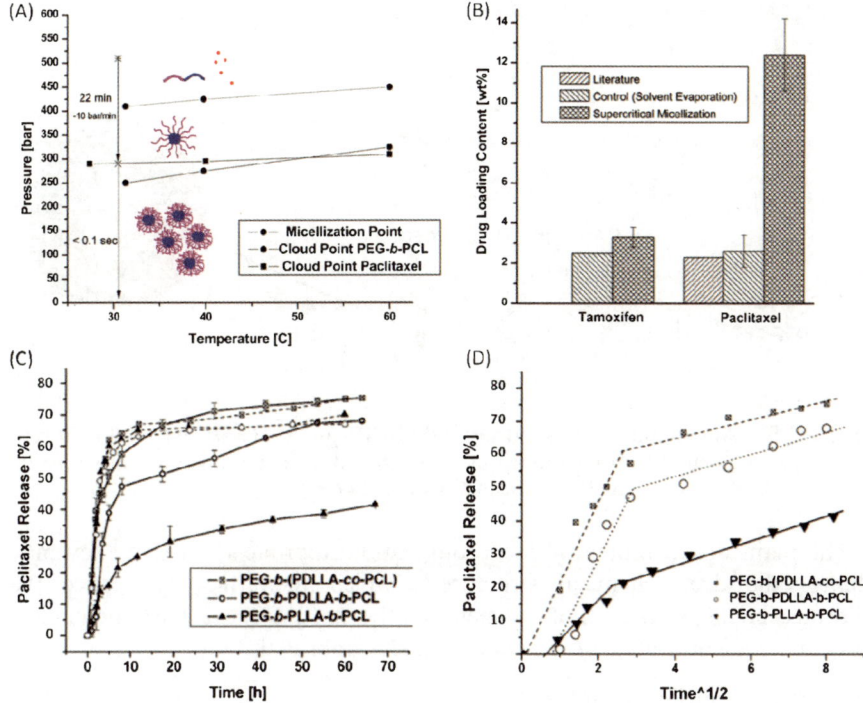

Figure 3.4 Nanoparticles prepared by a near-critical fluid micellization (NCM) method; (A) Micellization and cloud pressures of PEG-PCL and PTX in 70% dimethyl ether/30% trifluoromethane. (B) Drug loading contents of tamoxifen and PTX in PEG-PCL nanoparticles by the NCM process and conventional solution process. Adapted with permission from ref. 17. Copyright 2011 American Chemical Society. (C) Cumulative drug release as a function of time for a diblock and two triblocks copolymers. Solid lines represent micelles prepared by near-critical micelliazation while dashed lines represent micelles prepared conventionally by solvent evaporation. (D) Cumulative drug release plotted as a function of $t^{1/2}$ for experiments plotted in (C). Adapted with permission from ref. 18. Copyright 2012 American Chemical Society.

(PEG-*b*-PLAA) conjugated drugs eliminated any burst release.[25] Using labile linkers responsive to the tumor's extracellular or intracellular stimuli results in drug release triggered in a tumor extracellular environment.[26,27]

Our group demonstrated, using drugs as the hydrophobic part, directly making self-assembling amphiphilic prodrugs for fabricating burst-free carriers.[28] In this method, hydrophobic CPT molecules were conjugated to short oligomer chains of ethylene glycol (OEG) to form the amphiphilic phospholipid-mimicking prodrugs OEG-CPT or OEG-DiCPT (Figure 3.5). These prodrugs formed stable liposome-like nanocapsules with extremely high drug loading content but no burst release. Similar nanoparticles were prepared from an amphiphilic curcumin prodrug.[29]

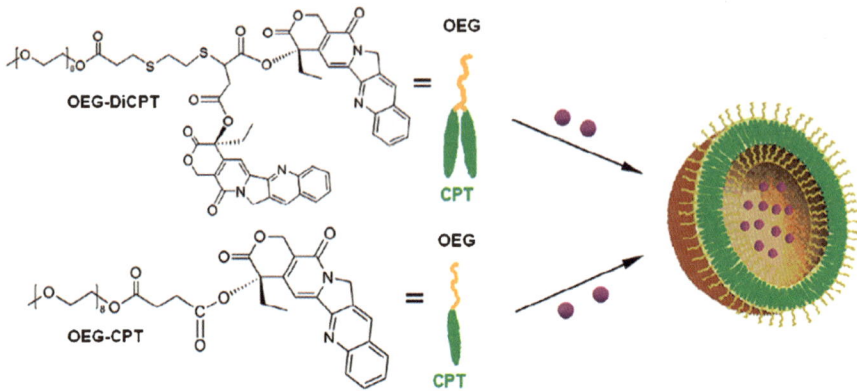

Figure 3.5   Amphiphilic CPT prodrugs (OEG-CPT and OEG-DiCPT) and their self-assembly into nanocapsules. Reprinted with permission from ref. 28. Copyright 2010 American Chemical Society.

The main disadvantage of such conjugation approaches is that they may change the drug chemical structure,[30] which in turn may reduce the pharmaceutical efficacy, not to mention the need for extensive preclinical tests and clinical trials before acquiring FDA approval.

(3)   Core- or shell-crosslinked micelles

The third approach aimed at reducing the burst release is crosslinking the core or the corona shell of micelles. For example, Wooley *et al.* developed methods for fabricating shell-crosslinked micelles.[31] In order to de-crosslink the shell to allow drug release at the target site, a linker labile in the presence of intracellular glutathione (GSH) was used.[32] As intended, such crosslinked shells inhibited drug diffusion from the micelles, and hence reduced burst release. However, such a crosslinked shell becomes more rigid and hence loses its ability to repel serum proteins or other biomacromolecules,[33] and thus may not continue to be stealthy in circulation.

Covalent crosslinking of the micelle hydrophobic core can therefore be a preferable approach.[34,35] For instance, crosslinked micelles consisting of PEG-*b*-poly(acryloyl carbonate)-*b*-poly(D,L-lactide) (PEG-PAC-PDLLA) had high stability and significantly inhibited PTX release at low micelle concentrations compared to the non-crosslinked controls.[36] Lavasanifar *et al.* applied click chemistry and developed hydrolysable core-crosslinked PEG-*b*-poly(α-propargyl carboxylate-ε-caprolactone) (PEG-PPCL) micelles that exhibited a lower degree of PTX burst release than equivalent non-crosslinked micelles.[37] When the crosslinked core had disulfide linkers, it was shown to hold the drug tightly but release it quickly once in the tumor cell, due to the cleavage of the crosslinkages by intracellular GSH.[38] Similarly, thiolated Pluronic copolymer (Plu-SH) was demonstrated to form core-crosslinked micelles that were reversible *via* dithiothreitol (DTT)-breakable disulfide bonds, which inhibited the premature release in an aqueous solution.[39]

3.2.1.2 Approaches to Increase Carrier Stability to Prevent Premature Release

A thermodynamically unstable carrier (an unstable carrier for short) may dissociate before reaching its target and thus prematurely release the drug. Such an unstable carrier may dissociate fast or slowly, referred to as micelle dissociation kinetics (some authors[25] use "kinetic stability"). We always prefer carriers that are thermodynamically stable until they reach their target. At a given temperature, micelles form at the polymer concentrations above the critical micelle concentration (CMC):[40]

$$C_{CMC} \approx \exp(-n\varepsilon_h/k_b T)$$

where $k_b T$ is the thermal energy and ε_h is the monomer effective interaction energy with the bulk solution (related to χ in polymer physics). Polymers with a low CMC suggest a high thermodynamic stability, and *vice versa*. Usually, the longer the hydrophobic blocks, the more stable the micelles they form.[41] Thermodynamic stability is particularly important because locally, in circulation, micelles may dissociate if the block copolymer concentration falls below the CMC. It seems intuitive that a drug-loaded micelle may have a CMC that is different from its virgin drug-free analog, but, to a first approximation, it is common to neglect this difference.

Once the copolymer concentration falls below its CMC, the micelle dissociation rate can vary, depending on cohesive forces among the core-forming blocks. Chain insertion/expulsion and micellar fusion/splitting are two mechanisms that can explain the overall dynamic exchange between monomers and micelles.[41] Monte Carlo simulation indicated that chain insertion/expulsion played a major role when the polymer concentration was low.[42] Because chain mobility plays a crucial role, the hydrophobic blocks with a relatively high glass transition temperature (T_g) make the micelles dissociate much more slowly than those with a low T_g.[43] Furthermore, the size of the hydrophobic block and the hydrophilic-to-hydrophobic block mass ratio were found to affect the rate of micelle dissociation, from size-exclusion chromatography (SEC) experiments. For simple PEG-PCL copolymers, micelles formed from PEG-PCL (5000:4000 and 5000:2500) dissociated slowly; however, micelles formed from the PEG-PCL (5000:1000) dissociated quickly into monomers.[44]

Even though there is evidence that some polymeric micelles can be stable in serum even *in vivo*,[45] the stability of micelles in the blood is far from understood. Quite different from carriers tested in water or in buffer solutions, micelles in blood circulation can be extremely diluted and encounter various blood components which may promote micelle dissociation. Burt *et al.* prepared radiolabeled PTX-loaded PEG-PDLLA micelles and found that PTX was rapidly released from the micelles, and the diblock copolymer was cleaved into its two polymer components in the blood.[46,47] Maysinger *et al.* conjugated fluorescein-5-carbonyl azide diacetate to PEG-PCL micelles and noticed that they were stable in buffer solutions but unstable in serum-containing culture media with or without cells.[48] Recently, Cheng *et al.* employed a fluorescence-resonance energy-transfer (FRET) technique to demonstrate that PEG-*b*-PDLLA micelles were not stable in the bloodstream due to the influence of α- and β-globulins rather than γ-globulin or serum albumin.[49] Based on those results, Cheng *et al.* summarized the possible mechanisms responsible for the micelle decomposition induced by serum proteins,[41] including protein adsorption,[48,49] protein penetration,[50,51] and drug extraction.[41] What exactly happens to the micelles after injection is poorly understood because it is hard to measure and estimate micelle concentration locally in the bloodstream.[52] Cheng *et al.* tracked unmodified copolymer micelles using the FRET imaging method, but unfortunately no direct evidence proved that the CMC was unchanged by incorporating a FRET pair.[53]

However, there is no doubt that, directionally, the lower the CMC, the higher the probability of micelle stability in the bloodstream. Therefore, the most common strategy to enhance the micelle stability is to reduce its CMC. Compared to liposomes, polymeric micelles usually have a much lower CMC, at a micromolar level, which imparts a higher stability. A further reduction of polymeric micelle CMC can be achieved by increasing the core-forming block hydrophobicity, molecular weight, or both.[40] One example is that of chemically modified Pluronics: Pluronic/PCL copolymeric nanospheres exhibited a lower CMC.[54,55] Another interesting finding is that stearic acid as side chains can keep micelles stable even in the presence of serum.[56] In the presence of serum albumin, α- and β-globulins, or γ globulins, the micelles from PEG-*b*-poly(*N*-hexyl stearate L-aspartamide) (PEG-*b*-PHSA) copolymers with nine stearic acid side chains still existed after two hours.

Crosslinking is a straightforward method to stabilize micelles. While the covalent crosslinking of the micelle core or shell can inhibit burst release from a stable micelle, it can also inhibit or prevent micelle dissociation. For instance, PEG-PCL micelles with cores crosslinked by radical polymerization of the double bonds introduced to the PCL blocks turned out to be more stable.[57] Biodegradable thermosensitive micelles with crosslinked cores formed from PEG-*b*-[*N*-(2-hydroxyethyl methacrylamide)-oligolactates] [PEG-*b*-p(HEMAm-Lac$_n$)] kept their integrity upon dilution and only degraded after cleavage of the ester bonds in the crosslinkers.[58]

The caveat, however, is that crosslinking reactions usually occur after core formation, which can alter the structure and properties of the encapsulated drug. To overcome this potential problem, our group developed stable core-surface crosslinked micelles (SCNs), shown in Figure 3.6, made from

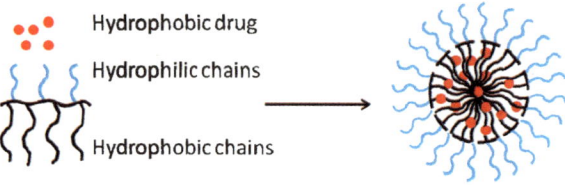

Figure 3.6 Formation of SCNs from amphiphilic brush polymers. Adapted with permission from ref. 59. Copyright 2004 American Chemical Society.

amphiphilic polymer brushes.[59] The key point is that the backbones of the polymer brushes acted as crosslinkages on the hydrophobic core surface, instead of chemical crosslinking, which substantially enhanced the micelle stability. Specifically, the resulting micelles had much lower CMCs than corresponding PEG-PCL block copolymers.

For the excretion of the nanocarriers from the body, crosslinked micelles must be able to break into small polymer chains. Toward this end, reversible crosslinking triggered by different stimuli like pH,[60] UV light,[61] or others[62] was later developed. Historically, pH-sensitivity was the first one used to trigger a desired carrier change because cancer or inflammation makes the extracellular pH at the disease site acidic.[63] For instance, micelles formed from the triblock copolymer PEG-*b*-poly[*N*-(3-aminopropyl)methacrylamide]-*b*-poly(*N*-isopropylacrylamide) (PEG-PAPMA-PNIPAM) were shell-crosslinked with terephthaldicarbaldehyde (TDA) at pH 9 *via* cleavable imine linkages.[60] However, at pH < 6 the hydrolytic cleavage of the imine crosslinkages occurred. Other examples[64,65] were inspired by crosslinking, using disulfide linkages that are sensitive to intracellular GSH ($\sim$0.5–10 mM as opposed to $\sim$20–40 μM in the bloodstream[66]). For example, micelles made of a PCL-*b*-poly[(2,4-dinitrophenyl)thioethyl ethylene phosphate]-*b*-PEG (PCL-*b*-PPE$_{DNPT}$-*b*-PEG) triblock copolymer, crosslinked with disulfide bonds, were found to be stable in circulation but quickly decomposed in intracellular fluid.[65]

Even if the micelle happens to be unstable, its decomposition rate can be reduced by choosing a stiff or bulky core. Toward this end, benzyl groups were introduced to increase the rigidity of hydrophobic cores.[67] Lavasanifar *et al.* synthesized benzyl carboxylate-substituted ε-CL monomers and prepared PEG-*b*-poly(α-benzyl carboxylate ε-caprolactone) (PEG-*b*-PBCL) copolymers.[67] For comparison, they also prepared PEG-*b*-poly(α-carboxyl-ε-caprolactone) (PEG-*b*-PCCL) by further catalytic debenzylation. Their results demonstrated that the stability of micelles with core structures containing aromatic groups (PEG-*b*-PBCL) was higher than that of the parent PEG-PCL micelles and of the PEG-*b*-PCCL micelles. The micelle decomposition rate can also be reduced by crystallizable hydrophobic blocks.[45,68] Another approach is to enhance ionic or hydrogen bonding interactions in the micelle core. For example, polyion complex (PIC) micelles with oppositely charged macromolecules, such as DNA or peptides, are resistant to enzymes in the bloodstream,[69] but they disassemble once the salt concentration rises above a certain threshold.[70] Hedrick *et al.* introduced urea functional groups,[71] while Zhu *et al.* introduced DNA base pairs[72] into block copolymers to show that hydrogen bonding can reduce micelle decomposition rates.

3.2.1.3 Approaches to Achieve Robust Intracellular Release

The chemical forces discussed above that make carriers retain drugs can conflict with the need for a rapid and complete release at the target site. Drugs become active only after liberation from their carriers.[73,74] DOX that was stably bonded to the nanoparticle core of poly(lactic-*co*-glycolic acid) (PLGA)[75] or P(Asp)[76] showed low or even no anticancer activity.[77] The rate of drug release is also very important because tumor cells have intrinsic and acquired drug-resistance mechanisms to remove intracellular drugs,[78,79] *e.g.* as a result of cell-membrane-associated multidrug resistance to efflux drugs[79,80] and cell-specific drug metabolism or detoxification.[81] Tumor cells can also sequestrate some weakly basic drugs in their lysosomes and use biomacromolecules to bind drugs to limit their access to their targets. Thus, it is only the intracellular drug molecules free to bind to their targets that are useful therapeutically. Such free drug concentration in the cytosol, herein referred to as the effective cytosolic drug concentration [D] (effective [D] for short) determines the overall therapeutic efficacy.

Drug carriers that reach tumor cells are generally internalized by endocytosis[82,83] and routed to endosomes and then acidic lysosomes, as shown in Figure 3.7. The internalized carrier can release the drug in one of two possible ways or both: (1) within the lysosome, followed by drug diffusion, as illustrated with the upper path in Figure 3.7, and (2) in the cytosol, following the carrier escape from the lysosome, as illustrated in the lower path in Figure 3.7. For a specific tumor cell, [D] is a function not only of the cellular uptake of the carrier but also of its drug release rate (see Eq. 1 on the figure). If either ends up being "too little, too late," it can prevent reaching an effective [D].

(1)   Intra-lysosome release

The intra-lysosome release mechanism (upper path in Figure 3.7) works for most carriers that can be endocytosed into endosomes/lysosomes. The pH in endosomes decreases progressively, typically near 6 in early endosomes, near 5 in late endosomes, and about 4–5 in lysosomes.[84] This acidic pH and the

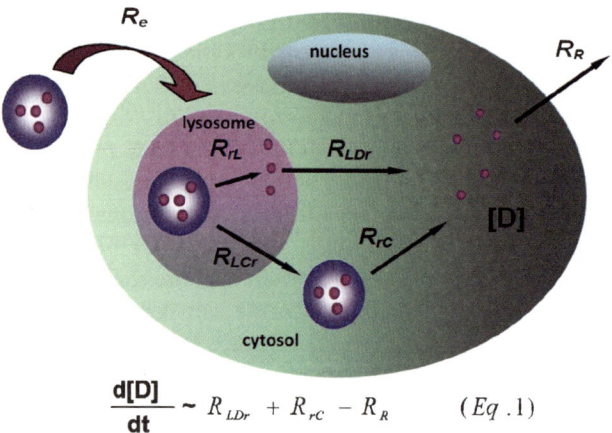

$$\frac{d[D]}{dt} \sim R_{LDr} + R_{rC} - R_R \qquad (Eq.1)$$

Figure 3.7   Cytosolic drug accumulation by drug delivery: [D], effective drug concentration in cytosol; R_e, endocytosis rate of the carrier; R_{rL}, drug release rate of the carrier in lysosomes; R_{rC}, drug release rate of the carrier in cytosol; R_{LDr}, lysosomal drug release rate; R_{LCr}, lysosomal-carrier escape rate; R_R, the overall rate of drug removal by P-gp pumps and drug consumption by other forms of drug resistance. Reprinted with permission from ref. 8. Copyright 2012 Elsevier.

special enzymes in lysosomes can trigger drug release from the carriers into lysosomes.[85] Because the harsh environment of lysosomes can easily degrade drugs sensitive to acid or these enzymes,[86,87] the drug must quickly diffuse out into the cytosol to avoid deactivation.

Polymer–drug conjugates, in which the drugs are conjugated to the polymer carriers *via* lysosomal pH-labile linkers, are the most popular design. Hydrazone and *cis*-aconityl are examples of such a linker.[88,89] Ulbrich *et al.* conjugated DOX to *N*-(2-hydroxypropyl)methacrylamide (HPMA) copolymers *via* this hydro-lytically labile spacer.[90] The results showed a fast DOX release from the polymer at intracellular pH 5, whereas at pH 7.4 the conjugates retained the drug. Recently, they synthesized new biodegradable star conjugates consisting of poly(amido amine) (PAMAM) dendrimer cores and HPMA grafts bearing DOX *via* hydrazone bonds.[89] The *in vitro* cytotoxicity and *in vivo* antitumor activity of all such conjugates were higher than those of classic conjugates. Another example is Wang *et al.*'s dual pH-responsive polymer–drug conjugate PPC-Hyd-DOX-DA, which could respond to the tumor extracellular pH gradients *via* amide bonds and the tumor intracellular pH gradients *via* hydrazone bonds.[91] Lysosomal degradable peptides [*e.g.* glycylphenylalanylleucylglycine (GFLG)], which are cleavable by lysosomal enzymes to release the drugs, are also used for drug conjugation.[74,92] For instance, DOX was conjugated to HPMA copolymers *via* GFLG peptides to form a cleavable HPMA-GFLG-DOX conjugate.[74]

Lysosomal pH has also been used to trigger drug release from pH-sensitive nanoparticles.[93,94] For example, pH-sensitive micelles composed of reducible poly(β-amino ester) (RPAE) cores dissociated rapidly in an acidic environment and at high levels of reducing reagents, inducing fast intracellular release.[94] Carriers with a core made from amine-containing hydrophobic polymers, such as polyhistidine (PHis), can be protonated and thus dissolve in acidic lysosomes, thereby releasing the drug.[95] Our group showed that a rapid cytoplasmic release from carriers could increase the anticancer activity of drugs.[96,97]

The additional advantage of such amine-containing polymers is that they may also have endosomal membrane-disruption activity induced by a "proton sponge" mechanism,[98] and thus disrupt the lysosomal membrane and further release the drug into the cytosol. Some specially designed polyacids, such as poly(propylacrylic acid) (PPAA),[99,100] were shown to disrupt endosomes at pH 6.5 or below, causing the cytosolic release of cargo molecules.

(2) Intra-cytosol release

An alternative to the carriers designed for intra-lysosome release discussed above is carriers designed for intra-cytosol release (lower path in Figure 3.7). Such intra-cytosol-release carriers retain the drug until escape from the endosome/lysosome[101] and then release the drugs into the cytosol, hence avoiding lysosomal drug retention and degradation. This is particularly important in small interfering RNA (siRNA) or gene delivery and thus various approaches have been explored to facilitate the endosomal release of DNA or RNA complexes.[102,103] In this approach the carriers must respond to the lysosomal environment for lysosomal escape and to the cytosolic environment for drug release.

Stealth carriers, such as HPMA[104] and pegylated particles, cannot diffuse through the lysosomal membrane and thus can be retained in the lysosomes for a long time. For instance, PEG-PCL particles were found confined in lysosomes.[105] Thus, they must be functionalized with lysosomal membrane-destabilizing polymers such as PPAA,[99,100] pH-dependent fusogenic peptides,[106,107] or cationic polymers such as polyethylenimine (PEI)[108] or histidine-rich peptides or polymers.[109] For cationic polymers or peptides, on the other hand, it is important first to mask their cationic charges (from primary and secondary amines) at physiological pH, so the carriers can be used for i.v. administration. However, once inside the tumor lysosome, the cationic charges are recovered to lyze the lysosomal membrane for escape. Such a "negative-to-positive charge-reversal" method makes the carrier stealthy in circulation, but enables endosomal lysis, once in lysosomes.[108,110]

Removal of a cleavable PEG layer can also allow lysosomal escape.[111] For instance, a PEG-cleavable lipid, *via* an acid-labile vinyl ether linker, was used for pegylation of (1,2-dioleoyl-*sn*-glycero-3-phosphoethanolamine) (DOPE) liposomes. At acidic lysosomal pH the vinyl ether linker hydrolyzed and the PEG layer was removed from the DOPE liposomes, enabling DOPE, which has excellent fusogenic capacity, to fuse with the lysosomal membrane for escape.[112] Disulfide linkages were also used to detach PEG and make the drug-loaded carriers quickly escape from the endosomes.[113] After the particles were internalized by cells and trapped by endosomes, the PEG layer was removed. The exposed particles interacted with the endosomal membrane, increased the endosomal pressure, or both, resulting in destruction of the endosomal membrane to enable effective endosomal escape.[113]

Most carriers reaching the cytoplasm have already experienced an initial burst release and are in a slow, diffusion-controlled drug release process, *e.g.* nanoparticles with cores made of solid glassy polymers such as PCL or polylactide (PLLA).[114] According to Eq. 1 (see Figure 3.7), such a slow drug release profile may not be able to lead to a high [D] lethal to cancer cells. Thus, carriers responding to cytosolic signals have been developed for faster drug release. The most common is a cytosolic redox signal resulting from an elevated intracellular GSH concentration (~ 10 mM) compared to that in the bloodstream (~ 2 μM).[115] GSH can effectively cleave the disulfide bonds to release conjugated drugs.[66,116] It is thus used to trigger decomposition of micelles with hydrophobic parts linked by disulfide bonds[117] or other carriers crosslinked[118] or gated[119,120] with disulfide linkers. It has also been observed that removal of the PEG corona could increase the drug release rate.[121,122]

3.2.2   *2S*: Stealthy in Circulation and Tumor Penetration *versus* Sticky to Tumor Cells

The second major material challenge is how to impart nanocarriers' stealth ability to circulate in blood for a long time and after extravasation to penetrate deep into the tumor, to the cells away from the blood vessels, but become effectively sticky upon interacting with tumor cells for fast cell internalization.

To be stealthy for a long circulation time in the blood compartments has been recognized as essential for a nanocarrier to achieve passive tumor targeting,[123,124] whereas transport in the tumor tissue after extravasation has been gradually realized in recent years.[125,126] Tumor resistance to anticancer drugs not only involves the cellular and genetic drug resistance mechanisms,[78,127] but also the physiological barriers of solid tumor tissues.[128,129] It is found that tumor drug distribution is not uniform. Drugs are rich in the areas surrounding the blood vessels and the concentration declines sharply away from the blood vessels, owing to the compact structure of tumor tissues.[130] Thus, the most aggressive tumor cells located in these hostile microenvironments (low pH and low pO_2) are actually exposed to few drugs.[7] Moreover, the exposure of those cancer cells to sublethal concentrations of anticancer drugs may facilitate the development of resistance.[7] Therefore, it is important for the nanocarrier to remain stealthy after extravasation for tumor penetration. It can be imagined that a nanocarrier strongly interacting with surrounding cells and matrix will be trapped there and cannot travel a long distance.

3.2.2.1 Approaches to Stealth Surfaces

(1) In circulation

The nanocarrier's stealth character hinges on many factors, including surface properties,[123] size,[131] and even shape.[132,133] In circulation, those with molecular weights below the renal threshold (*e.g.* 40 kDa for PEG) or sizes below 5 nm are rapidly cleared from the blood by glomerular filtration,[86] while those with diameters above 200 nm will be scavenged by the RES, mainly the liver and spleen.[131,134]

Most stealth carriers capable of avoiding opsonization[135] and interaction with the mononuclear phagocyte system (MPS)[131] are made from HPMA,[136,137] PEG, or polysaccharides[138] (*e.g.* heparin[139]). Nanoparticles coated with a layer of these polymers become stealthy by both hydration and steric hindrance.[140] For example, pegylation of particles or liposomes is well established,[135,141] and the DOX-loaded stealth liposome named Doxil® was approved by the FDA for cancer therapy.[142] Huang *et al.* reported that, on 100 nm liposomes pegylated with 1,2-distearoyl-*sn*-glycero-3-phosphoethanolamine-PEG$_{2000}$ (DSPE-PEG$_{2000}$), PEG chains were arranged in a mushroom configuration at a DSPE-PEG fraction less than 4 mol% but in a brush configuration at a DSPE-PEG content greater than 8 mol%.[124] The high density of PEG chains on the liposome surface with the brush configuration was the key to reduce liposome liver sequestration.[124] Discher *et al.* incorporated the PEG brushes onto polymersomes and obtained polymersomes having a blood circulation time two-fold longer than pegylated liposomes.[143] Dai *et al.* pegylated single-wall carbon nanotubes (SWNT) and found that, with the increase of linear PEG chain length from 2 kDa to 5 kDa, the blood circulation time of pegylated SWNTs was significantly extended, but a further increase of the PEG chain length showed no significant effect.[144] Although pegylation reduces the recognition of the carriers by the MPS system and thereby extends their blood circulation time, the "accelerated blood clearance (ABC)" phenomenon was observed upon repeated injection of pegylated liposomes[145,146] due to IgM bound to pegylated liposomes secreted into the bloodstream after the first dose.[147] Such an immune reaction against the pegylated liposomes occurred in the spleen at least 2–3 days after the first administration.[145,146]

The carrier shape is also recognized as an important parameter that can substantially affect the blood circulation time. In fact, Mitragotri *et al.* reported that the particle shape, not size, played a dominant role in phagocytosis of polystyrene (PS) particles of various sizes and shapes: the rod-like particles entered the cells much faster.[148] Discher *et al.* found that flexible worm-like micelles efficiently evaded the RES and circulated in the blood for a week,[149,150] much longer than spherical micelles. Dai *et al.* found that carbon nanotubes pegylated with long PEG chains exhibited a long blood circulation time ($t_{1/2}$ = 22.1 h) upon intravenous injection into mice.[151] All these studies suggest that particle phagocytosis can be inhibited by minimizing its size-normalized curvature.[148,152] Thus, particle shape is an important variable to make it remain stealthy in circulation long enough for enhanced tumor accumulation.[149,150,153]

(2) In tumor tissue

Solid tumors are characteristic of poorly structured blood vessels,[154] a stiff extracellular matrix (ECM),[155–157] tightly packed cells,[158] high interstitial fluid pressure (IFP),[159,160] and drug metabolism and binding.[126] Together they impose strong diffusion barriers to nanocarriers, and even small molecules (Figure 3.8).[125,161]

For instance, it seems intuitive that as long as a nanocarrier extravasates from tumor blood capillaries and releases the carried small molecular drugs, the drug molecules will diffuse deep into the tumor tissue. Actually, free drugs, either hydrophobic or carrying positive charges, cannot migrate far from the nanocarrier due to their avid binding.[162] Diffusion of larger macromolecules, such as bovine serum albumin (BSA, 68 kDa, 9 nm in diameter) and immunoglobulin G (IgG, 150 kDa, 11 nm in diameter), in the tumor ECM is also hindered compared to that in buffered saline.[163] After extravasation, dextrans with a molecular weight between 40 and 70 kDa (and a diameter of 11.2–14.6 nm) were observed to be concentrated near the vascular surface.[130] Apparently, nanocarriers, which are larger in size than BSA and dextrans, are likely to face greater difficulties in tumor penetration.[161] Chan *et al.* systematically examined the effect of nanoparticle size on tumor penetration using sub-100 nm pegylated gold nanoparticles.[164] As expected, larger nanoparticles appeared to stay near the vasculature while smaller nanoparticles (20 nm in diameter) rapidly diffused into the tumor matrix (Figure 3.9). Similarly, Lee *et al.* demonstrated that PEG-PCL micelles with a mean diameter of 25 nm diffused further away from the blood vessels compared to those with diameters of 60 nm, which mainly remained in the perivascular

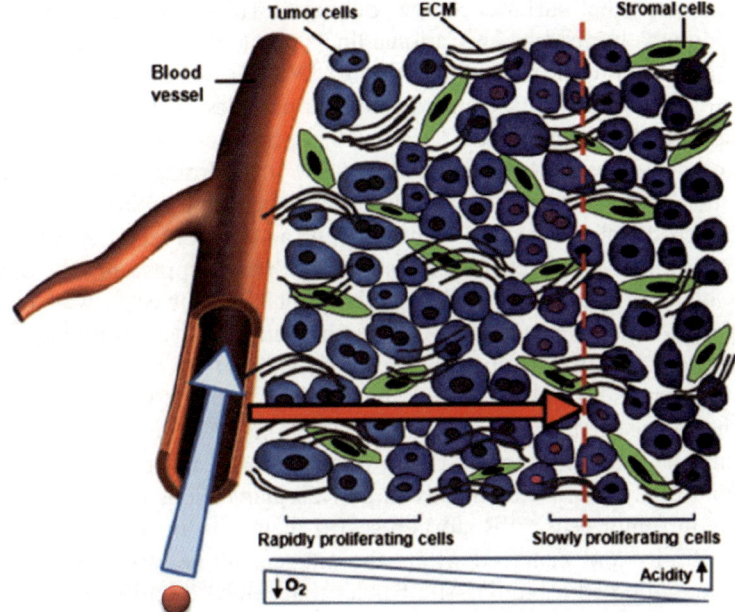

Nanocarrier

Figure 3.8 Scheme of solid tumor tissue which is characteristic of stiff ECM and compact tumor cells. Adapted with permission from ref.125. Copyright 2012 Elsevier.

regions.[165] Furthermore, Kataoka *et al*,[166] and Liang *et al*.[167] also proved that small-sized nanocarriers were essential for improved diffusion.[168] However, small-sized nanocarriers possess a high probability to be fast cleared during the circulation, as mentioned above.

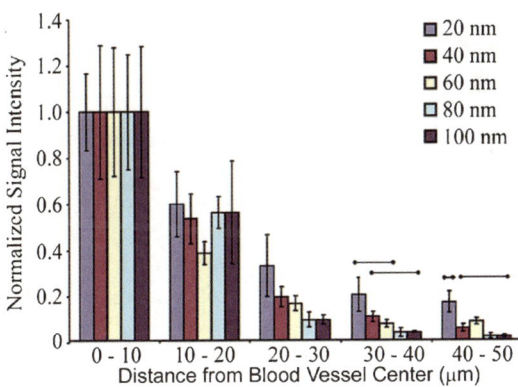

Figure 3.9 Size-dependent penetration of nanoparticles within tumor tissues. Reprinted with permission from ref. 164. Copyright 2009 American Chemical Society.

Besides size, the surface charge of nanocarriers also influences their penetration into tumor tissues. Cationic liposomes (150 nm) were observed not able to travel far into the tumor interstitium.[169] Recently, Forbes *et al.* compared the penetration of oppositely charged gold nanoparticles (+30 *vs.* −36 mV, 6 nm) into cylindroidal cell aggregates.[170] Cationic nanoparticles were taken up by the proliferating cells on the periphery of the cylindroids, whereas anionic nanoparticles were better at penetrating the extracellular matrix and entered hypoxic necrotic cells in the core of the mass. As a matter of fact, the extracellular matrix presents as an effective electrostatic bandpass, suppressing the diffusive motion of both positively and negatively charged objects, which allows uncharged particles to easily diffuse through while effectively trapping charged particles (Figure 3.10).[171] Jain *et al.* demonstrated that the optimal particles for delivery to tumors should be neutral after exiting the blood vessels.[172]

Another issue that needs addressing is affinity.[173] The affinity plays an important role in antibody-based tumor targeting nanocarriers. It was visualized that the antibody distributed mostly in perivascular regions rather than homogeneously in tumor cells.[174] Reports revealed there was an inverse relationship between affinity and penetration, *i.e.* the antigen–antibody interaction in the tumor tissue imposed a binding-site barrier that retarded antibody penetration and caused a heterogeneous distribution.[175–177] The higher the affinity of binding and the higher antigen density caused fewer free molecules to be able to penetrate farther into the tumor interstitium.[175,176] Increasing the antibody dose gave better penetration and more uniform distribution.[175]

Therefore, it is clear that to deliver a sufficient drug concentration to the tumor center region lacking vascular perfusion, where the most aggressive and resistant cells reside, the nanocarrier should not release the carried drug after

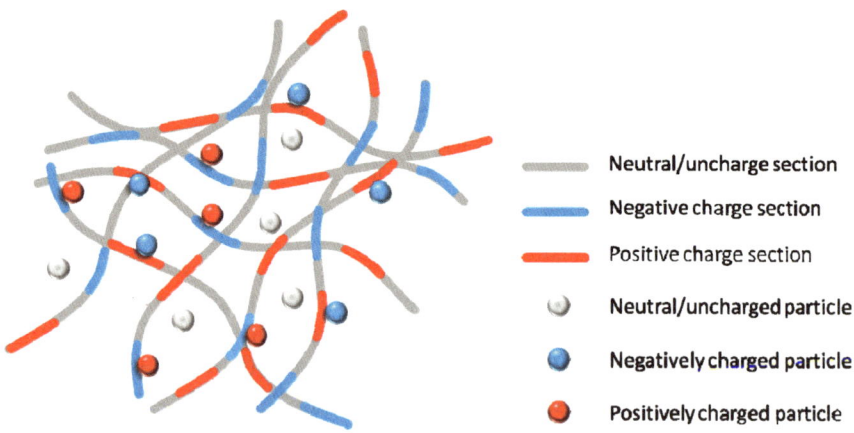

Figure 3.10   Scheme of the ECM exerts the filtering function in tumor tissue. Charged particles (red, blue) are trapped in the respective region of opposite charge (blue, red), while neutral particles (gray) can diffuse nearly unhindered.

extravasation but should further diffuse deep into the tumor. This requires the nanocarrier to remain slippery and have as small a size as possible. Thus, it is better for the nanocarrier to be neutral and not to present any binding groups (including targeting groups) until reaching the center of the tumor.

3.2.2.2 Approaches to Becoming Sticky to Tumor Cells for Cellular Uptake

After reaching the targeted region the nanocarrier should efficiently enter the cells for drug release. Now the same properties that impart stealth to the nanocarrier prevent it from cellular uptake by tumor cells. Nanocarriers that are negatively charged will be repelled from the cell membrane due to the electrostatic repulsion. The PEG corona of pegylated polymeric micelles or liposomes retards their interaction with cell membranes due to steric hindrance. Thus, once in the tumor, the carrier must become cell-binding or sticky to targeting tumor cells for fast cellular uptake.[178] The challenge is how to reconcile these two opposite requirements, stealth for circulation and diffusion *versus* sticky for targeting. For instance, it is well known that positively charged carriers reliably stick to cell membranes due to electrostatic adsorption triggering fast cellular uptake, but positively charged carriers are not suitable for *in vivo* applications because they are systemically toxic[179] and have short circulation times.[180]

One strategy to convert a carrier from stealth circulation to sticky targeting is to equip it with PEG groups that are cleavable upon encountering a tumor-specific stimulus. Once the PEG chains are removed, the bare particle can be adsorbed onto the cell membrane. Toward this end, Thompson *et al.* prepared acid labile PEG-conjugated vinyl ether lipids to stabilize fusogenic DOPE liposomes.[181] At lower pH, the PEG layer was removed by the acid-catalyzed hydrolysis of the vinyl ether bond, triggering membrane fusion. Similarly, Harashima *et al.* connected PEG to the lipid through a matrix of metalloproteinase (MMP)-cleavable peptide.[182,183] MMP is overexpressed in tumor-tissue angiogenesis, invasion, and metastasis[184] and thus the peptide can be degraded quickly in tumors. They prepared a multifunctional envelope-type nano device (MEND) using the PEG-peptide lipid and found that pDNA expression was dependent on the MMP expression level in the host cell.

Positive charges can promote carrier adsorption on the negatively charged membrane and hence trigger adsorption-mediated endocytosis. Thus, an alternative is to use tumor extracellular acidity to impart positive charges to the carrier by a "charge-reversal" technique (illustrated in Figure 3.11). Amine-containing carriers, such as PCL-*b*-PEI,[108] poly(L-lysine) (PLL),[185] and PAMAM dendrimers,[110] were amidized to acid-labile β-carboxylic acid amides to make them negatively charged at physiological pH. Once in weakly acidic tumor extracellular fluid, the amides hydrolyzed and regenerated the amines with cationic charges, which led to fast cellular uptake (Figure 3.11A). In yet another example, a pH-responsive layer becomes positively charged at tumor extracellular acidity but collapses, forming a middle layer, at neutral pH (Figure 3.11B).[16] Bae *et al.* reported tumor extracellular pH-triggered TAT-presenting micelles. The TAT moieties were anchored to a PEG micelle corona and shielded at pH >7.0 by their electrostatic complexation with poly(methacryloyl sulfadimethoxine) (anionic PSD)-PEG (PSD-*b*-PEG) diblock copolymer. At pH 6.6, however, PSD turned to a nonionized form and fell off the TAT, exposing it and enabling the micelle a fast cellular uptake.[186] Another design was to anchor TAT onto the PEG corona through a pH-sensitive PHis spacer. At pH 7.4, the PHis was water insoluble, which kept the TAT moieties buried in the PEG corona. At pH lower than 7.2, however, ionization of the PHis spacer made it water soluble, which stretched it, exposing the TAT on the corona surface.[187]

The most common approach enabling a carrier to become sticky to the cell membrane is to decorate it with a ligand whose receptors are overexpressed on the cancer-cell membrane. The ligand–receptor binding enables receptor-mediated endocytosis, promoting cellular uptake.[11,188] Only a few ligands are needed for rapid internalization.[189] More ligand groups can theoretically increase uptake, but a high surface ligand density may make the carrier less stealthy as a result of opsonization-mediated clearance.[190] Many examples of targeting ligands include folic acid,[191] peptides,[192–194] antibodies,[195–197] transferrin,[198] aptamers,[199,200] and other moieties[201] that have been tested and subsequently reviewed.[2,202]

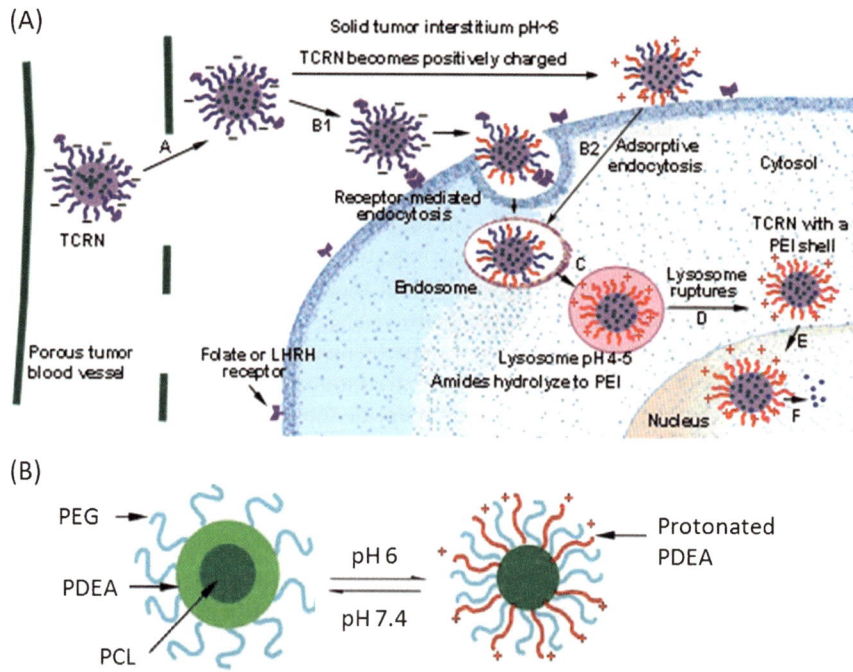

Figure 3.11   (A) The charge-reversal concept for drug delivery. Reprinted with permission from ref. 14. Copyright 2010 Elsevier; (B) The pH-responsive three-layered nanoparticles (3LNPs). Reprinted with permission from ref. 16. Copyright 2008 American Institute of Chemical Engineers.

3.3 The Material Excipientability and Production Process Scale-Up Ability

The *2R2S* capability for nanocarriers discussed in the previous sections determines the adsorption, distribution, metabolism, and excretion (ADME) of the carried drug. Such a nanocarrier simultaneously having *2R2S* capability can deliver a high cytosolic drug concentration and give rise to high therapeutic efficacy. However, this is not sufficient for it to be translational.[203–205] The nanocarrier itself should also have proper ADME. According to Choi and Frangioni, safety and clearance (renal or hepatic) and a proper stealth surface should be included among the basic criteria for clinical translation of formulation/materials administered to humans,[205] "from the benchtop to the bedside" translation. Thus, a nanocarrier must meet the requirements for the pharmaceutical excipient for i.v. uses. For simplicity, this ability of the nanocarrier material(s) to be used or approved to be an excipient, herein denoted as excipientability, is the second element for a nanocarrier to be translational (see Figure 3.2). It goes without saying that the production of the nanocarrier and the resulting nanomedicine should be able to be scaled-up and establish the required GMP, or scale-up ability, for short. Some of these important points of the two key elements are summarized as follows.

1) Safety. The nanocarrier itself should have proper ADME and no nanotoxicity, and should be nontoxic and easy to excrete completely from the body *via* the liver (into bile) or the kidneys (into urine) or both. This is because retention of polymers or nanosized materials in the body, even inert polymers like polyvinylpyrrolidone (PVP),[206–208] can cause health problems. The threshold for rapid renal excretion is about 5.5 nm in hydrodynamic diameter. This corresponds to a molecular weight of about ~ 45 kDa for HPMA[209] and 40 kDa for PEG.[86]

2) Approval. In order to expedite and increase the probability of the approval success, the carrier should have a clear and simple structure with known degradation products. An even better case would be that it is made of FDA-approved building blocks.

3) Production scale-up. This involves the feasibility of making large volumes of consistently reproducible quality to establish GMP. For instance, because the molecular weight of a polymer–drug conjugate strongly affects its pharmacokinetics, the polymer itself must have consistently low polydispersity and reproducible average molecular weight from batch to batch. The same applies to drug-loaded micelles made of block copolymers, such as PEG-PCL, in addition to reproducible particle size, particle-size distribution, and drug-loading efficiency and content. As the micelle structure becomes more and more complicated, the number of quality control parameters drastically increases,[14,210] which makes it more and more difficult to produce an acceptably consistent formulation. Also, although not crucial to clinical success, it is also worth considering a high, ideally close to 100%, drug-loading

efficiency to simplify the manufacturing process and minimize losses of these very expensive anticancer drugs.

4) High drug-loading content. In current commercial formulations, the drug-loading content tends to be on the low side.[211–213] High drug-loading contents are needed to minimize the body's exposure to excipient carrier matter, even if it is biocompatible and relatively benign. For instance, PEG-containing liposomal carriers may induce acute immune toxicity, manifested in hypersensitivity reactions (HSRs).[214,215]

3.4   Challenges of Rational Design for Translational Nanomedicine

With the above analysis in mind, it is clear that the key to translational nanomedicine is to develop nanocarriers with optimal *2R2S* capability, excipientability, and scale-up ability.

As for the nanocarrier *2R2S* capability, we still do not have ones that can *fully and simultaneously* achieve the *2R2S* capability, despite a large volume of the scientific literature on each topic separately, or on various subsets of them, giving rise to unsatisfied therapeutic efficacy and side effects. As a consequence, a particular problem of those systems is that a large majority doses of the drugs are still sequestrated in the liver or spleen, even though the tumor drug accumulations are indeed enhanced compared to free drugs.[133,216] For instance, the PF-PTX micelles[217] and IT-101 CPT conjugates[218] give drug accumulation in tumors much better than Taxol® and CPT, respectively, but the total amounts of drugs accumulated in the liver were still about 4.5 and 3.5 times of those in tumors. In many cases, only a few percent of the injected drugs were in the tumors. Thus, for many nanomedicine systems, liver toxicity is the killer for further developments. Other necessities are how to achieve effective cellular uptake of the nanocarriers once in the tumor and robust intracellular release. Delayed or insufficient intracellular release directly leads to lower cytotoxicity than the free drugs.[219,220]

The material excipientability of nanocarriers and the production scale-up ability of the nanocarriers and their nanomedicine systems are equally important. For instance, a large variety of inorganic nanomaterials and sophisticated polymeric nanostructures have been proposed and investigated as nanocarriers for cancer drug delivery. These studies provide useful proof-of-concepts and rich insights into various aspects of cancer drug delivery essential to the design of nanocarriers towards *2R2S* capability, but those aimed at clinical applications must comprehensively design and characterize their materials, nanosize effects, and scale-up ability. Of the three, the material is the basic concern for a translational nanocarrier. If the material used for the nanocarrier is not proper for *in vivo* clinical uses (for instance, inherently toxic or non-clearable from the body), the resulting nanocarrier, even with perfect nanosize effects and *2R2S* capability, would not be able, or take an impractically long time, to be translated into clinics. Thus, except for proof-of-concepts, it is better to look into these issues early at the bench in order for a successful nanocarrier to move forward quickly.

3.5 Conclusion

The challenge to develop truly translational nanocarriers and nanomedicine is to use excipientable materials and processes of scale-up ability to produce nanocarriers with optimal *2R2S* capability. While research aimed at proof-of-concepts remains important, it is important to increasingly focus on comprehensive approaches or systems that include *all* the three key elements as early as possible in the innovation chain, to speed up developments of translational nanomedicine.

References

1. R. Tong, D. A. Christian, L. Tang, H. Cabral, J. R. Baker, Jr., K. Kataoka, D. E. Discher and J. Cheng, *MRS Bull.*, 2009, **34**, 422–431.
2. F. Danhier, O. Feron and V. Preat, *J. Controlled Release*, 2010, **148**, 135–146.
3. M. E. R. O'Brien, N. Wigler, M. Inbar, R. Rosso, E. Grischke, A. Santoro, R. Catane, D. G. Kieback, P. Tomczak, S. P. Ackland, F. Orlandi, L. Mellars, L. Alland, C. Tendler and C. B. C. S. Grp, *Ann. Oncol.*, 2004, **15**, 440–449.
4. W. J. Gradishar, S. Tjulandin, N. Davidson, H. Shaw, N. Desai, P. Bhar, M. Hawkins and J. O'Shaughnessy, *J. Clin. Oncol.*, 2005, **23**, 7794–7803.
5. V. P. Torchilin, *Eur. J. Pharm. Sci.*, 2000, **11**, S81–S91.
6. H. Maeda, J. Wu, T. Sawa, Y. Matsumura and K. Hori, *J. Controlled Release*, 2000, **65**, 271–284.
7. C. Wong, T. Stylianopoulos, J. Cui, J. Martin, V. P. Chauhan, W. Jiang, Z. Popovic, R. K. Jain, M. G. Bawendi and D. Fukumura, *Proc. Natl. Acad. Sci. U. S. A.*, 2011, **108**, 2426–2431.
8. Q. Sun, M. Radosz and Y. Shen, *J. Controlled Release*, 2012, **164**, 156–169.
9. M. Ye, S. Kim and K. Park, *J. Controlled Release*, 2010, **146**, 241–260.
10. F. Mohamed and C. F. van der Walle, *J. Pharm. Sci.*, 2008, **97**, 71–87.
11. D. Peer, J. M. Karp, S. Hong, O. C. FaroKhzad, R. Margalit and R. Langer, *Nat. Nanotechnol.*, 2007, **2**, 751–760.
12. G. Shazly, T. Nawroth and P. Langguth, *Dissol. Techn.*, 2008, **15**, 7–10.
13. B. Narasimhan and R. Langer, *J. Controlled Release*, 1997, **47**, 13–20.
14. Z. L. Tyrrell, Y. Shen and M. Radosz, *Prog. Polym. Sci.*, 2010, **35**, 1128–1143.
15. C. O. Rangel-Yagui, A. Pessoa and L. C. Tavares, *J. Pharm. Pharm. Sci.*, 2005, **8**, 147–163.
16. Y. Shen, Y. Zhan, J. Tang, P. Xu, P. A. Johnson, M. Radosz, E. A. Van Kirk and W. J. Murdoch, *AIChE J.*, 2008, **54**, 2979–2989.

17. Z. L. Tyrrell, Y. Shen and M. Radosz, *J. Phys. Chem. C*, 2011, **115**, 11951–11956.
18. Z. L. Tyrrell, Y. Shen and M. Radosz, *Macromolecules*, 2012, **45**, 4809–4817.
19. M. Yokoyama, G. S. Kwon, T. Okano, Y. Sakurai, T. Seto and K. Kataoka, *Bioconjugate Chem.*, 1992, **3**, 295–301.
20. M. Yokoyama, S. Inoue, K. Kataoka, N. Yui and Y. Sakurai, *Makromol. Chem. Rapid Commun.*, 1987, **8**, 431–435.
21. A. Ponta and Y. Bae, *Pharm. Res.*, 2010, **27**, 2330–2342.
22. Y. Bae, A. W. G. Alani, N. C. Rockich, T. S. Z. C. Lai and G. S. Kwon, *Pharm. Res.*, 2010, **27**, 2421–2432.
23. S. Aryal, C.-M. J. Hu and L. Zhang, *ACS Nano*, 2010, **4**, 251–258.
24. R. Tong and J. Cheng, *J. Am. Chem. Soc.*, 2009, **131**, 4744–4754.
25. A. Lavasanifar, J. Samuel and G. S. Kwon, *Adv. Drug Delivery Rev.*, 2002, **54**, 169–190.
26. C. Wei, J. Guo and C. Wang, *Macromol. Rapid Commun.*, 2011, **32**, 451–455.
27. L. Wong, M. Kavallaris and V. Bulmus, *Polym. Chem.*, 2011, **2**, 385–393.
28. Y. Shen, E. Jin, B. Zhang, C. J. Murphy, M. Sui, J. Zhao, J. Wang, J. Tang, M. Fan, E. Van Kirk and W. J. Murdoch, *J. Am. Chem. Soc.*, 2010, **132**, 4259–4265.
29. Y. Q. Shen, H. D. Tang, C. J. Murphy, B. Zhang, M. H. Sui, E. A. Van Kirk, X. W. Feng and W. J. Murdoch, *Nanomedicine (London, U. K.)*, 2010, **5**, 855–865.
30. H. S. Yoo, E. A. Lee and T. G. Park, *J. Controlled Release*, 2002, **82**, 17–27.
31. M. J. Joralemon, R. K. O'Reilly, C. J. Hawker and K. L. Wooley, *J. Am. Chem. Soc.*, 2005, **127**, 16892–16899.
32. Y. T. Li, B. S. Lokitz, S. P. Armes and C. L. McCormick, *Macromolecules*, 2006, **39**, 2726–2728.
33. X.-B. Xiong, A. Falamarzian, S. M. Garg and A. Lavasanifar, *J. Controlled Release*, 2011, **155**, 248–261.
34. M. Iijima, Y. Nagasaki, T. Okada, M. Kato and K. Kataoka, *Macromolecules*, 1999, **32**, 1140–1146.
35. X. Jiang, J. Zhang, Y. Zhou, J. Xu and S. Liu, *J. Polym. Sci., Part A: Polym. Chem.*, 2008, **46**, 860–871.
36. J. Xiong, F. Meng, C. Wang, R. Cheng, Z. Liu and Z. Zhong, *J. Mater. Chem.*, 2011, **21**, 5786–5794.
37. S. M. Garg, X.-B. Xiong, C. Lu and A. Lavasanifar, *Macromolecules*, 2011, **44**, 2058–2066.
38. F. Meng, W. E. Hennink and Z. Zhong, *Biomaterials*, 2009, **30**, 2180–2198.
39. N. Abdullah Al, H. Lee, Y. S. Lee, K. D. Lee and S. Y. Park, *Macromol. Biosci.*, 2011, **11**, 1264–1271.
40. D. E. Discher and F. Ahmed, *Annu. Rev. Biomed. Eng.*, 2006, **8**, 323–341.

41. S. Kim, Y. Shi, J. Y. Kim, K. Park and J.-X. Cheng, *Exp. Opin. Drug Delivery*, 2010, **7**, 49–62.
42. T. Haliloglu, I. Bahar, B. Erman and W. L. Mattice, *Macromolecules*, 1996, **29**, 4764–4771.
43. N. Rapoport, *Prog. Polym. Sci.*, 2007, **32**, 962–990.
44. K. K. Jette, D. Law, E. A. Schmitt and G. S. Kwon, *Pharm. Res.*, 2004, **21**, 1184–1191.
45. J. Liu, F. Zeng and C. Allen, *Eur. J. Pharm. Biopharm.*, 2007, **65**, 309–319.
46. X. C. Zhang, J. K. Jackson and H. M. Burt, *Int. J. Pharm.*, 1996, **132**, 195–206.
47. H. M. Burt, X. C. Zhang, P. Toleikis, L. Embree and W. L. Hunter, *Colloids Surf., B*, 1999, **16**, 161–171.
48. R. Savic, T. Azzam, A. Eisenberg and D. Maysinger, *Langmuir*, 2006, **22**, 3570–3578.
49. H. Chen, S. Kim, W. He, H. Wang, P. S. Low, K. Park and J.-X. Cheng, *Langmuir*, 2008, **24**, 5213–5217.
50. S. M. Li, H. Garreau, B. Pauvert, J. McGrath, A. Toniolo and M. Vert, *Biomacromolecules*, 2002, **3**, 525–530.
51. C. Chen, C. H. Yu, Y. C. Cheng, P. H. F. Yu and M. K. Cheung, *Biomaterials*, 2006, **27**, 4804–4814.
52. Y. H. Bae and H. Yin, *J. Controlled Release*, 2008, **131**, 2–4.
53. H. Chen, S. Kim, L. Li, S. Wang, K. Park and J.-X. Cheng, *Proc. Natl. Acad. Sci. U. S. A.*, 2008, **105**, 6596–6601.
54. J. C. Ha, S. Y. Kim and Y. M. Lee, *J. Controlled Release*, 1999, **62**, 381–392.
55. S. Y. Kim, J. C. Ha and Y. M. Lee, *J. Controlled Release*, 2000, **65**, 345–358.
56. T. A. Diezi, Y. Bae and G. S. Kwon, *Mol. Pharmaceutics*, 2010, **7**, 1355–1360.
57. X. T. Shuai, T. Merdan, A. K. Schaper, F. Xi and T. Kissel, *Bioconjugate Chem.*, 2004, **15**, 441–448.
58. C. J. Rijcken, C. J. Snel, R. M. Schiffelers, C. F. van Nostrum and W. E. Hennink, *Biomaterials*, 2007, **28**, 5581–5593.
59. P. S. Xu, H. D. Tang, S. Y. Li, J. Ren, E. Van Kirk, W. J. Murdoch, M. Radosz and Y. Q. Shen, *Biomacromolecules*, 2004, **5**, 1736–1744.
60. X. Xu, J. D. Flores and C. L. McCormick, *Macromolecules*, 2011, **44**, 1327–1334.
61. J. Jiang, B. Qi, M. Lepage and Y. Zhao, *Macromolecules*, 2007, **40**, 790–792.
62. V. Torchilin, *Eur. J. Pharm. Biopharm.*, 2009, **71**, 431–444.
63. M. Stubbs, P. M. J. McSheehy and J. R. Griffiths, *Adv. Enzyme Regul.*, 1999, **39**, 13–30.
64. T. Xing, B. Lai, X. Ye and L. Yan, *Macromol. Biosci.*, 2011, **11**, 962–969.

65. Y.-C. Wang, Y. Li, T.-M. Sun, M.-H. Xiong, J. Wu, Y.-Y. Yang and J. Wang, *Macromol. Rapid Commun.*, 2010, **31**, 1201–1206.
66. G. Saito, J. A. Swanson and K. D. Lee, *Adv. Drug Delivery Rev.*, 2003, **55**, 199–215.
67. A. Mahmud, X.-B. Xiong and A. Lavasanifar, *Macromolecules*, 2006, **39**, 9419–9428.
68. C. Allen, D. Maysinger and A. Eisenberg, *Colloids Surf., B*, 1999, **16**, 3–27.
69. K. Kataoka, A. Harada and Y. Nagasaki, *Adv. Drug Delivery Rev.*, 2001, **47**, 113–131.
70. A. V. Kabanov, T. K. Bronich, V. A. Kabanov, K. Yu and A. Eisenberg, *Macromolecules*, 1996, **29**, 6797–6802.
71. S. H. Kim, J. P. K. Tan, F. Nederberg, K. Fukushima, J. Colson, C. Yang, A. Nelson, Y.-Y. Yang and J. L. Hedrick, *Biomaterials*, 2010, **31**, 8063–8071.
72. D. Wang, Y. Su, C. Jin, B. Zhu, Y. Pang, L. Zhu, J. Liu, C. Tu, D. Yan and X. Zhu, *Biomacromolecules*, 2011, **12**, 1370–1379.
73. J. Kopecek, P. Kopeckova, T. Minko, Z. R. Lu and C. M. Peterson, *J. Controlled Release*, 2001, **74**, 147–158.
74. A. Malugin, P. Kopeckova and J. Kopecek, *J. Controlled Release*, 2007, **124**, 6–10.
75. H. S. Yoo, K. H. Lee, J. E. Oh and T. G. Park, *J. Controlled Release*, 2000, **68**, 419–431.
76. M. Yokoyama, S. Fukushima, R. Uehara, K. Okamoto, K. Kataoka, Y. Sakurai and T. Okano, *J. Controlled Release*, 1998, **50**, 79–92.
77. M. Shahin and A. Lavasanifar, *Int. J. Pharm.*, 2010, **389**, 213–222.
78. R. Agarwal and S. B. Kaye, *Nat. Rev. Cancer*, 2003, **3**, 502–516.
79. M. M. Gottesman, *Annu. Rev. Med.*, 2002, **53**, 615–627.
80. E. V. Batrakova and A. V. Kabanov, *J. Controlled Release*, 2008, **130**, 98–106.
81. M. Michael and M. M. Doherty, *J. Clin. Oncol.*, 2005, **23**, 205–229.
82. A. M. Kaufmann and J. P. Krise, *J. Pharm. Sci.*, 2007, **96**, 729–746.
83. G. Sahay, D. Y. Alakhova and A. V. Kabanov, *J. Controlled Release*, 2010, **145**, 182–195.
84. R. M. Steinman, I. S. Mellman, W. A. Muller and Z. A. Cohn, *J. Cell Biol.*, 1983, **96**, 1–27.
85. S. Ganta, H. Devalapally, A. Shahiwala and M. Amiji, *J. Controlled Release*, 2008, **126**, 187–204.
86. K. D. Jensen, A. Nori, M. Tijerina, P. Kopeckova and J. Kopecek, *J. Controlled Release*, 2003, **87**, 89–105.
87. V. P. Torchilin, *Annu. Rev. Biomed. Eng.*, 2006, **8**, 343–375.
88. B. Rihova, T. Etrych, M. Sirova, L. Kovar, O. Hovorka, M. Kovar, A. Benda and K. Ulbrich, *Mol. Pharmaceutics*, 2010, **7**, 1027–1040.
89. T. Etrych, L. Kovar, J. Strohalm, P. Chytil, B. Rihova and K. Ulbrich, *J. Controlled Release*, 2011, **154**, 241–248.

90. K. Ulbrich, T. Etrych, P. Chytil, M. Jelinkova and B. Rihova, *J. Controlled Release*, 2003, **87**, 33–47.

91. J.-Z. Du, X.-J. Du, C.-Q. Mao and J. Wang, *J. Am. Chem. Soc.*, 2011, **133**, 17560–17563.

92. Y. Shiose, H. Kuga, H. Ohki, M. Ikeda, F. Yamashita and M. Hashida, *Bioconjugate Chem.*, 2009, **20**, 60–70.

93. X. Huang, Y. Xiao and M. Lang, *J. Colloid Interface Sci.*, 2011, **364**, 92–99.

94. J. Chen, X. Qiu, J. Ouyang, J. Kong, W. Zhong and M. M. Q. Xing, *Biomacromolecules*, 2011, **12**, 3601–3611.

95. E. S. Lee, K. Na and Y. H. Bae, *J. Controlled Release*, 2005, **103**, 405–418.

96. P. S. Xu, E. A. Van Kirk, W. J. Murdoch, Y. H. Zhan, D. D. Isaak, M. Radosz and Y. Q. Shen, *Biomacromolecules*, 2006, **7**, 829–835.

97. P. S. Xu, E. A. Van Kirk, S. Y. Li, W. J. Murdoch, J. Ren, M. D. Hussain, M. Radosz and Y. Q. Shen, *Colloids Surf., B*, 2006, **48**, 50–57.

98. M. Belting, S. Sandgren and A. Wittrup, *Adv. Drug Delivery Rev.*, 2005, **57**, 505–527.

99. T. R. Kyriakides, C. Y. Cheung, N. Murthy, P. Bornstein, P. S. Stayton and A. S. Hoffman, *J. Controlled Release*, 2002, **78**, 295–303.

100. R. A. Jones, C. Y. Cheung, F. E. Black, J. K. Zia, P. S. Stayton, A. S. Hoffman and M. R. Wilson, *Biochem. J.*, 2003, **372**, 65–75.

101. A. K. Varkouhi, M. Scholte, G. Storm and H. J. Haisma, *J. Controlled Release*, 2011, **151**, 220–228.

102. R. F. Minchin and S. Yang, *Expert Opin. Drug Delivery*, 2010, **7**, 331–339.

103. J. G. Huang, T. Leshuk and F. X. Gu, *Nano Today*, 2011, **6**, 478–492.

104. A. Nori and J. Kopecek, *Adv. Drug Delivery Rev.*, 2005, **57**, 609–636.

105. R. Savic, L. B. Luo, A. Eisenberg and D. Maysinger, *Science*, 2003, **300**, 615–618.

106. T. Wang, S. Yang, V. A. Petrenko and V. P. Torchilin, *Mol. Pharmaceutics*, 2010, **7**, 1149–1158.

107. H. Hatakeyama, E. Ito, H. Akita, M. Oishi, Y. Nagasaki, S. Futaki and H. Harashima, *J. Controlled Release*, 2009, **139**, 127–132.

108. P. Xu, E. A. Van Kirk, Y. Zhan, W. J. Murdoch, M. Radosz and Y. Shen, *Angew. Chem. Int. Ed.*, 2007, **46**, 4999–5002.

109. C. Pichon, C. Goncalves and P. Midoux, *Adv. Drug Delivery Rev.*, 2001, **53**, 75–94.

110. Y. Shen, Z. Zhuo, M. Sui, J. Tang, P. Xu, E. A. Van Kirk, W. J. Murdoch, M. Fan and M. Radosz, *Nanomedicine (London, U. K.)*, 2010, **5**, 1205–1217.

111. B. Romberg, W. E. Hennink and G. Storm, *Pharm. Res.*, 2008, **25**, 55–71.

112. J. A. Boomer, M. M. Qualls, H. D. Inerowicz, R. H. Haynes, V. S. Patri, J.-M. Kim and D. H. Thompson, *Bioconjugate Chem.*, 2009, **20**, 47–59.

113. S. Takae, K. Miyata, M. Oba, T. Ishii, N. Nishiyama, K. Itaka, Y. Yamasaki, H. Koyama and K. Kataoka, *J. Am. Chem. Soc.*, 2008, **130**, 6001–6009.
114. Z. P. Zhang and S. S. Feng, *Biomacromolecules*, 2006, **7**, 1139–1146.
115. D. P. Jones, J. L. Carlson, P. S. Samiec, P. Sternberg, V. C. Mody, R. L. Reed and L. A. S. Brown, *Clin. Chim. Acta*, 1998, **275**, 175–184.
116. Y. E. Kurtoglu, R. S. Navath, B. Wang, S. Kannan, R. Romero and R. M. Kannan, *Biomaterials*, 2009, **30**, 2112–2121.
117. J.-H. Ryu, R. Roy, J. Ventura and S. Thayumanavan, *Langmuir*, 2010, **26**, 7086–7092.
118. R. Cheng, F. Feng, F. Meng, C. Deng, J. Feijen and Z. Zhong, *J. Controlled Release*, 2011, **152**, 2–12.
119. A. M. Sauer, A. Schlossbauer, N. Ruthardt, V. Cauda, T. Bein and C. Braeuchle, *Nano Lett.*, 2010, **10**, 3684–3691.
120. H. Kim, S. Kim, C. Park, H. Lee, H. J. Park and C. Kim, *Adv. Mater.*, 2010, **22**, 4280–4283.
121. H.-Y. Wen, H.-Q. Dong, W.-J. Xie, Y.-Y. Li, K. Wang, G. M. Pauletti and D.-L. Shi, *Chem. Commun.*, 2011, **47**, 3550–3552.
122. H. Sun, B. Guo, R. Cheng, F. Meng, H. Liu and Z. Zhong, *Biomaterials*, 2009, **30**, 6358–6366.
123. N. T. Huynh, E. Roger, N. Lautram, J.-P. Benoit and C. Passirani, *Nanomedicine (London, U. K.)*, 2010, **5**, 1415–1433.
124. S.-D. Li and L. Huang, *J. Controlled Release*, 2010, **145**, 178–181.
125. M. Yu and I. F. Tannock, *Cancer Cell*, 2012, **21**, 327–329.
126. A. I. Minchinton and I. F. Tannock, *Nat. Rev. Cancer*, 2006, **6**, 583–592.
127. G. D. Wang, E. Reed and Q. Q. Li, *Oncol. Rep.*, 2004, **12**, 955–965.
128. S. H. Jang, M. G. Wientjes, D. Lu and J. L. S. Au, *Pharm. Res.*, 2003, **20**, 1337–1350.
129. R. K. Jain, *Science*, 2005, **307**, 58–62.
130. M. R. Dreher, W. G. Liu, C. R. Michelich, M. W. Dewhirst, F. Yuan and A. Chilkoti, *J. Natl. Cancer Inst.*, 2006, **98**, 335–344.
131. S.-D. Li and L. Huang, *Mol. Pharmaceutics*, 2008, **5**, 496–504.
132. F. Alexis, E. Pridgen, L. K. Molnar and O. C. Farokhzad, *Mol. Pharmaceutics*, 2008, **5**, 505–515.
133. P. Decuzzi, B. Godin, T. Tanaka, S. Y. Lee, C. Chiappini, X. Liu and M. Ferrari, *J. Controlled Release*, 2010, **141**, 320–327.
134. D. C. Litzinger, A. M. J. Buiting, N. Vanrooijen and L. Huang, *Biochim. Biophys. Acta, Biomembr.*, 1994, **1190**, 99–107.
135. K. Knop, R. Hoogenboom, D. Fischer and U. S. Schubert, *Angew. Chem. Int. Ed.*, 2010, **49**, 6288 6308.
136. K. Ulbrich and V. Subr, *Adv. Drug Delivery Rev.*, 2010, **62**, 150–166.
137. M. Talelli, C. J. F. Rijcken, C. F. van Nostrum, G. Storm and W. E. Hennink, *Adv. Drug Delivery Rev.*, 2010, **62**, 231–239.
138. M. P. Patel, R. R. Patel and J. K. Patel, *J. Pharm. Pharm. Sci.*, 2010, **13**, 536–557.

139. Y.-I. Chung, J. C. Kim, Y. H. Kim, G. Tae, S.-Y. Lee, K. Kim and I. C. Kwon, *J. Controlled Release*, 2010, **143**, 374–382.
140. M. Wang and M. Thanou, *Pharmacol. Res.*, 2010, **62**, 90–99.
141. K. Park, *J. Controlled Release*, 2010, **142**, 147–148.
142. J. T. Thigpen, C. A. Aghajanian, D. S. Alberts, S. M. Campos, A. N. Gordon, M. Markman, D. S. McMeekin, B. J. Monk and P. G. Rose, *Gynecol. Oncol.*, 2005, **96**, 10–18.
143. P. J. Photos, L. Bacakova, B. Discher, F. S. Bates and D. E. Discher, *J. Controlled Release*, 2003, **90**, 323–334.
144. Z. Liu, C. Davis, W. Cai, L. He, X. Chen and H. Dai, *Proc. Natl. Acad. Sci. U. S. A.*, 2008, **105**, 1410–1415.
145. T. Ishida, M. Ichihara, X. Wang and H. Kiwada, *J. Controlled Release*, 2006, **115**, 243–250.
146. T. Tagami, K. Nakamura, T. Shimizu, N. Yamazaki, T. Ishida and H. Kiwada, *J. Controlled Release*, 2010, **142**, 160–166.
147. T. Ishida, M. Ichihara, X. Wang, K. Yamamoto, J. Kimura, E. Majima and H. Kiwada, *J. Controlled Release*, 2006, **112**, 15–25.
148. J. A. Champion and S. Mitragotri, *Proc. Natl. Acad. Sci. U. S. A.*, 2006, **103**, 4930–4934.
149. Y. Geng, P. Dalhaimer, S. Cai, R. Tsai, M. Tewari, T. Minko and D. E. Discher, *Nat. Nanotechnol.*, 2007, **2**, 249–255.
150. D. A. Christian, S. Cai, O. B. Garbuzenko, T. Harada, A. L. Zajac, T. Minko and D. E. Discher, *Mol. Pharmaceutics*, 2009, **6**, 1343–1352.
151. G. Prencipe, S. M. Tabakman, K. Welsher, Z. Liu, A. P. Goodwin, L. Zhang, J. Henry and H. Dai, *J. Am. Chem. Soc.*, 2009, **131**, 4783–4787.
152. J. A. Champion and S. Mitragotri, *Pharm. Res.*, 2009, **26**, 244–249.
153. G. Sharma, D. T. Valenta, Y. Altman, S. Harvey, H. Xie, S. Mitragotri and J. W. Smith, *J. Controlled Release*, 2010, **147**, 408–412.
154. R. K. Jain, *Cancer Res.*, 1990, **50**, S814–S819.
155. R. K. Jain, *Adv. Drug Delivery Rev.*, 2001, **46**, 149–168.
156. J. Choi, K. Credit, K. Henderson, R. Deverkadra, Z. He, H. Wiig, H. Vanpelt and M. F. Flessner, *Clin. Cancer Res.*, 2006, **12**, 1906–1912.
157. G. Alexandrakis, E. B. Brown, R. T. Tong, T. D. McKee, R. B. Campbell, Y. Boucher and R. K. Jain, *Nat. Med.*, 2004, **10**, 203–207.
158. M. F. Flessner, J. Choi, K. Credit, R. Deverkadra and K. Henderson, *Clin. Cancer Res.*, 2005, **11**, 3117–3125.
159. Y. Boucher, L. T. Baxter and R. K. Jain, *Cancer Res.*, 1990, **50**, 4478–4484.
160. C. H. Heldin, K. Rubin, K. Pietras and A. Ostman, *Nat. Rev. Cancer*, 2004, **4**, 806–813.
161. H. Holback and Y. Yeo, *Pharm. Res.*, 2011, **28**, 1819–1830.
162. A. J. Primeau, A. Rendon, D. Hedley, L. Lilge and I. F. Tannock, *Clin. Cancer Res.*, 2005, **11**, 8782–8788.

163. P. A. Netti, D. A. Berk, M. A. Swartz, A. J. Grodzinsky and R. K. Jain, *Cancer Res.*, 2000, **60**, 2497–2503.
164. S. D. Perrault, C. Walkey, T. Jennings, H. C. Fischer and W. C. W. Chan, *Nano Lett.*, 2009, **9**, 1909–1915.
165. H. Lee, B. Hoang, H. Fonge, R. M. Reilly and C. Allen, *Pharm. Res.*, 2010, **27**, 2343–2355.
166. H. Cabral, Y. Matsumoto, K. Mizuno, Q. Chen, M. Murakami, M. Kimura, Y. Terada, M. R. Kano, K. Miyazono, M. Uesaka, N. Nishiyama and K. Kataoka, *Nat. Nanotechnol.*, 2011, **6**, 815–823.
167. N. Tang, G. Du, N. Wang, C. Liu, H. Hang and W. Liang, *J. Natl. Cancer Inst.*, 2007, **99**, 1004–1015.
168. S. Ramanujan, A. Pluen, T. D. McKee, E. B. Brown, Y. Boucher and R. K. Jain, *Biophys. J.*, 2002, **83**, 1650–1660.
169. R. B. Campbell, D. Fukumura, E. B. Brown, L. M. Mazzola, Y. Izumi, R. K. Jain, V. P. Torchilin and L. L. Munn, *Cancer Res.*, 2002, **62**, 6831–6836.
170. B. Kim, G. Han, B. J. Toley, C.-K. Kim, V. M. Rotello and N. S. Forbes, *Nat. Nanotechnol.*, 2010, **5**, 465–472.
171. O. Lieleg, R. M. Baumgaertel and A. R. Bausch, *Biophys. J.*, 2009, **97**, 1569–1577.
172. T. Stylianopoulos, M.-Z. Poh, N. Insin, M. G. Bawendi, D. Fukumura, L. L. Munn and R. K. Jain, *Biophys. J.*, 2010, **99**, 1342–1349.
173. S. I. Rudnick and G. P. Adams, *Cancer Biother. Radiopharm.*, 2009, **24**, 155–161.
174. J. H. E. Baker, K. E. Lindquist, L. Huxham, A. H. Kyle, J. T. Sy and A. I. Minchinton, *Clin. Cancer Res.*, 2008, **14**, 2171–2179.
175. K. Fujimori, D. G. Covell, J. E. Fletcher and J. N. Weinstein, *J. Nucl. Med.*, 1990, **31**, 1191–1198.
176. M. Juweid, R. Neumann, C. Paik, M. J. Perezbacete, J. Sato, W. Vanosdol and J. N. Weinstein, *Cancer Res.*, 1992, **52**, 5144–5153.
177. W. Vanosdol, K. Fujimori and J. N. Weinstein, *Cancer Res.*, 1991, **51**, 4776–4784.
178. E. Gullotti and Y. Yeo, *Mol. Pharmaceutics*, 2009, **6**, 1041–1051.
179. V. Mishra, U. Gupta and N. K. Jain, *J. Biomater. Sci., Polym. Ed.*, 2009, **20**, 141–166.
180. N. Malik, R. Wiwattanapatapee, R. Klopsch, K. Lorenz, H. Frey, J. W. Weener, E. W. Meijer, W. Paulus and R. Duncan, *J. Controlled Release*, 2000, **68**, 299–302.
181. J. Shin, P. Shum and D. H. Thompson, *J. Controlled Release*, 2003, **91**, 187–200.
182. H. Hatakeyama, H. Akita, K. Kogure, M. Oishi, Y. Nagasaki, Y. Kihira, M. Ueno, H. Kobayashi, H. Kikuchi and H. Harashima, *Gene Ther.*, 2007, **14**, 68–77.

183. H. Hatakeyama, H. Akita, E. Ito, Y. Hayashi, M. Oishi, Y. Nagasaki, R. Danev, K. Nagayama, N. Kaji, H. Kikuchi, Y. Baba and H. Harashima, *Biomaterials*, 2011, **32**, 4306–4316.
184. R. Roy, B. Zhang and M. A. Moses, *Exp. Cell Res.*, 2006, **312**, 608–622.
185. Z. X. Zhou, Y. Q. Shen, J. B. Tang, M. H. Fan, E. A. Van Kirk, W. J. Murdoch and M. Radosz, *Adv. Funct. Mater.*, 2009, **19**, 3580–3589.
186. V. A. Sethuraman and Y. H. Bae, *J. Controlled Release*, 2007, **118**, 216–224.
187. E. S. Lee, Z. Gao, D. Kim, K. Park, I. C. Kwon and Y. H. Bae, *J. Controlled Release*, 2008, **129**, 228–236.
188. N. M. Zaki and N. Tirelli, *Exp. Opin. Drug Delivery*, 2010, **7**, 895–913.
189. P. Rai, C. Padala, V. Poon, A. Saraph, S. Basha, S. Kate, K. Tao, J. Mogridge and R. S. Kane, *Nat. Biotechnol.*, 2006, **24**, 582–586.
190. M. Ferrari, *Nat. Nanotechnol.*, 2008, **3**, 131–132.
191. M. A. Phillips, M. L. Gran and N. A. Peppas, *Nano Today*, 2010, **5**, 143–159.
192. S. Zhu, L. Qian, M. Hong, L. Zhang, Y. Pei and Y. Jiang, *Adv. Mater.*, 2011, **23**, H84–H89.
193. X.-B. Xiong, H. Uludag and A. Lavasanifar, *Biomaterials*, 2010, **31**, 5886–5893.
194. C. Zhan, B. Gu, C. Xie, J. Li, Y. Liu and W. Lu, *J. Controlled Release*, 2010, **143**, 136–142.
195. M. V. Pasquetto, L. Vecchia, D. Covini, R. Digilio and C. Scotti, *J. Immunother.*, 2011, **34**, 611–628.
196. J. Mathew and E. A. Perez, *Curr. Opin. Oncol.*, 2011, **23**, 594–600.
197. M. Lopus, *Cancer Lett.*, 2011, **307**, 113–118.
198. T. Kakudo, S. Chaki, S. Futaki, I. Nakase, K. Akaji, T. Kawakami, K. Maruyama, H. Kamiya and H. Harashima, *Biochemistry*, 2004, **43**, 5618–5628.
199. W. Tan, H. Wang, Y. Chen, X. Zhang, H. Zhu, C. Yang, R. Yang and C. Liu, *Trends Biotechnol.*, 2011, **29**, 634–640.
200. T. Chen, M. I. Shukoor, Y. Chen, Q. Yuan, Z. Zhu, Z. Zhao, B. Gulbakan and W. Tan, *Nanoscale*, 2011, **3**, 546–556.
201. D. J. Yoon, C. T. Liu, D. S. Quinlan, P. M. Nafisi and D. T. Kamei, *Ann. Biomed. Eng.*, 2011, **39**, 1235–1251.
202. G. Trapani, N. Denora, A. Trapani and V. Laquintana, *J. Drug Targeting*, 2012, **20**, 1–22.
203. E. Lavik and H. von Recum, *ACS Nano*, 2011, **5**, 3419–3424.
204. S. T. Stern, J. B. Hall, L. L. Yu, L. J. Wood, G. F. Paciotti, L. Tamarkin, S. E. Long and S. E. McNeil, *J. Controlled Release*, 2010, **146**, 164–174.
205. H. S. Choi and J. V. Frangioni, *Mol. Imaging*, 2010, **9**, 291–310.
206. P. Dunn, T. T. Kuo, L. Y. Shih, P. N. Wang, C. F. Sun and M. W. J. Chang, *Am. J. Hematol.*, 1998, **57**, 68–71.
207. T. T. Kuo, S. Hu, C. L. Huang, H. L. Chan, M. J. W. Chang, P. Dunn and Y. J. Chen, *Am. J. Surg. Pathol.*, 1997, **21**, 1361–1367.

208. P. Schneider, T. A. Korolenko and U. Busch, *Microsc. Res. Techn.*, 1997, **36**, 253–275.
209. L. W. Seymour, R. Duncan, J. Strohalm and J. Kopecek, *J. Biomed. Mater. Res.*, 1987, **21**, 1341–1358.
210. M. Irfan and M. Seiler, *Ind. Eng. Chem. Res.*, 2010, **49**, 1169–1196.
211. K. M. Huh, S. C. Lee, Y. W. Cho, J. W. Lee, J. H. Jeong and K. Park, *J. Controlled Release*, 2005, **101**, 59–68.
212. D. M. Vail, L. D. Kravis, A. J. Cooley, R. Chun and E. G. MacEwen, *Cancer Chemother. Pharmacol.*, 1997, **39**, 410–416.
213. S. Y. Kim and Y. M. Lee, *Biomaterials*, 2001, **22**, 1697–1704.
214. J. Szebeni, *Toxicology*, 2005, **216**, 106–121.
215. J. Szebeni, L. Baranyi, S. Savay, J. Milosevits, R. Bunger, P. Laverman, J. M. Metselaar, G. Storm, A. Chanan-Khan, L. Liebes, F. M. Muggia, R. Cohen, Y. Barenholz and C. R. Alving, *J. Liposome Res.*, 2002, **12**, 165–172.
216. C. Zhang, G. Qu, Y. Sun, X. Wu, Z. Yao, Q. Guo, Q. Ding, S. Yuan, Z. Shen, Q. Ping and H. Zhou, *Biomaterials*, 2008, **29**, 1233–1241.
217. W. Zhang, Y. Shi, Y. Chen, J. Hao, X. Sha and X. Fang, *Biomaterials*, 2011, **32**, 5934–5944.
218. T. Schluep, J. J. Cheng, K. T. Khin and M. E. Davis, *Cancer Chemother. Pharmacol.*, 2006, **57**, 654–662.
219. J. Hu, Y. Su, H. Zhang, T. Xu and Y. Cheng, *Biomaterials*, 2011, **32**, 9950–9959.
220. D. Ding, Z. Zhu, Q. Liu, J. Wang, Y. Hu, X. Jiang and B. Liu, *Eur. J. Pharm. Biopharm.*, 2011, **79**, 142–149.

CHAPTER 4
Functional Polymers for Gene Delivery

XUAN ZENG, REN-XI ZHUO AND
XIAN-ZHENG ZHANG*

Key Laboratory of Biomedical Polymers of Ministry of Education &
Department of Chemistry, Wuhan University, Wuhan 430072, P. R. China
*E-mail: xz-zhang@whu.edu.cn

4.1 Introduction

Gene therapy is an innovative approach for devastating inherited or acquired disease treatment through delivering therapeutic genes to targeted cells and replacing the disorder genes, where conventional therapy, such as radiotherapy or chemotherapy, has met strong resistance. This therapeutic process can be administrated *in vivo* or *ex vivo* (Figure 4.1). However, the negatively charged cell membrane inhibits the entry of naked DNA due to its electronegativity, and the unprotected DNA will be rapidly degraded by nucleases present in plasma, so that most gene transfer is carried out using a powerful gene delivery vehicle. Over the past two decades, gene therapy has made great progress, especially in the development of effective gene vectors.[1] Some of these vectors have been successfully used in animal models and are currently being tested in clinical trials to treat maladies such as cardiovascular disease, cystic fibrosis, Parkinson's disease, and various cancers. The gene delivery system is mainly categorized to be viral and nonviral vectors. Viral vectors are biological systems derived from naturally evolved viruses capable of transferring their genetic material into the host cells, and thus are very

RSC Polymer Chemistry Series No. 3
Functional Polymers for Nanomedicine
Edited by Youqing Shen
© The Royal Society of Chemistry 2013
Published by the Royal Society of Chemistry, www.rsc.org

effective in achieving high efficiency for both gene delivery and expression. Nevertheless, the limitations associated with viral vectors, including immunogenic responses, risk of tumorigenicity, complicated preparation, and high cost, have encouraged researchers to investigate and develop safe and efficient nonviral gene delivery vehicles.[2]

The advantages of nonviral systems are obvious. Besides their lesser toxicity and lower immune responses than viral vectors, no integration into the host genome occurs during nonviral vector-mediated gene delivery. Moreover, nonviral methods are not limited by the size of the gene cargo, are stable to storage, are easier to produce on a large scale, and can offer remarkable structural and chemical versatility. Nonviral gene vectors can be broadly categorized into polymers, liposomes, peptides, and organic/inorganic nanoparticles.[3] Especially, cationic polymer vectors have been considered as the most common DNA condensing agents. They interact with negatively charged plasmid DNA through electrostatic interactions and package them into nanoscale polyplexes, which can protect DNA from enzymatic degradation and transport the gene into target cells through an endocytic pathway (Figure 4.2).

In order to achieve maximum expression of a therapeutic gene carried by vectors, multiple hurdles must be overcome.[4] These include (1) protection of the complexes from *in vivo* degradation; (2) efficient transfection of the complexes into the target cell; (3) protection/prevention from nuclease

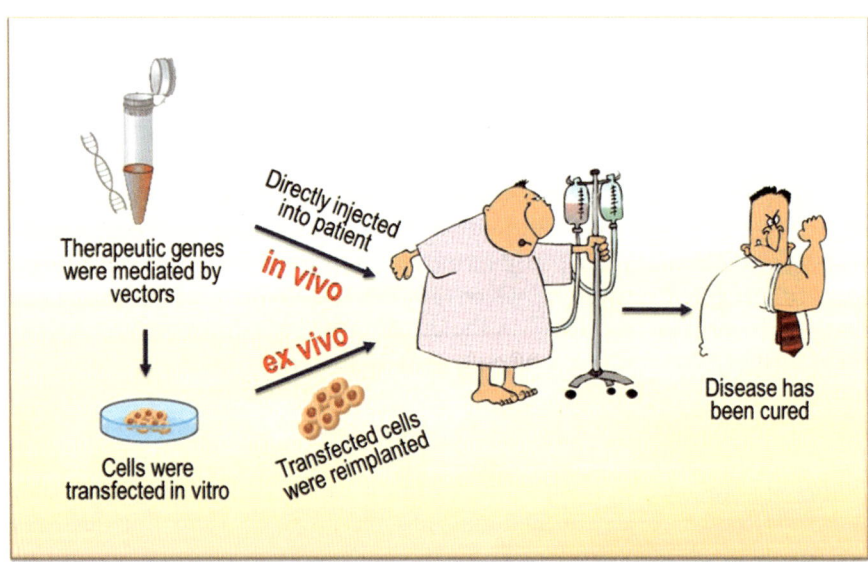

Figure 4.1   *In vivo* and *ex vivo* gene therapy. (1) During the *in vivo* process, vector-mediated therapeutic genes are injected into the patient and transfected directly. (2) During the *ex vivo* process, therapeutic genes are first inserted into cells *in vitro*. The transfected cells are then expanded and reimplanted into the patient.

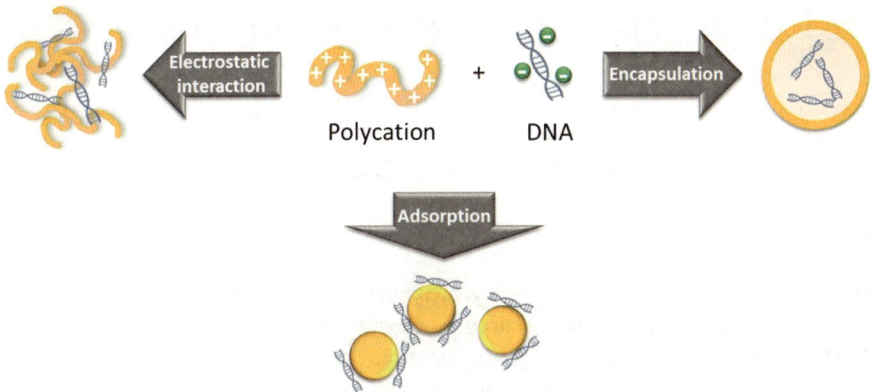

Figure 4.2 Self-assembled formations of polycation/DNA complexes.

degradation within endosomes; (4) efficient entry into the cell nucleus; (5) efficient gene expression; and (6) rapid clearance of the vector without any toxicity. Consequently, a tremendous amount of research in the past decade has focused on designing cationic compounds that can form complexes with DNA and can avoid both *in vitro* and *in vivo* barriers for gene delivery. Here, we review the current progress of the most significant nonviral gene delivery systems, and introduce novel design strategies for enhancing the safety and efficacy of nonviral vectors in gene delivery and tumor targeting.

4.2 Polyethylenimine-Based Gene Vectors

Polyethylenimine (PEI) has been known as the most promising gene carrier and a gold standard among nonviral vectors used for *in vitro* and *in vivo* transfection, due to its relatively high and efficient transfection performance in various cell lines and tissues.[1] PEI has a number of attractive characteristics beneficial to gene delivery. It can condense DNA or RNA to form nanosized compactable particles through electrostatic interaction, attributed to its large number of protonable amino groups, which results in a high cationic charge density at physiological pH. In addition, the "proton sponge" nature of PEI leads to osmotic swelling and rupture of endosomes, so that complex particles can prevent lysosomal degradation and escape into the cytoplasm. Despite these advantages, there are still some thorny problems associated with PEI-mediated gene delivery, especially high cytotoxicity and lacking of cell specificity, seriously limiting further applications.

It was reported that there were at least two types of cytotoxicity in PEI-mediated gene delivery: immediate and delayed toxicities.[4] Before being internalized by cells, free PEIs can cause immediate cell death by membrane destabilization. In addition, after the internalization, free PEIs produced by dissociation of PEI/DNA complexes can induce cell apoptosis related to delayed toxicity. Many characteristics of PEI-derived vectors and polyplexes

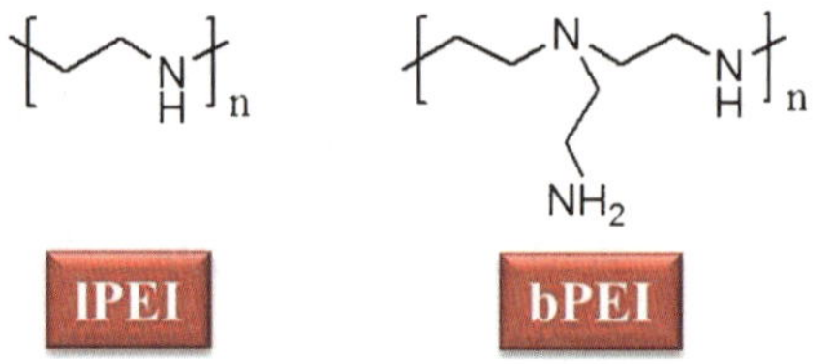

Figure 4.3 Chemical structures of linear PEI (lPEI) and branched PEI (bPEI).

have strong connections with the cytotoxicity or even exacerbate it, including molecular weight, molecular structure, degree of branching, cationic charge density, buffer capacity, polyplex particle size, polyplex zeta potential, polyplex concentration, the transfection time, *etc.*

High molecular weight (HMW) PEI, such as 25 kDa PEI (both branched and linear structures; Figure 4.3), shows high transfection efficiency. Nonetheless, the HMW PEI is nonbiodegradable and thus cannot be cleared from the circulation, which leads to accumulative toxicity inside the body and other side effects. In contrast, low molecular weight (LMW) PEI (less than 2 kDa) has demonstrated low cytotoxicity but cannot be used as a gene vector due to the poor transfection efficiency. Therefore, various methods have been adopted to overcome PEI's drawbacks, and extensive work was focused on modification of LMW PEI. Because the advantages of HMW PEI and LMW PEI are well complementary to each other, we believe that appropriate combinations may generate advanced gene delivery systems retaining the advantages while avoiding the shortcomings of each.

4.2.1 Low-Toxicity Polyethylenimine

Currently, in order to reduce the toxicity and enhance the *in vivo* gene delivery efficiency of PEI-based delivery systems, the design of biocompatible and biodegradable PEI vectors presents two main trends: (1) surface decoration of PEIs with hydrophilic polymers and biodegradable polymers by covalent or noncovalent attachment in order to shield positive charges and enhance serum stability; (2) synthesis of HMW crosslinked PEI compounds by LMW PEIs *via* the incorporation of reducible disulfide linkages or ester conjugation, and these crosslinked PEIs could be biodegraded into LMW PEIs again in the physiological environment (Figure 4.4).

Various biocompatible polymers have been employed to modify PEI molecules, such as poly(ethylene glycol) (PEG), natural glucose polymers, proteins, peptides, *etc.* The most common method is conjugating PEG to PEI molecules, as PEGylation is a well-established technique that can mask the complex from the host's immune system and prolong the circulation time of complexes in the bloodstream. PEGylated PEIs have been studied widely as potential gene delivery systems. Kissel's group have focused on synthesizing PEG-*g*-PEI conjugates by grafting linear PEG (550 Da, 2 kDa, 5 kDa,

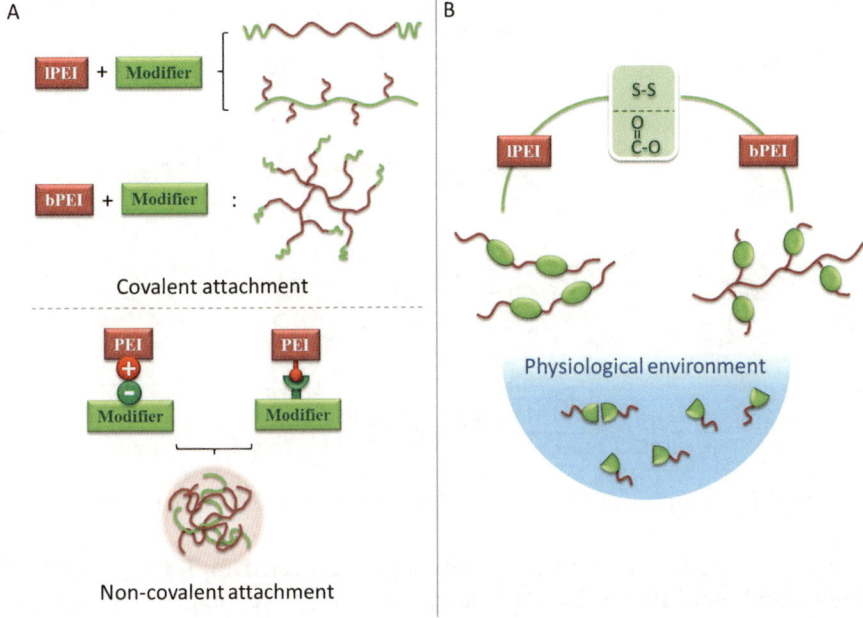

Figure 4.4 Schematic illustration of decorating strategies to obtain low-toxicity PEI-based vectors: (A) biocompatible PEI derivatives; (B) biodegradable crosslinked PEIs.

20 kDa) onto branched PEI (25 kDa); the results have shown favorable transfection efficiency and reduced cytotoxicity in small interfering RNA (siRNA), plasmid DNA (pDNA), and messenger RNA (mRNA) delivery (Scheme 4.1A).[5,6] Merkel *et al.* reported that *in vitro* complement activation was prominently caused by PEI 25 kDa, whereas the PEGylated versions of PEI 25 kDa showed no significant activity.[7] It was shown that PEGylation of polycations with 20 kDa PEG or higher molecular weight may be favorable.

Chen *et al.* synthesized a PEG-*g*-PEI copolymer by grafting 8 kDa PEG onto 25 kDa PEI.[8] The PEG-*g*-PEI/pEGFP-C1 nanoparticles displayed low cytotoxicity, good solubility, and compatibility with serum. They found that biocompatibility was guaranteed by dense PEG shells, which endowed the nanoparticles with water solubility and prevented their interaction with serum protein in the culture medium. Beyerle *et al.* investigated the side effects of PEG-*g*-PEI/siGFP polyplexes in mice lungs.[9] The results showed that hydrophilic modifications, with high PEG-grafting degrees, induced less proinflammatory effects without depleting macrophages and disrupting the epithelial/endothelial barrier in the lungs, while showing only a minor oxidative stress response.

The length and density of PEG chains conjugated to PEI also have an effect on transfection efficiency. Weber and co-workers grafted 25 kDa PEI with 20 kDa or 2 kDa PEG to form biocompatible PEI-*g*-PEG gene vectors. Then

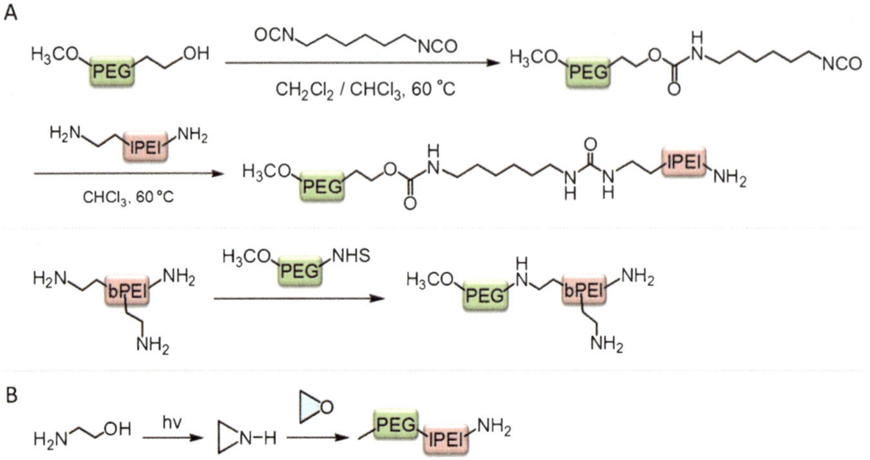

A

B

Scheme 4.1   Synthesis of (A) PEI-*g*-PEG and (B) PEI-*co*-PEG copolymers.

siRNA was mediated by these vectors and transfected to T lymphocytes.[10] They found that PEI-*g*-PEG (20 kDa)/siRNA polyplexes had low toxicity and could knockdown GAPDH expression and inhibit HIV replication. However, siRNA gene knockdown was less effective when a high density of shorter PEG chains (2 kDa) was conjugated to PEI, as they hindered uptake or release of active siRNA from internal vesicles.

Tsai *et al.* synthesized a single-monomer-derived linear-like PEI-*co*-PEG for efficient siRNA delivery and silencing *in vitro* and *in vivo* (Scheme 4.1B).[11] This copolymer was only synthesized from a single monomer by intensive synchrotron X-ray irradiation in the absence of catalyst and organic solvent. They found that the incorporation of PEG segments into the copolymer not only solved the cytotoxicity problems, but also improved the efficiency of siRNA release compared to either linear PEI or Lipofectamine 2,000.

Besides PEGylation, several other modifications to PEI molecules have been developed to reduce toxicity. Natural glucose polymers [*e.g.* dextran, β-cyclodextrin (β-CD)] are a series of natural cyclic oligosaccharides which have low immunogenicity and toxicity in humans. These natural polymers can enhance the absorption and resistance to nucleases through binding and interacting with oligonucleotides; thus, they have been incorporated into cationic polymers as new gene delivery vectors.[12] Liu *et al.* employed β-CD to crosslink LMW PEI (600 Da) and then coupled the product with ligand YC21 (epidermal growth factor receptor targeted oligopeptide) to form tumor-targeted gene vectors (Scheme 4.2A).[13] The results showed that these vectors possessed lower cytotoxicity and higher efficient gene delivery ability to the EGFR-positive liver cancer cells *in vitro* and achieved favorable therapeutic effects in the inhibition of tumor growth *in vivo*. Tang's group also reported a series of β-CD crosslinked PEIs, grafted with folic acid, TAT-R8, or NLS peptide.[14–16] These compounds exhibited higher targeting and transfection efficiency as well as lower cytotoxicity compared to 25 kDa PEI in various tumor cell lines and different mouse models. Xiao and co-workers synthesized (dextran–hexamethylene diisocyanate)-*g*-polyethylenimine [(Dex-HMDI)-*g*-PEI] as a siRNA vector (Scheme 4.2B).[17] This dextran-linked compound showed reduced cytotoxicity and significant knockdown ability compared to 25 kDa PEI.

Luo *et al.* developed a "hydroxylation camouflage" strategy to promote serum-tolerant capability of polycation-based gene delivery systems (Scheme 4.2C).[18] In this design, a 25 kDa PEI surface was coated with abundant hydroxyl groups *via* a catalyst-free ring-opening aminolysis reaction with 5-ethyl-5-(hydroxymethyl)-1,3-dioxan-2-one (EHDO). The results indicated that the serum-tolerant capability largely depended on the surface composition and substitution degree. The hydroxyl-enriched "skin" would render PEI-*g*-EHDO with remarkably improved biocompatibility and stronger resistance against the serum-associated detrimental effects such as protein

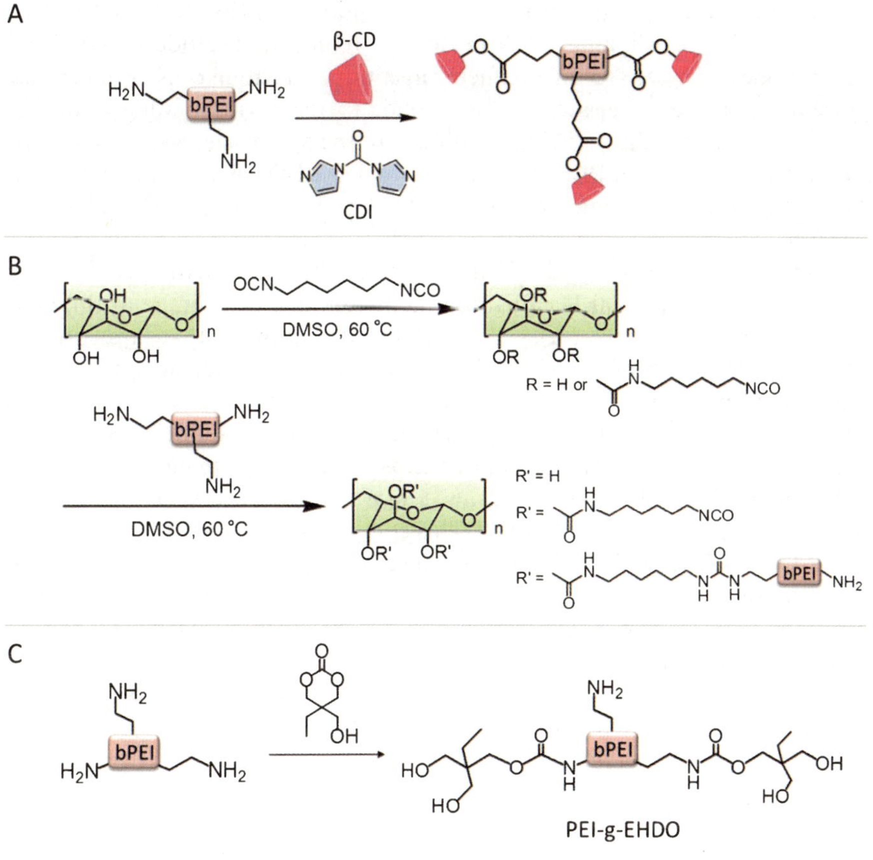

Scheme 4.2 Synthesis of (A) β-CD-modified PEI, (B) dextran-modified PEI copolymers, and (C) hydroxyl-enriched PEI.

adsorption, particle aggregation, and polycation–protein exchange. The transfection efficiency of PEI-*g*-EHDO was 30-fold higher than 25 kDa PEI in the presence of serum.

Peptide-modified PEIs are also advantageous because they are easily metabolized and nontoxic to cells, making different combinations possible. Dey *et al.* designed histidine-based peptide-linked PEI polymers for transfection studies.[19] The results demonstrated that these peptide-linked PEI complexes significantly reduced the toxic effects of the 25 kDa PEI. L-Carnosine derivatives of PEI showed significantly higher transfection efficiency in primary human cells with respect to the widely used transfection agent, Lipofectamine. PEI linked with the dipeptides L-carnosine or Boc-L-carnosine appeared to be the most promising gene delivery agents among all the complexes tested, based on their reduced production of reactive oxygen species and markedly high transfection efficiency.

Generally, based on different degradation mechanisms, crosslinked PEIs could be divided into three types: reduction, hydrolysis, and acidic hydrolysis. As for reduction type, the disulfide bond-containing PEIs (SS-PEIs) can be synthesized by crosslinking LMW PEI with reducible disulfide crosslinkers. Glutathione as a water-soluble reducing agent, found within cells at millimolar concentrations, can degrade disulfide bonds to the corresponding thiols. In consequence, under the reducing conditions of the cytoplasm, SS-PEI could be fragmentized into small PEIs, so that the condensed DNA would release from the complexes easily and rapidly, reducing cytotoxicity and enhancing transgene expression.

Peng and co-workers synthesized a series of disulfide-crosslinked PEIs by the reaction between 800 Da PEI and methylthiirane or thiirane at different molar ratios (Scheme 4.3A).[20,21] The results showed that transfection efficiency was dependent on disulfide content and molecular weight. The compounds with an adequate thiolation degree between 2.6 and 4.5 have relatively lower cytotoxicity and higher gene transfection efficiency than 25 KDa PEI. Wei *et al.* synthesized bioreducible SS-PEIs by chemical coupling of 3′-dithiobispropanoic acid (DTPA) and LMW PEI (800 Da) *via* an EDC/NHS activation reaction (Scheme 4.3B).[22] This compound was employed to transfer hTERT siRNA into HepG2 cells and revealed relatively low cytotoxicity *in vitro* and at an appropriate dose had no adverse effect on liver and kidney functions *in vivo*. Sun *et al.* synthesized the SS-PEIs by Michael addition between cystamine bisacrylamide and LMW branched 800 Da PEI (Scheme 4.3C).[23] *In vitro* transfection experiments showed that SS-PEI/DNA binary complexes demonstrated comparable transfection efficiency, but lower cytotoxicity, in comparison with that of 25 kDa PEI. Deng *et al.* synthesized disulfide-containing poly(β-amino amine)s and then used them to crosslink LMW PEI (1800 Da) through Michael addition to obtain SS-PBAA-PEIs as the final gene carriers.[24] *In vitro* transfection showed that the SS-PBAA-PEIs had comparable transfection efficiencies and lower cytotoxicities compared with 25 kDa PEI. Confocal laser scanning microscopy confirmed that these

SS-PBAA-PEIs could transport and trigger the release of DNA in the cytoplasm by bioreducible degradation. Koo *et al.* synthesized biodegradable b-PEIS [branched poly(ethylenimine sulfide)] by crosslinking linear PEIs (Scheme 4.3D).[25] These b-PEIS derivatives showed high transfection efficiencies and almost no toxicity. Owing to the biocompatibility of b-PEIS, a greater amount of DNA could be applied to the cells in order to improve the transfection efficiency. The disulfide-based b-PEIS was degraded in the cytosol and finally excreted from the cells during the transfection process.

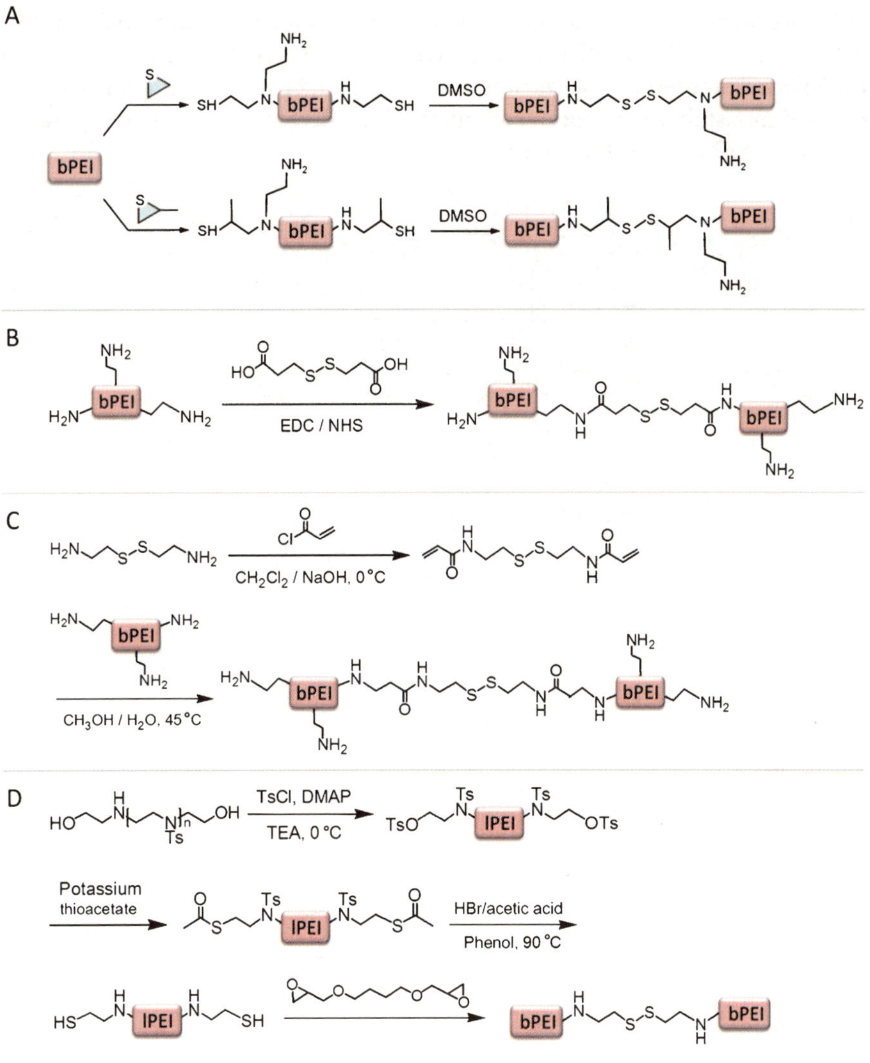

Scheme 4.3 Synthesis of disulfide bond-containing PEI derivatives.

In addition to disulfide linkages, the hydrolysis capability of ester (including acid-labile ester) or amide bonds also has been widely used to create biodegradable PEI-based gene vectors. These vectors can be degraded into nontoxic LMW PEIs without inducing toxicity or immune response. However, common hydrolysis degradation is unspecific and hard to control, while acidic hydrolysis would damage the nucleic acids due to the low pH of late endosomes (or lysosomes). Endres *et al.* synthesized amphiphilic PEG-PCL-PEI triblock copolymers as self-assemble multifunctional carriers (Scheme 4.4A).[26] The synthesis was in three steps: (1) PEG-PCL was produced by ring-opening polymerization initiated by the PEG hydroxyl endgroup, using Sn(Oct)$_2$ as a catalyst; (2) a double bond was attached onto the PEG-PCL chain-end by esterification of the PEG-PCL hydroxyl group with acryloyl chloride; (3) 2,500 kDa PEI was coupled onto the PEG-PCL-linker copolymer by Michael addition. The authors found that increasing the hydrophilicity resulted in a stability increase, combined with a decrease in cytotoxicity due to effective PEG charge shielding.

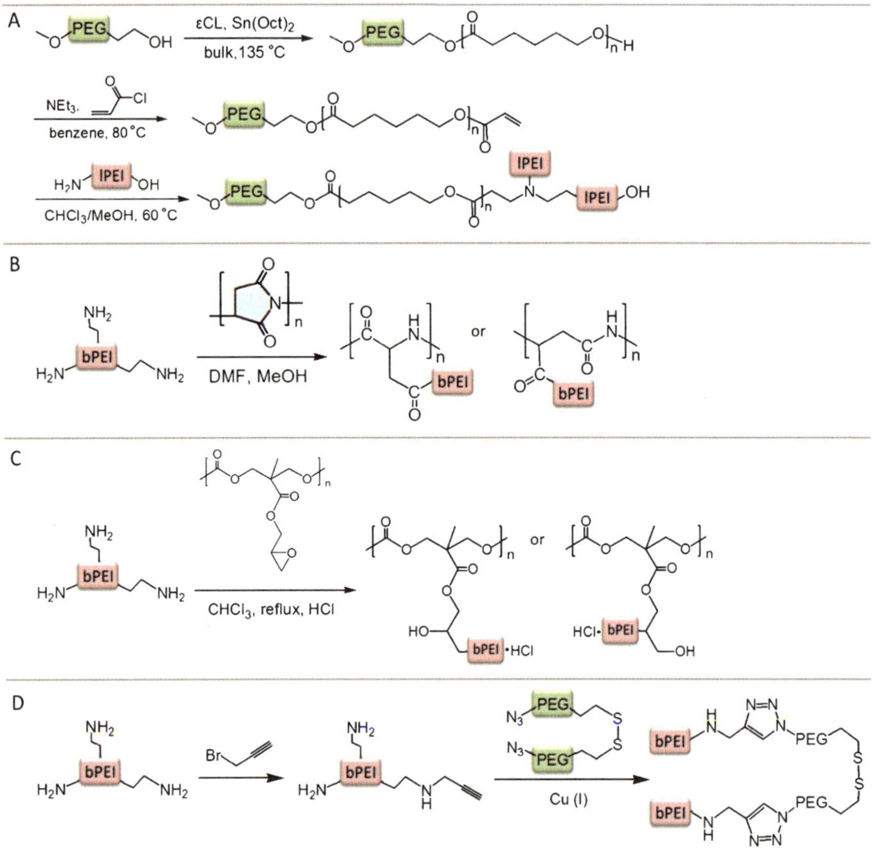

Scheme 4.4   Synthesis of biodegradable PEI derivatives.

Huang *et al.* synthesized a biodegradable PEI-based vector by grafting branched 800 Da PEI onto poly(L-succinimide) (PSI) backbones (Scheme 4.4B).[27] This compound was considered to ultimately degrade into amino acids *in vivo* and help release condensed plasmid DNA in the nucleolus. They further conjugated methoxy-PEG to this biodegradable vector, which exhibited remarkably lower cytotoxicity and higher transfection efficiency compared to 25 kDa PEI.[28] In the study of He *et al.*, biodegradable backbone polycarbonates, poly[(5-methyl-5-allyloxycarbonyl-trimethylene carbonate)-*co*-(5,5-dimethyl-trimethylene carbonate)], were grafted by 1800 kDa PEI (Scheme 4.4C).[29] This polycation presented better degradability, lower cytotoxicity, and much higher gene transfection efficacy in comparison with 25 kDa PEI.

Recently, more focus has been placed on the combination of both methods in one gene delivery system, thereby resulting in better performance. Zhao *et al.* synthesized a biodegradable disulfide-containing PEG-PEI by click chemistry (Scheme 4.4D).[30] PEI (2 kDa) and short chain length PEG (tetraethylene glycol, TEG) were crosslinked to a HMW PEG-PEI copolymer (~22 kDa). This PEG-PEI polymer is about 22-fold less toxic than 25 kDa PEI in the L929 cell line. After coupling of small PEG chains and crosslinking by disulfide bridges, the transfection efficiency was increased approximately six-fold in comparison to 2 kDa PEI. This click cluster crosslinked disulfide-containing PEG-PEI copolymer could be an attractive cationic polymer for nonviral gene delivery.

4.2.2 Cell-Targeted Polyethylenimine

To develop *in vivo* suitable gene carriers, the important goal is to increase the specific uptake efficiency and to decrease the nonspecific uptake efficiency. In order to promote cell specificity, efforts have been made to combine or even exchange the nonspecific electrostatic polyplex–cell surface interaction with a specific receptor-mediated cellular uptake, by the incorporation of cell binding ligands into the transfection complexes (Figure 4.5). A variety of ligands has been successfully coupled to PEI, including galactose for hepatocyte targeting, mannose for enhanced uptake by dendritic cells, epidermal growth factor for enhanced uptake by epithelial cells, integrin-binding peptides and anti-CD3 antibodies for gene delivery to CD3-expressing cells (Table 4.1).

Much work has been directed to realize this aim by coupling cell-targeting ligands, such as arginine-glycine-aspartic acid (RGD) peptides and folic acid, onto cationic polymer vehicles in order to target integrin receptors on the cell surface. Tian and co-workers modified the hyperbranched PEI with a hydrophobic poly(γ-benzyl L-glutamate) segment (PBLG).[31,32] The biocompatible PBLG grafting onto PEI shielded the toxicity of the PEI and condensed the DNA into small particles (~100 nm). The compound PEI-PBLG exhibited much higher transfection efficiency than that of 25 kDa PEI. Recently, the same group conjugated RGD peptides onto natural anionic polymer

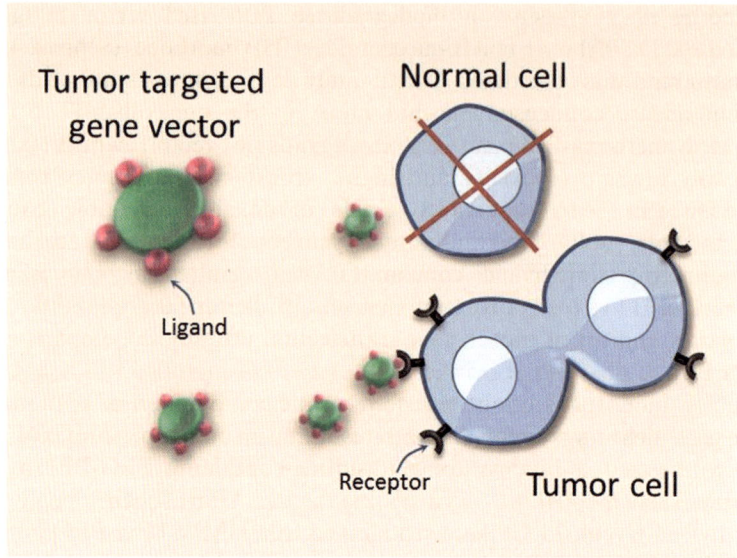

Figure 4.5   Tumor targeted gene transfection.

hyaluronic acid (HA), and then coated the cationic surface of PEI-PBLG/
DNA complexes *via* electrostatic interaction.[33] The resulting HA-RGD/PEI-
PBLG/DNA ternary complexes showed higher cell viability and transfection
efficiency compared with PEI-PBLG/DNA, owing to the RGD target bonding
affinity for integrin on HeLa cells.

HA is a biodegradable, non-immunogenic, non-cytotoxic, and negatively
charged polysaccharide. It has several advantages, including the negligible
nonspecific interaction with serum components due to the polyanionic
characteristics and the highly efficient target specific delivery to the liver
tissues with HA receptors. Yao *et al.* synthesized HA-*g*-PEI copolymer by an
imine reaction between periodate-oxidized HA and PEI.[34] This compound

Table 4.1   Ligands used to target tumors and enhance transfection efficiency
of PEI-based vectors.

Ligand	Target	Ref.
RGD	HepG2, HeLa	33
Hyaluronic acid	Liver and kidney cancer cells	34
Folate	SKOV3, HeLa	35,36
Tetanus toxin fragment	Neuronal cell lines	37
Rabies virus glycoprotein	Mouse neuroblastoma Neuro2a cells	38
Chondrocyte-homing peptide	Chondrocyte cells	39
Avidin	HepG2	40–42
Transferrin	HepG2, HeLa	43
N-Acetylglucosamine	293FT, HeLa	44
Phenylboronic acid	HepG2	45

exhibited lower cytotoxicity and higher transfection efficiency in HepG2 cells compared to 25 kDa PEI. Moreover, these HA-modified complexes were obviously accumulated in tumor after i.v. administration, which indicated that HA could assist PEI targeting to the tumor.

One of the best-characterized targeting ligands for tumor treatment is folate, since folate receptors (FRs) exhibit limited expression on healthy cells, but are overexpressed in certain cancer cells. Folate-modified PEI has previously been shown to promote target-specific gene delivery both *in vitro* and *in vivo*, showing superior performance compared to PEI. Liang *et al.* found that when folate-PEG-PEI was used for the delivery of therapeutic genes, by direct injection into tumors of glioma-xenografted rats, a significant antitumor effect was achieved.[35] Zhang *et al.* synthesized folate/PEG-modified PEI as a delivery system for tumor targeting transfer of mcDNA.[36] These folate-labeled polyplexes containing mcDNA exhibited strong tumor targeting capability and high levels of gene expression both *in vitro* and *in vivo*.

Efforts have been made in order to attain neuron-specific complexes. Oliveira *et al.* developed a PEI-based multi-component gene vector targeted to peripheral nervous system cells.[37] A thiolated PEI/DNA complex was used as the complex core, and a PEG-modified tetanus toxin fragment was grafted onto the core as the targeting moiety. The results demonstrated that these ternary vectors were able to transfect primary cultures of dorsal root ganglion dissociated neurons in a targeted manner and elicit the expression of a relevant neurotrophic factor. Hwang *et al.* synthesized rabies virus glycoprotein (RVG)-labeled SS-PEI as neuron-specific miR-124a delivery to neuron *in vivo*.[38] This compound showed low toxicity and high transfection efficiency in acetylcholine receptor-positive Neuro2a cells.

Gene therapy is a promising method for osteoarthritis and cartilage injury. However, specifically delivering target genes into chondrocytes is a great challenge because of their non-vascularity and the dense extracellular matrix of cartilage. Pi *et al.* constructed a cartilage-targeting gene delivery system by conjugating chondrocyte-homing peptide to PEI.[39] This system showed apparently cartilage-targeting property in an *in vivo* assay, so could be used as a specific cartilage-targeting vector for cartilage disorders.

To improve the gene delivery efficiency and safety of nonviral vectors in liver cells, Zeng *et al.* synthesized a series of biotinylated PEI/avidin bioconjugates that presented less toxic and higher gene expression in HepG2 cells due to the biocompatibility of avidin and the specific interactions between avidin and HepG2 cells.[40,41] They also confirmed the optimum conditions for *in vitro* gene transfection of these bioconjugates in HepG2 cells.[42] Tumor cells with a high rate of proliferation usually overexpress transferrin receptors, so they could act as ideal targets in antitumor gene delivery. Recently, this group developed a biotinylated transferrin/avidin/biotinylated disulfide-containing PEI bioconjugate (TABP-SS) mediated p53 gene delivery system attributed to a "avidin-biotin bridge".[43] TABP-SS exhibited much lower cytotoxicity and higher transfection efficacy in HepG2 and HeLa cells due to the specific interactions between transferrin ligands and their receptors on tumor cells.

N-Acetylglucosamine (GlcNAc)-conjugated agents can be targeted to the vimentin- and desmin-expressing cells and tissues. Kim *et al.* synthesized GlcNAc grafted PEI conjugates that showed reduced cytotoxicity and significant transfection efficiency in vimentin-expressing 293FT and HeLa cells.[44] This gene-delivery system could be used to target various vimentin-expressing cells such as fibroblasts and tumor cells.

Boronic acid was found to selectively bind oligo- or polysaccharides such as heparin and glycoproteins. Peng *et al.* synthesized phenylboronic acid-modified PEI by coupling 1800 Da PEI with 4-(bromomethyl)phenylboronic acid molecules.[45] This compound showed obviously higher transfection efficiency in HepG2 cells compared to 25 kDa PEI, attributed to the interaction between boronic acid and carbohydrate on the cell surface.

4.2.3 Other Polyethylenimine Derivatives

In addition to improved low-toxicity and tumor-targeting capability, various efforts have been made to enhance the transfection efficiency of PEI. Yang *et al.* reported that polyurethane–short-branch polyethylenimine (PU-PEI) exhibited high transfection efficiency with relatively low cytotoxicity *in vitro* and *in vivo*.[46] As a therapeutic delivery vehicle, PU-PEI-mediated miR145 delivery to glioblastomas-CD133+ significantly inhibited their tumorigenic and cancer stem cell-like abilities.

Swami *et al.* reacted butane-1,4-diol diglycidyl ether (BDE)-crosslinked PEI (25 kDa) nanoparticles with varying proportions of a novel linker, 2-(*N*-1-tritylimidazol-4-yl)-*N*-(6-glycidyloxyhexyl)acetamide, to yield PN-*g*-imidazolyl nanoparticles with improved transfection efficiency.[47] Recently, Goyal *et al.* showed the effect of imidazole grafting through an epoxy end-based linker on BDE-crosslinked PEI nanoparticles.[48] It was found that introduction of imidazole groups resulted in enhancement of the transfection efficiency of crosslinked nanoparticles, which was higher than PEI and commercial transfection reagents *in vitro* and *in vivo*. They also synthesized a series of linear PEI nanoparticles by crosslinking with BDE that exhibited excellent transfection efficiency in comparison with 25 kDa PEI and Lipofectamine.[49]

Morris *et al.* synthesized arginine-modified oligo(alkylaminosiloxane)-conjugated PEI by simple reproducible reaction methods.[50] These nanoparticles were found to exhibit higher cell viability and gene transfection efficiency than 25 kDa PEI in KB cell lines. They found that the enhanced transfection was due to the existence of a uniformly spaced arginine moiety *via* oligoalkylaminosiloxane arms, which could promote the cellular uptake by multiple pathways and subsequent entry into the nucleus.

Some studies have characterized lyophilized complexes and have investigated the applicability of dry powder aerosols for pulmonary gene delivery. Pfeifer *et al.* investigated PEI/pDNA dry powder aerosols as novel gene

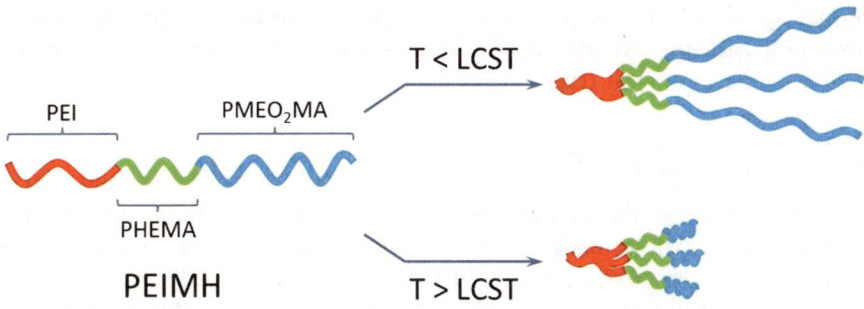

Figure 4.6 Thermoresponsive PEI derivatives as gene vectors.

vector formulations for gene transfer *in vitro* and murine lungs *in vivo*.[51] They found that *in vivo* experiments resulted in up to three-fold higher gene expression in the lung with lactose as lyoprotectant compared to sucrose or trehalose.

Yang *et al.* synthesized thermoresponsive diblock copolymers, poly[2-(2-methoxyethoxy)ethyl methacrylate]-*b*-poly(2-hydroxyethyl methacrylate), with low polydispersity by atomic transfer radical polymerization, which were then grafted with 1200 Da PEI to form PEIMH copolymer vectors (Figure 4.6).[52] The lower critical solution temperature (LCST) of PEIMH was dependent upon the grafting number of the PEI. When the temperature was elevated above the LCST, PEIMH could condense DNA more efficiently due to the shielding effect of collapsed PMEO2MA chains. Meanwhile, the contracted PMEO2MA chains led to more exposure of surface positive charges of the PEIMH/pDNA complexes, which was favorable for gene transport. Moreover, PEIMH showed superior transfection efficiency and considerably lower cytotoxicity compared to 25 kDa PEI.

4.3 Chitosan-Based Gene Vectors

Natural cationic polymers, such as chitosan, which are biocompatible, biodegradable, and have low toxicity, have been widely used as alternative gene vectors and much attention has been paid to them. Chitosan is composed of D-glucosamine and *N*-acetyl-D-glucosamine linked by β-(1,4)-glycosidic linkages. The above positive characteristics as well as multiple functionalization make chitosan a good gene carrier candidate and a series of experiments confirmed that chitosan could effectively combine with pDNA and protect it from nuclease degradation. In addition, the amino group of chitosan could bind with the negatively charged cell-membrane surface by electrostatic attraction and hydrogen bonding, which could increase the cell uptake of pDNA. However, the application of chitosan is significantly limited by its poor solubility (the amino groups on chitosan are only partially protonated at a physiological pH of 7.4), poor stability of the polyplex at physiological pH, low cell specificity, and low transfection efficiency, which need to be overcome before its use in clinical trials.

4.3.1   PEI-Modified Chitosans

In order to improve the transfection efficiency of chitosan, many efforts have been made and various modifications have been carried out. For example, Gao *et al.* linked chitosan with 1.8 kDa PEI by 1,1′-carbonyldiimidazole to form a chitosan-linked PEI copolymer (Scheme 4.5A).[53] In HepG2, A549, and HeLa cells, this compound exhibited lower cytotoxicity and better long-term transfection ability compared with 25 kDa PEI. Furthermore, it was used as a gene carrier to deliver the therapeutic gene CCL22 into peripheral hepatoma cells (H22). When these gene-altered cells were inoculated in mice, the tumor growth rate was significantly decreased. Ping *et al.* synthesized water-soluble chitosan-*g*-(PEI-β-CD) cationic copolymers *via* reductive amination between oxidized chitosan and LMW PEI (423 Da, 600 Da)-modified β-CD

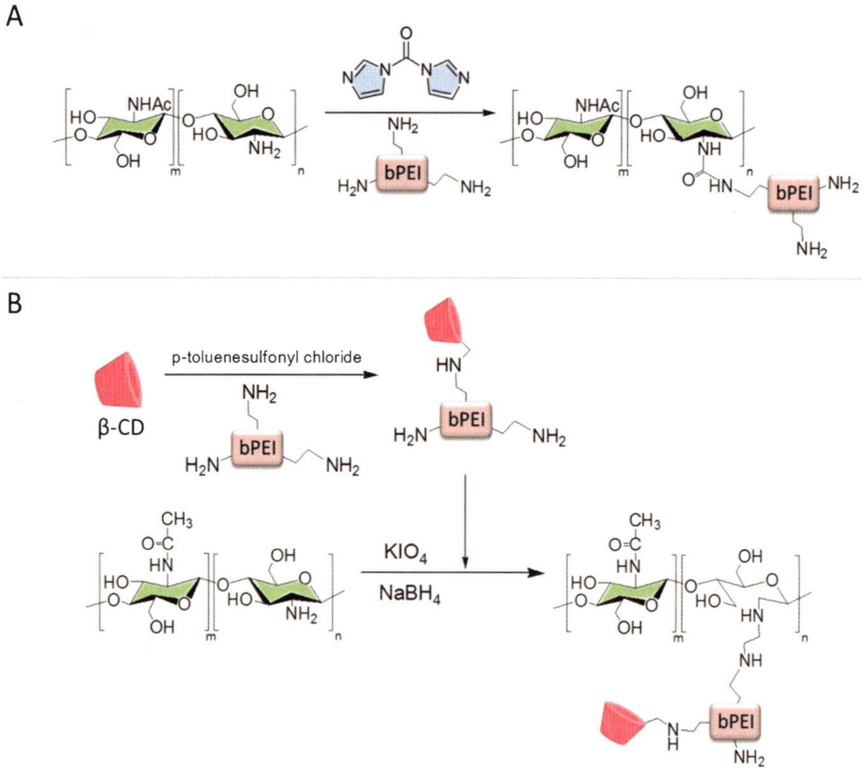

Scheme 4.5   Synthesis of PEI-modified chitosans.

(Scheme 4.5B).[54] These compounds exhibited good ability to condense plasmid DNA and siRNA, much lower cytotoxicity, and higher transfection efficiency than 25 kDa PEI in HEK293, L929, and COS7 cell lines.

4.3.2 Cell-Targeted Chitosans

Inefficient release of the polymer/DNA complexes from endocytic vesicles into the cytoplasm is one of the primary causes of poor gene delivery. Therefore, various specific ligands, such as galactose, transferrin, folate, RGD, mannose, *etc.*, have been incorporated into chitosan to enhance the cell-specific delivery. Lu *et al.* synthesized lactobionic acid-modified *N*-succinyl-chitosan-*g*-PEI copolymers as gene vectors with hepatocyte targeting properties.[55] These compounds showed higher gene transfection efficiency and cell specificity compared to 25 kDa PEI in HepG2 cells, which therefore has the potential to be a safe and efficient gene vector. Recently, Mohammadi *et al.* prepared chitosan–DNA nanoparticles by a complex coacervation process.[56] Fibronectin attachment protein of *Mycobacterium bovis* (FAP-B) was added to the chitosan–DNA nanoparticles *via* electrostatic attraction as a ligand for attachment to its specific receptors on the surface of epithelial cells. This compound showed 10-fold higher transfection efficiency than a nonspecific chitosan control in alveolar epithelial cells (A549), and higher cell viability compared to Turbofect control. Han *et al.* developed a cyclic RGD peptide-labeled chitosan nanoparticle by a thiolation reaction for tumor-targeted delivery of siRNA.[57] The cyclic RGD provided conformation stability and improved binding selectivity for the $\alpha_v\beta_3$ integrin overexpressed tumor cells. They found that this targeted delivery system-mediated gene silencing significantly enhanced antitumor therapeutic efficacy compared with a nontargeted delivery system in preclinical ovarian cancer models. Du *et al.* synthesized folic acid-conjugated stearic acid-grafted chitosan micelles by a 1-ethyl-3-(3-dimethylaminopropyl)carbodiimide coupling reaction for specific receptor-mediated gene delivery.[58] They found that folate conjugation increased the intracellular uptake of complexes in folate receptor-positive SKOV3 cells *via* folate receptor-mediated endocytosis, and resulted in higher transfection efficiency.

4.3.3 Other Chitosan Derivatives

Most of the modifications mentioned above were made *via* an amidation reaction, which could consume the amino group of chitosans and influence the cationic property of chitosan-mediated gene delivery, even resulting in higher cytotoxicity. As a recent alternative, thiolated chitosan has been designed and used as a gene vector. The free thiol groups on its side chains can form disulfide bonds with mucin glycoproteins on cell membranes, thereby promoting cellular uptake of the thiolated chitosan/pDNA complexes (Scheme 4.6A).[59] Li *et al.* designed a glutathione- and PMPEG-modified chitosan copolymer through this unique structural design (Scheme 4.6B).[60] A series of well-defined PEG brush-like PMPEG living chains with dithioester residues was prepared by employing the reversible addition–fragmentation chain transfer (RAFT) method, which then was grafted onto the allyl-chitosan *via* the radical coupling method. The resulting conjugate as a novel gene delivery vector showed excellent transfection efficiency and less cytotoxicity in mouse embryonic fibroblast cells (NIH3T3) compared to pristine chitosan. *N*,*N*,*N*-Trimethylated-chitosan (TMC) is a partially quaternized chitosan derivative which, compared to chitosan, has improved solubility in aqueous solution at neutral pH, safety, and effectiveness. Varkouhi *et al.* reported that the gene-silencing activity of the siRNA complexes based on thiolated TMC was substantially higher compared to the non-thiolated TMC and Lipofectamine-based complexes (Scheme 4.6C).[61] Zhao *et al.* synthesized a TMC–cysteine conjugate by conjugating L-cysteine hydrochloride onto TMC, which exhibited significant higher *in vivo* transfection efficiency than Lipofectamine 2000 (Scheme 4.6D).[62]

Polypeptides can bind the negatively charged backbone of a DNA chain, not only promoting its condensation but also favoring the interaction of the nanoparticle with the cell membrane and the consequent internalizations. Therefore, combining chitosan with polypeptides would allow higher transfection efficiency and low cytotoxicity. To assist the intracellular release of siRNA and enhance its effectiveness in gene silencing, Liao *et al.* reported an approach for the enhancement of the efficiencies of cellular uptake and gene silencing through the inclusion of a negatively charged poly(γ-glutamic acid) (γ-PGA) into the formulation of chitosan/siRNA complexes.[63] γ-PGA is a water-soluble, biodegradable, and nontoxic naturally occurring peptide. The results demonstrated that γ-PGA played an important role in improving the cellular uptake of chitosan/siRNA complexes, expediting their intracellular unpackaging, and the release of siRNA, thus significantly enhancing the efficiency of gene silencing and prolonging the duration of its action. Opanasopit and co-workers demonstrated that ternary complexes, resulting from association of chitosan with poly-L-arginine/DNA complexes, were significantly more efficient in mediating transfection than the corresponding chitosan/DNA complexes and had low cytotoxicity.[64]

NOVAFECT chitosans are ultrapure chitosan oligomers that were recently marketed as carriers for nonviral gene therapy. Klausner *et al.* designed chitosan–DNA nanoparticles based on NOVAFECT.[65] *In vitro* transfection studies demonstrated the ability of oligomeric chitosan–DNA nanoparticles to effectively transfect COS-7 cells. In rat corneas, injection of a select formulation of oligomeric chitosan–DNA nanoparticles into the stroma showed that gene expression was 5.4 times greater than PEI/DNA complexes. It was a promising approach for the treatment of acquired and inherited corneal diseases that otherwise lead to blindness.

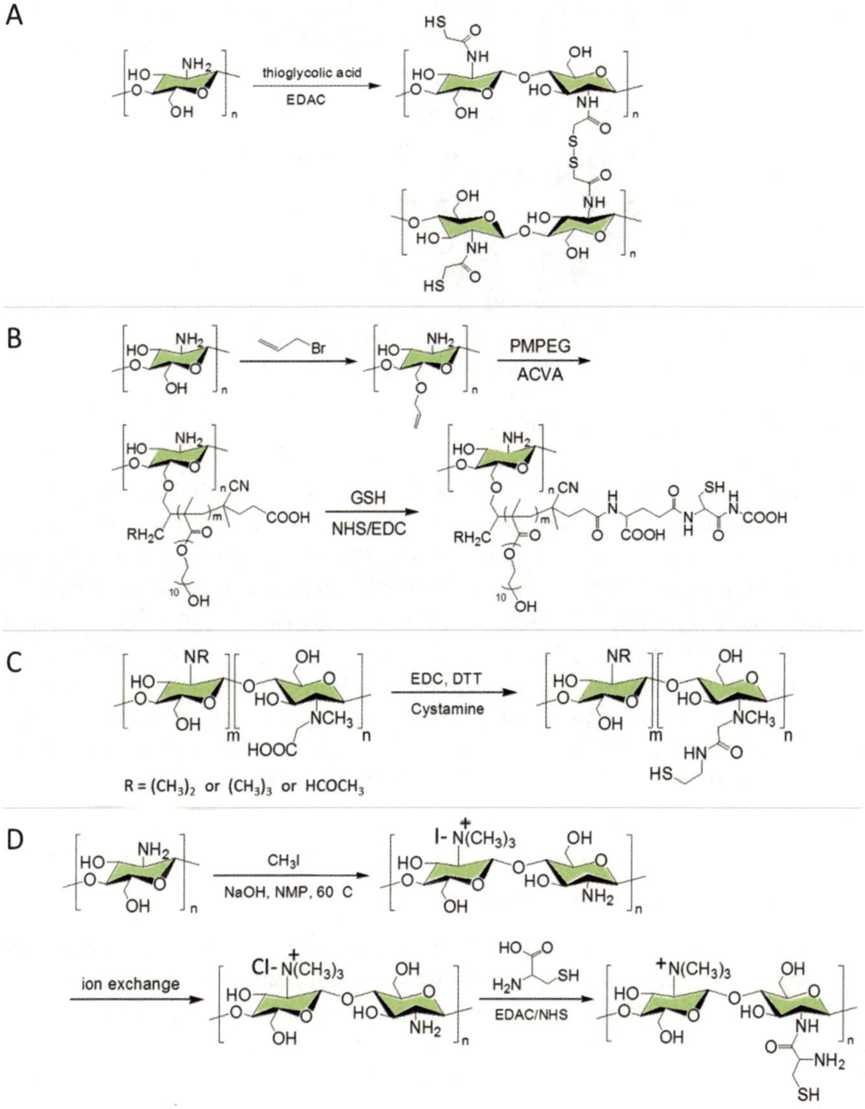

Scheme 4.6 Synthesis of thiolated chitosans.

4.4 Dendrimer-Based Gene Vectors

Dendrimers are highly branched, monodisperse macromolecules with symmetrical nanometer-sized architecture, which are prepared by multistep synthetic procedures. They consist of a central core, branching units, and terminal functional groups (Figure 4.7). This type of architecture induces the formation of nanocavities, the environment of which determines their solubilizing or encapsulating properties, while the external groups primarily

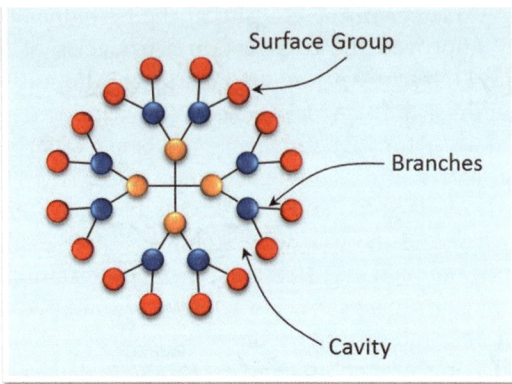

Figure 4.7   Structure of dendrimers.

characterize their solubility and chemical behavior. As nonviral gene delivery vectors, dendrimers have a couple of unique advantages. Their nanometer size enables them to mimic cell organelles, easing the transport of therapeutic genes from the cytoplasm to the nucleus. The intense proton sponge effect helps the dendrimer/DNA complex escape from the lysosome, thereby avoiding enzymatic degradation. Therefore, dendrimers display high transfection efficiency *in vitro*. However, *in vivo*, the sensitivity of the size and dispersal stability of these types of polyplexes to buffer systems and serum results in high toxicity, low transfection efficiency, poor pharmacokinetics, and poor biological distribution. To solve these problems, various surface modifications to dendrimers have been developed.

4.4.1   Polyamidoamine Dendrimers

Polyamidoamine (PAMAM), based on an ethylenediamine or ammonia core with four and three branching points, was the first complete dendrimer family to be synthesized, characterized, and commercialized. PAMAM dendrimers have been widely studied for gene delivery processes *in vitro* and *in vivo*, and their transfection capability appears to depend on the generation and number of primary amino groups on the surface. However, high cytotoxicity and relatively low transfection efficiency in the presence of serum limited their clinical application. Thus, tremendous research efforts have been made to develop PAMAM dendrimer-based gene vectors with low generation, low toxicity, and high transfection efficiency.

Biodegradable PAMAM dendrimers have been developed to reduce the cytotoxicity and improve the release of DNA inside the cell. Xue *et al.* synthesized bioreducible PAMAMs containing repeating disulfide linkages in the main chains and dendritic PAMAMs in the side chains *via* repetitive Michael addition and amidation (Scheme 4.7A).[66] These dendrimers showed low cytotoxicity. In comparison to the original disulfide-containing PAMAM with an aminoethyl side chain, the grafting of the bioreducible PAMAM with dendrimer greatly improved the transfection efficiencies of 293T and HeLa cells with foreign DNA at various N/P ratios. Alternatively, Wu *et al.* synthesized a serum-resistant PAMAM-based polypeptide dendrimer through peptide bond linkages for gene delivery (Scheme 4.7B).[67] The obtained compound, ALA, had nontoxic low-generation G2 PAMAM as its central core, polyglutamate as its star-shaped backbone branches, and G1 PAMAM as its branch grafts and peripheral terminals. ALA exhibited negligible cytotoxicity, significant high transfection efficiency, exciting serum tolerant capacity, and sustained transfection characteristics compared with commercial G5 and Lipofectamine, and proved itself a promising and challenging potential gene delivery carrier in safe and efficient *in vivo* gene delivery.

In addition to reducing cytotoxicity, various efforts have focused on incorporating cell targeting capability into PAMAM-based dendriplexes. Recently, Yu *et al.* reported a synthetic strategy *via* click chemistry for site-specific epidermal growth factor-PEG functionalization of PAMAM dendrons

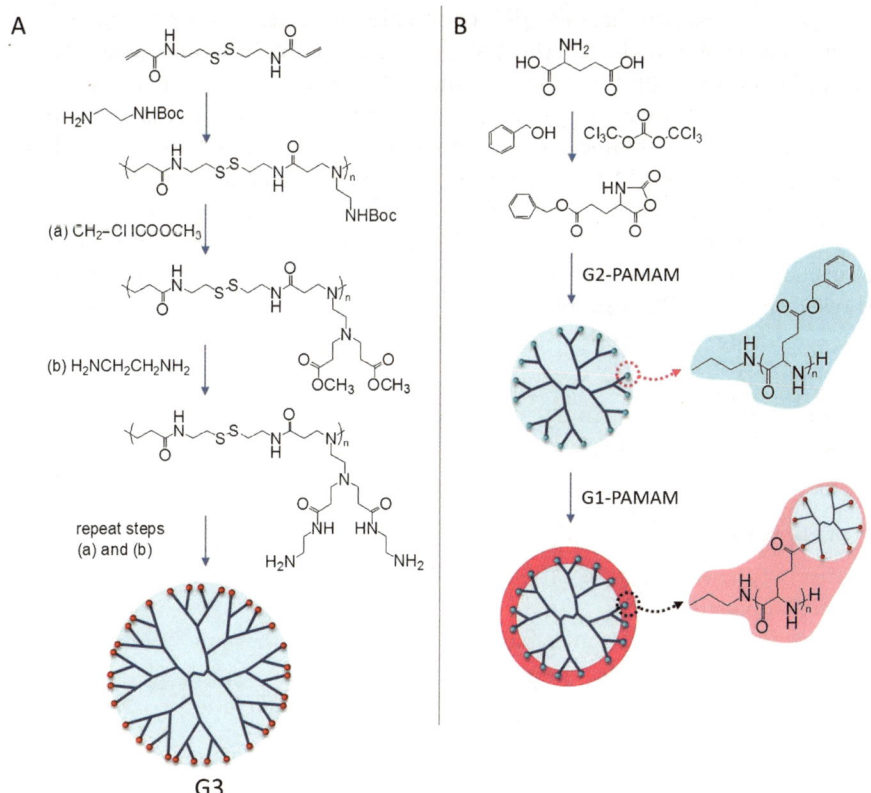

Scheme 4.7 Synthesis of (A) bioreducible PAMAM and (B) serum-resistant PAMAM.

for targeted gene delivery (Scheme 4.8).[68] PAMAM-pentaethylenehexamine (PEHA) dendron polyplexes displayed the best gene transfer ability. Conjugation of a PAMAM-PEHA dendron with a PEG spacer was conducted *via* a click reaction, which was performed before amidation with PEHA. The resultant PEG-PAMAM-PEHA copolymer was then coupled with an EGF ligand. This compound showed 10-fold higher transfection efficiency in HuH-7 hepatocellular carcinoma cells compared to the ligand-free ones. Zhang *et al.* reported that heparin and heparin-biotin were introduced into PAMAM/DNA complexes *via* electrostatic interactions to form self-assembled PAMAM/ DNA/heparin and PAMAM/DNA/heparin-biotin terplexes, respectively.[69] The results indicated that these terplexes exhibited decreased cytotoxicity after incorporation of heparin and heparin-biotin. Compared with PAMAM/ DNA and PAMAM/DNA/heparin complexes, the terplexes exhibited much higher cellular uptake into HeLa cells due to the specific interactions between biotin and biotin receptors on HeLa cells, which led to the enhanced transfection activity.

The blood–brain barrier (BBB) exerts a neuroprotective function, as it hinders the delivery of diagnostic and therapeutic agents to the brain. To overcome this natural barrier, an intracerebral gene delivery strategy has been extensively explored during the last two decades. A 29 amino acid peptide derived from the rabies virus glycoprotein (RVG29) was exploited as a ligand for efficient brain-targeting gene delivery due to its specific binding ability to the nicotinic acetylcholine receptor (nAchR) on neuronal cells. Liu and co-workers modified PAMAM with RVG29 through bifunctional PEG, and then complexed it with DNA, yielding PAMAM-PEG-RVG29/DNA nanoparticles (NPs).[70] PAMAM-PEG-RVG29/DNA NPs showed higher BBB crossing efficiency than PAMAM/DNA NPs in an *in vitro* BBB model. *In vivo* imaging showed that the NPs were preferably accumulated in the brain. Moreover, the

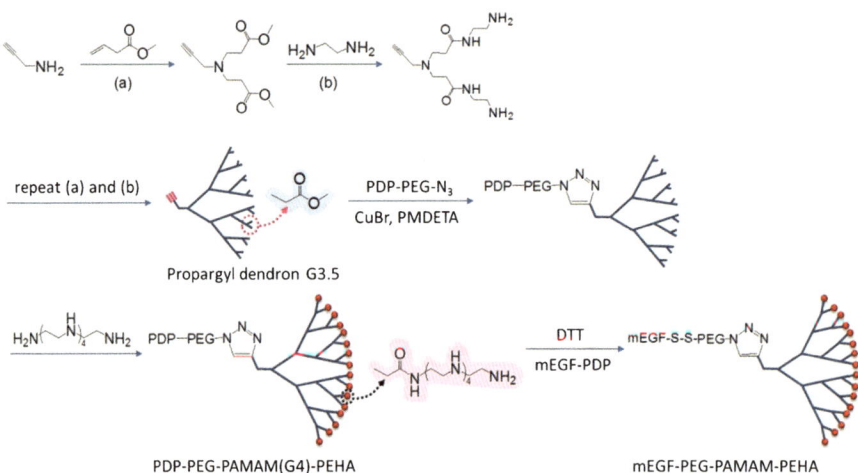

Scheme 4.8   Synthesis of ligand-modified PAMAM dendrimers *via* click chemistry.

report gene expression of the PAMAM-PEG-RVG29/DNA NPs was observed in the brain, and significantly higher than unmodified NPs.

Huang *et al.* developed a PAMAM-based glioma-targeting delivery system by conjugating a chlorotoxin ligand to PAMAM *via* bifunctional PEG.[71] The results showed that *in vivo* distribution and gene expression of this compound were significantly broader and higher in glioma than with unmodified PAMAM. In the study of Huang *et al.*, angiopep-2, which can target the low-density lipoprotein receptor-related protein-1 (LRP1) expressed on BCECs and glial cells, was exploited as the targeting ligand to conjugate PAMAM *via* bifunctional PEG to form a glial cell targeting gene vector.[72] The results showed that *in vivo* biodistribution of PAMAM-PEG-angiopep/DNA NPs in the brain, especially in the tumor site, was much more obvious due to the enhanced permeability and retention (EPR) effect and angiopep-2 targeting ability. This compound showed low cytotoxicity after *in vitro* transfection. Ke *et al.* synthesized PAMAM-PEG-angiopep for brain-targeting gene delivery both *in vitro* and *in vivo*.[73] The cellular uptake of the angiopep-modified NPs was in competition with angiopep-2, receptor-associated protein, and lactoferrin, indicating that LRP1-mediated endocytosis may be the main mechanism of cellular internalization of angiopep-modified NPs. Also the angiopep-modified NPs showed higher efficiency in BBB than unmodified NPs in an *in vitro* BBB model, and accumulated in brain more *in vivo*.

4.4.2 Polypropylenimine Dendrimers

As an alternative to PAMAM dendrimers, various polypropylenimine dendrimer-based compounds have been synthesized and investigated for gene transfer. Dufès' group have focused on the synthesis of gene vectors based on polypropylenimine dendrimers. They demonstrated that the conjugation of transferrin (TF) to the G3 polypropylenimine dendrimer led to an improved tumor specificity of gene expression and sustained tumor regression after intravenous administration without visible secondary effects to the mice.[74] Recently, they further found that TF-modified polypropylenimine dendrimer complexed to a plasmid DNA encoding p73 led to an enhanced anti-proliferative activity *in vitro* and *in vivo* compared to the unmodified dendriplex.[75] On the basis that amino acids such as arginine, lysine, and leucine are involved in enhancing DNA transportation into cells, the group grafted these amino acids onto polypropylenimine dendrimers.[76] These amino acid-modified dendrimers showed enhanced anti-proliferative activity *in vitro* and improved tumor gene expression *in vivo* compared to unmodified dendrimer.

Minko's group investigated the ability of superparamagnetic iron oxide (SPIO) nanoparticles and polypropylenimine G5 dendrimers to cooperatively provoke siRNA complexation in order to develop a targeted, multifunctional siRNA delivery system for cancer therapy.[77] A PEG coating and a cancer-specific targeting moiety (LHRH peptide) have been incorporated into SPIO-PPI G5-siRNA complexes to enhance serum stability and selective internalization by cancer cells. Such a modification of siRNA nanoparticles enhanced its internalization into cancer cells and increased the efficiency of targeted gene suppression *in vitro*. Moreover, the developed siRNA delivery system was capable of sufficiently enhancing *in vivo* antitumor activity of an anticancer drug (cisplatin). Recently, they constructed a nanocarrier-based delivery system by taking advantage of the lessons learned from the problems in the delivery of DNA.[78] In this study, siRNA nanoparticles were first formulated with polypropylenimine dendrimers. To provide lateral and steric stability to withstand the aggressive environment in the blood stream, the formed siRNA nanoparticles were caged with dithiol-containing crosslinker molecules followed by coating them with PEG. A synthetic analog of a luteinizing hormone-releasing hormone (LHRH) peptide was conjugated to the distal end of the PEG polymer to direct the siRNA nanoparticles specifically to the cancer cells. The results demonstrated that this layer-by-layer modification and targeting approach conferred the siRNA nanoparticles with stability in plasma and intracellular bioavailability, and provided for their specific uptake by tumor cells, accumulation of siRNA in the cytoplasm of cancer cells, and efficient gene silencing. In addition, *in vivo* body distribution data confirmed high specificity of the proposed targeting delivery approach, which created the basis for the prevention of adverse side effects of the treatment on healthy organs. Hao *et al.* modified G3 cationic dendritic polymeric polypropylenimine by pluronic P123 and investigated its use for gene delivery.[79] The transfection efficiency of SPC-A1 cells using P123-PPI/DNA nanoparticles was much higher than the transfection utilizing PPI/DNA nanoparticles. The addition of free P123 during the preparation of P123-PPI/DNA nanoparticles could significantly enhance the transfection efficiency in the presence of 10% fetal bovine serum.

4.4.3   Poly(L-lysine) Dendrimers

Peptide dendrimers are constructed by peptide bond linkages from natural or unnatural amino acids. They share common features with proteins such as globular structure, water solubility, biocompatibility, and biodegradability. Poly(L-lysine) dendrimers are a new type of polypeptide dendrimer, which have been explored as gene and drug delivery systems in recent years. Luo *et al.* reported the synthesis and characterization of different generations of dendritic poly(L-lysine) vectors and their use for *in vitro* gene transfection (Scheme 4.9).[80] They found that higher generations tended to produce greater positive potentials, indicating a stronger potency of the complexes to interact with negatively charged cell membranes. *In vitro* and *in vivo* cytotoxicity results showed good biocompatibility of the dendrimers. *In vitro* gene transfection revealed higher efficiency of G5 than other dendrimers and insensitive variation to the presence of serum. Al-Jamal *et al.* reported that G6 PLL dendrimer has the ability to accumulate and persist in solid tumor sites after systemic administration and exhibit antiangiogenic activity in the absence of cytotoxicity.[81]

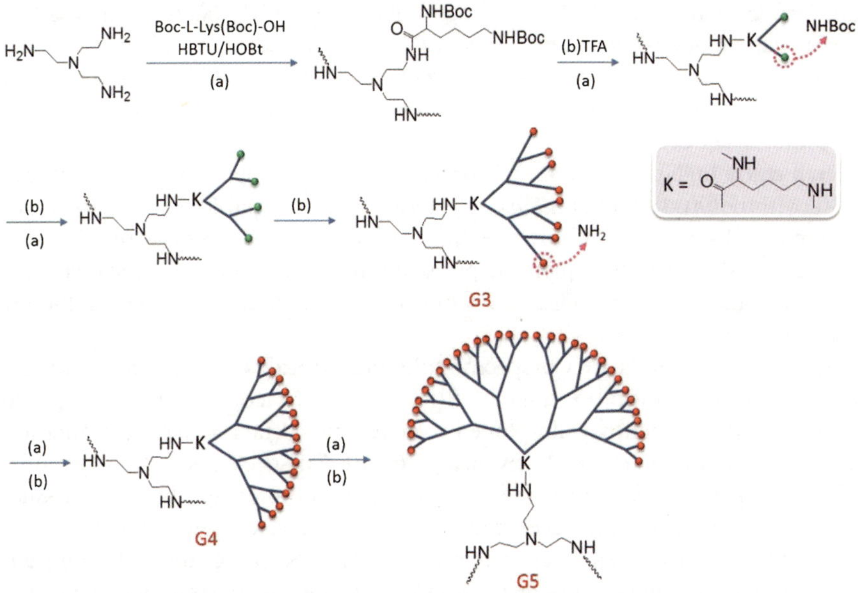

Scheme 4.9 Synthesis of L-lysine-based peptide dendrimers.

4.5 Polypeptide Gene Vectors

Among various nonviral gene delivery methods, using natural or artificial polypeptides with certain biological functions is considered as one promising approach. Unlike other vectors, polypeptides have low cytotoxicity, greater biodegradability, and also serve different functions. For example, lysine- and/ or arginine-rich cationic peptides can condense DNA into compact particles, TAT-based cell-penetrating peptides can disrupt the endosomal membrane, nuclear localization signal (NLS) peptides can traffic DNA to the nucleus, and RGD-rich peptides can target polyplexes to specific receptors. These properties may all be part of a single peptide sequence or a combination of peptides chemically conjugated to form a vector capable of packaging and targeting DNA for efficient delivery.

4.5.1 Normal Peptide-Based Vectors

Polypeptide-based vectors have great potential for gene delivery, as numerous new functional peptides will be continually discovered, and they have the ability to achieve these goals alone or in combination with other systems. The approach of introducing ligands that lead gene vectors to target caveolae-mediated endocytosis on a nanoparticle surface might serve as a promising strategy for effective gene transfection. Recently, in an attempt to enhance the possibility of caveolae-mediated endocytosis, Liu *et al.* demonstrated a peptide-targeted gene vector for highly efficient receptor-mediated intracellular delivery (Scheme 4.10).[82] A cyclic Asn-Gly-Arg (cNGR) peptide was used to target gene-loaded poly(lactic acid)-PEG nanoparticles into human umbilical vein endothelial cells overexpressing CD13. The cNGR peptide could specifically mediate the fast and efficient internalization of these nanoparticles into the target cells.

In a study by Numata *et al.*, silk-based block copolymers were bioengineered both with poly(L-lysine) domains, to interact with plasmid DNA, and RGD, to enhance cell-binding and transfection efficiency.[83] Samples with 30-lysine residues and 11 RGD sequences, prepared at an N/P ratio of 2, showed the highest transfection efficiency in HeLa cells and human embryonic kidney cells.

Kang *et al.* found a PKCα-specific substrate peptide that exhibited a higher phosphorylation ratio in tumor cells and tissues, compared with normal tissues.[84] Moreover, they developed a novel tumor-targeted gene regulation system responding to PKCα by using the PKCα-specific substrate peptide. This system could distinguish between normal and tumor cells but is not tissue-specific.[85] Recently, they developed a hepatoma-targeted gene delivery system by combining a human liver cell-specific bionanocapsule and a tumor cell-specific gene regulation polymer, which responds to hyperactivated protein kinase C (PKC)α in hepatoma cells.[86]

Chen *et al.* synthesized arginine-rich amphiphilic lipopeptides with hydrophobic aliphatic tails ($C_{12}GR_8GDS$ and $C_{18}GR_8GDS$) as functional gene vectors (Scheme 4.11).[87] They found that these lipopeptides exhibited very low cytotoxicity even at high concentration, and could be specifically recognized by cancer cells due to the incorporation of RGD sequences (specifically recognized by integrins $\alpha_v\beta_3$ and $\alpha_v\beta_5$ overexpressed on cancer cells).

Disulfide-linked bioreducible polymers are more efficient for the intracellular delivery of DNA. This increased efficiency is likely due to enhanced decomplexation as a result of degradation of the disulfide bonds in the reductive environment of the cytoplasm. Hyun *et al.* synthesized a reducible poly(oligo-D-arginine) gene carrier by connecting nine-arginine oligopeptides *via* internal disulfide linkages, and then mediated the heme oxygenase-1 (HO-1) gene for the treatment of ischemia/reperfusion (I/R)-induced brain stroke.[88] This polypeptide carrier showed markedly enhanced gene transfection and lower toxicity than 25 kDa PEI *in vitro* and *in vivo*. As the sequence of primary

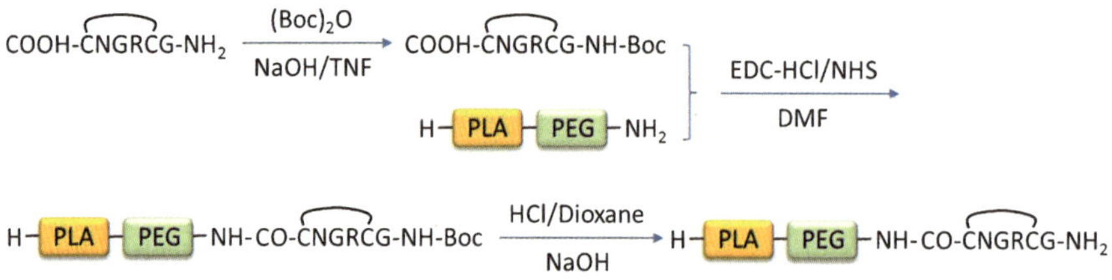

Scheme 4.10 Synthesis of the cNGR-PEG-PLA peptide gene vector.

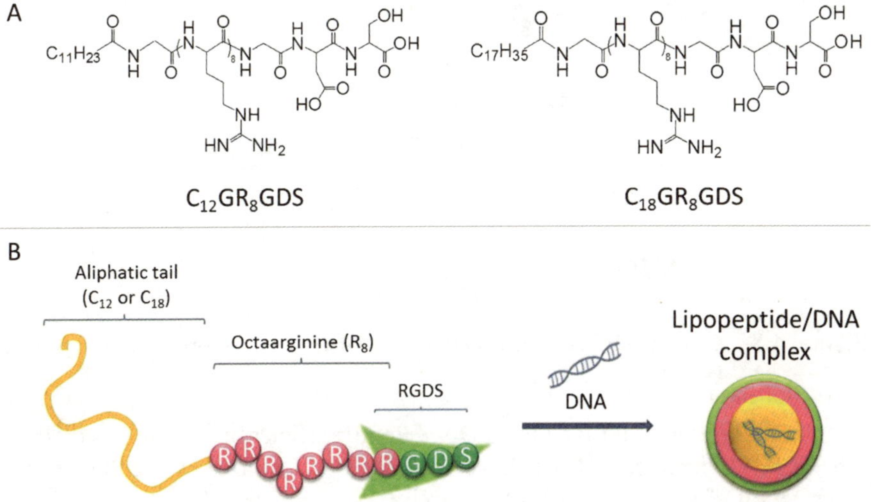

Scheme 4.11 Tumor-targeted arginine-rich amphiphilic lipopeptide gene vectors: (A) chemical structure of arginine-rich peptides; (B) complex formed by lipopeptide and DNA.

cardiomyocyte (PCM) (a 20-amino acid peptide) showed high selectivity for cardiomyocytes, Nam *et al.* developed a cardiomyocyte-targeted Fas siRNA delivery system using PCM-specific peptide-modified bioreducible poly(CBA-DAH) (PCD).[89] The impact of PCM conjugation on cellular uptake and transfection efficiency was greater in H9C2 rat cardiomyocytes than in NIH 3T3 cells. Fas siRNA/PCM-polymer polyplexes exhibited significant Fas gene silencing in rat cardiomyocytes under hypoxic conditions, leading to inhibition of cardiomyocyte apoptosis.

Harris *et al.* developed electrostatically adsorbed poly(glutamic acid)-based peptide coatings to alter the exterior composition of a core gene delivery particle and thereby affect the tissue specificity of the gene delivery function *in vivo*.[90] They found that, with all coating formulations tested, the coatings reduced the potential toxicity associated with uncoated cationic gene delivery nanoparticles following systemic injection. Particles coated with a low peptide:DNA weight ratio (w/w) form large 2 μm sized particles that could facilitate specific gene delivery to the liver. The same particles coated at a higher w/w form small 200 nm particles that can facilitate specific gene delivery to the spleen and bone marrow. Thus, variations in nanoparticle peptide coating density could alter the tissue specificity of gene delivery *in vivo*.

4.5.2 Cell-Penetrating Peptides

Cell-penetrating peptides (CPPs) are cationic polypeptides that can bind to plasmid DNA *via* charge interaction and condensation, and which can shuttle nucleic acids through cell membranes (Figure 4.8). CPP-based delivery systems

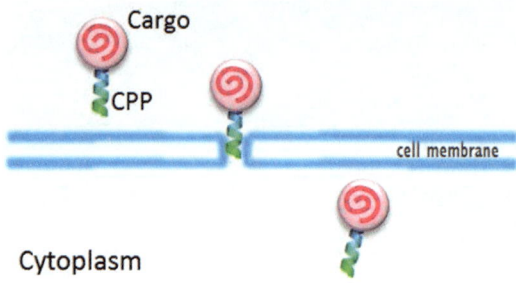

Figure 4.8   Cell membrane translocation mediated by a cell-penetrating peptide (CPP).

represent a strategy that facilitates DNA import efficiently and nonspecifically into cells. Therefore, CPPs have been widely utilized to enhance the transfection efficiency in current research.[91]

The most important classes of CPP are isolated from viruses, such as TAT (transactivated transcription, an arginine-rich CPP) derived from HIV-1. The HIV-1 TAT peptide has been successfully used for intracellular gene delivery. Yamano *et al.* demonstrated that TAT incorporated with histidine and cysteine residues and combined with the cationic lipid transfection reagent FuGENE HD results in very efficient gene transfer in a medium supplemented with serum across a range of cell lines.[92] Govindarajan *et al.* conjugated a cell-penetrating peptide (TAT-Mu) with a targeting ligand (HER2 antibody mimetic-affibody), resulting in specific TF-knockdown *in vitro* and regression of tumors in a xenograft mouse model.[93]

Another CPP, KALA (a cationic endosomolytic and fusogenic peptide), has also been investigated as a gene vector. Chen *et al.* studied the effect of KALA on the substrate-mediated gene delivery.[94] A fast degrading cholic acid functionalized star poly(DL-lactide) was used to fabricate calcium phosphate(Ca-P)/DNA/KALA co-precipitate-deposited films. They found that this system could rapidly release Ca-P/DNA/KALA to mediate efficient gene expression, and gene expression was significantly enhanced by the addition of KALA.

Owing to their intracellular permeability, protein transduction domains (PTDs) have been widely used to deliver proteins and peptides to mammalian cells. However, their performance in gene delivery has been relatively poor. To improve the efficiency of PTD-mediated gene delivery, Min *et al.* synthesized a new peptide, KALA-Antp (K-Antp), which contains the sequences for PTD of the third α-helix of the antennapedia (Antp) homeodomain and the fusogenic peptide KALA.[95] In this configuration, Antp is designed to provide the cell permeation capacity and nuclear localization signal, while the KALA moiety promotes cellular entry of the peptide–DNA complex. An optimal K-Antp/DNA formula was nearly 400- to 600-fold more efficient than Antp or polylysine-Antp (L-Antp) in gene delivery, and comparable or superior to a commercial liposome.

4.5.3   Nuclear Localization Signal

Nonviral gene therapy is challenged by inefficient delivery at the level of intracellular processing. Several barriers have been described and studied, including failure to escape from vesicular structures, lysosomal degradation, enzymatic degradation in the cytosol, entrapment in the highly viscous and crowded cytosol, lack of transport towards the nucleus and uptake into the nucleus, and finally inefficiency of transcription and/or translation. Capecchi *et al.* found that transport into the nucleus is the major bottleneck for successful nonviral gene delivery.[96] The nuclear localization signal (NLS), which is recognized by nuclear transport proteins (*e.g.*, importin-α which directly binds the NLS signal; and importin-β, which is the mediator of the actual import process through the nuclear pores) could overcome the nuclear membrane barrier and promote the nuclear translocation (Figure 4.9). Among them, the extensively studied NLS sequence is the SV40 large-T antigen (especially in its shortest form, Pro-Lys-Lys-Lys-Arg-Lys-Val).

   Much effort has been made to mimic this process by incorporating NLSs in gene delivery complexes. Dean *et al.* first demonstrated that plasmid nuclear import could be facilitated by the insertion into the plasmid of nuclear DNA-targeting sequences.[97] Recently, Wang and co-workers introduced an iodine atom to a nuclear localization signal (PKKKRKV) by chemically attaching 2-iodobenzoic acid to the peptide to obtain NLS-I peptide for cell targeting and nuclear transport.[98] They found that cell internalization and nuclear accumulation of NLS-I was markedly increased compared to NLS in MCF-7 cells, and gene expression by PEI1800/DNA/NLS-I complexes exhibited much enhanced efficiency (up to 130-fold). This study demonstrates an

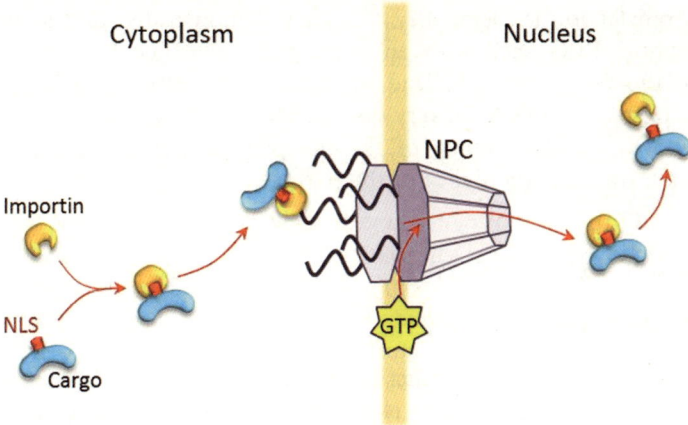

Figure 4.9   Nuclear localization signal-mediated nuclear import.

alternative method to construct a nonviral delivery system for targeted gene transfer into breast cancer cells.

4.5.4 Asp-Based Peptides

Kataoka's group found that polyplex micelles formed with plasmid DNA and PEG-*b*-poly{*N*-[*N*-(2-aminoethyl)-2-aminoethyl]aspartamide}   [PAsp(DET)] exhibit effective endosomal escaping properties based on di-protonation of diamine side chains with decreasing pH, which improves their transfection efficiency and thus are promising candidates for local *in vivo* gene transfer.[99] Recently, they improved the design of PEG-PAsp(DET)-based gene vectors by incorporating a cholesterol moiety into the terminus of the PAsp(DET) segment in the block copolymer.[100] PEG-PAsp(DET)-Chole micelles could be formed over the stoichiometric charge ratio due to self-association of cholesterol, and increased transfection efficiency at low N/P ratios. Furthermore, cholesterol introduction led to increased stability of polyplex micelles in the blood, which resulted in significant suppression of subcutaneous pancreatic tumor growth by intravenous injection of polyplex micelles loading sFlt-1 pDNA. They also succeeded in bone regeneration by introducing osteogenic factor-expressing genes for bone defects in the mouse skull.[101] Recently, the same group improved the safety and transfection efficiency of this polyplex micelle system by adding an anionic polycarbohydrate, chondroitin sulfate (CS).[102] The results showed that the addition of CS markedly reduced tissue damage and subsequent inflammatory responses in the skeletal muscle and lungs of mice following *in vivo* gene delivery with the polyplex micelles, which led to prolonged transgene expression in the target organs. This combination of polyplex micelles and CS holds great promise for safe and efficient gene introduction in clinical settings.

Lai *et al.* synthesized pluronic P85-based cationomers comprising poly{*N*-[*N*-(2-aminoethyl)-2-aminoethyl]aspartamide} (P[Asp(DET)]) cationic copolymers for transfection (Scheme 4.12).[103] They demonstrated that substituting the hydrophilic PEG shell with an amphiphilic pluronic P85 layer on the surface of the polyplex could lead to higher transfection efficiency of the P[Asp(DET)]-based pDNA delivery system. The amphiphilic nature of the P85 shell promoted cellular uptake of the polyplex particles and subsequently increased the transfection ability of the particles.

4.6   Other Gene Vectors

4.6.1   Lipid-Based Vectors

For gene carriers, cationic liposomes have recently emerged as leading nonviral vectors in worldwide gene therapy clinical trials. However, cytotoxic effects or apoptosis are often observed, which is mostly dependent on the cationic lipid used. In order to reduce the cytotoxicity, Li *et al.* synthesized a new cationic

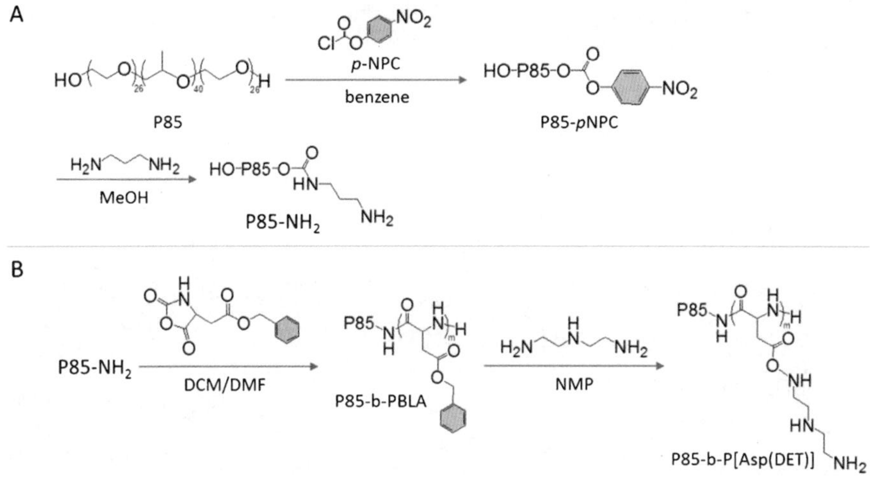

Scheme 4.12 Synthesis of (A) pluronic (P85) amine and (B) P85-*b*-P[Asp(DET)] copolymers.

liposome for pulmonary gene transfection using 6-lauroxyhexyl lysinate (LHLN).[104] When administered by intratracheal instillation into rat lungs for *in vivo* evaluation, LHLN-liposome/DNA complexes exhibited higher pulmonary gene transfection efficiency than Lipofectamine 2000/DNA complexes.

Hu *et al.* developed pegylated immuno-lipopolyplex (PILP) gene delivery systems by employing PEI/DNA complexes, as well as anionic liposomes composed of POPC, (DSPE)-PEG2000, and (DSPE)-PEG2000-biotin, and by using streptavidin-monoclonal antibody conjugating through the biotin group located at the distal end of the PEG spacer as the targeting antibody.[105] These complexes showed high efficiency in gene delivery to liver cancer cells but with no significant cytotoxicity. Intravenous administration of the PILP resulted in tumor- and liver-targeted gene expression of the reporter genes EGFP and luciferase, as opposed to the lung-targeted gene expression obtained with PEI/DNA complexes, causing no cytokine production and liver injury. PILP are promising gene delivery systems which may be used to target liver cancer.

Shirazi *et al.* have presented the efficient synthesis of a series of degradable multivalent cationic lipids (CMVL$_n$, n = 2–5) containing a disulfide bond spacer between the headgroup and lipophilic tails (Scheme 4.13).[106] This spacer is designed to be cleaved in the reducing milieu of the cytoplasm and thus decrease lipid toxicity. The CMVLs (n = 3–5) exhibit reduced cytotoxicity and transfect mammalian cells with efficiencies comparable to those of highly efficient nondegradable analogs and benchmark commercial reagents such as Lipofectamine 2000.

Huang *et al.* have synthesized two novel protonated cyclen and imidazolium salt-based cationic lipids, which differ only in their hydrophobic region (cholesterol or diosgenin).[107] Cationic liposomes were easily prepared from each of these lipids individually or from the mixtures of each cationic lipid and

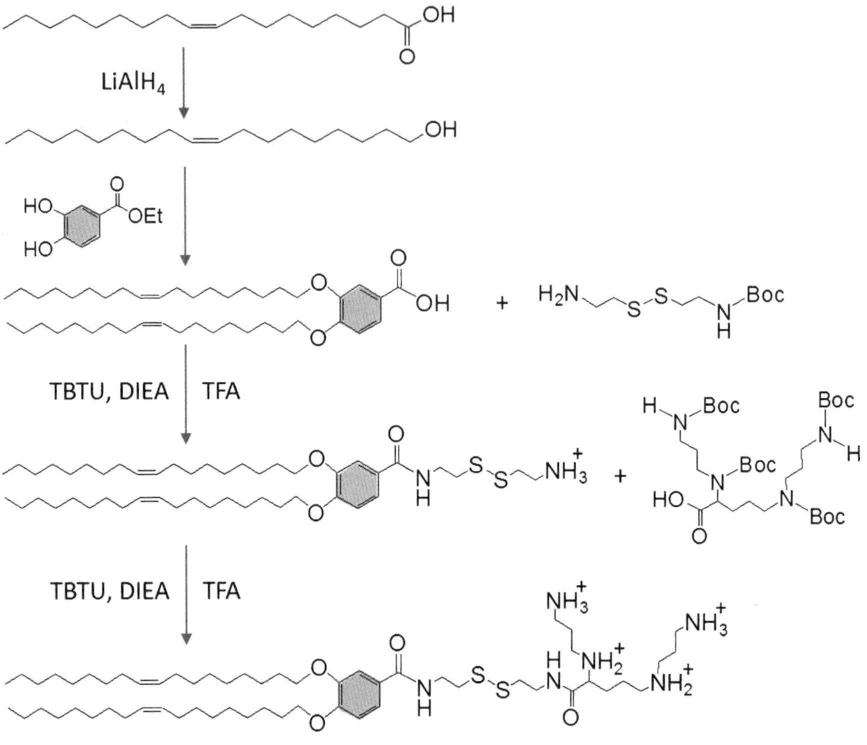

Scheme 4.13 Synthesis of reductively degradable multivalent cationic lipid (CMVL$_4$).

dioleoylphosphatidylethanolamine (DOPE). The results showed that two cationic lipids could induce effective gene transfection in HEK293 cells in association with DOPE. The gene transfection efficiencies of two cationic lipids were dramatically increased in the presence of calcium ion (Ca^{2+}), and their gene transfection abilities were maintained in the presence of 10% serum.

4.6.2 Polyallylamine

Polyallylamine, another synthetic cationic polymer, possesses a high density of primary amino groups which are suitable for binding and packaging negatively charged DNA. However, the utility of polyallylamine for gene delivery application is limited by its cytotoxicity of a too strong polycationic character. Different chemical modifications have been used to decrease the cytotoxicity and enhance the transfection efficiency of polyallylamine. For example, Pathak *et al.* demonstrated that polyallylamine modified with imidazole and PEG-bis(phosphate) could reduce the positive surface charge to decrease the

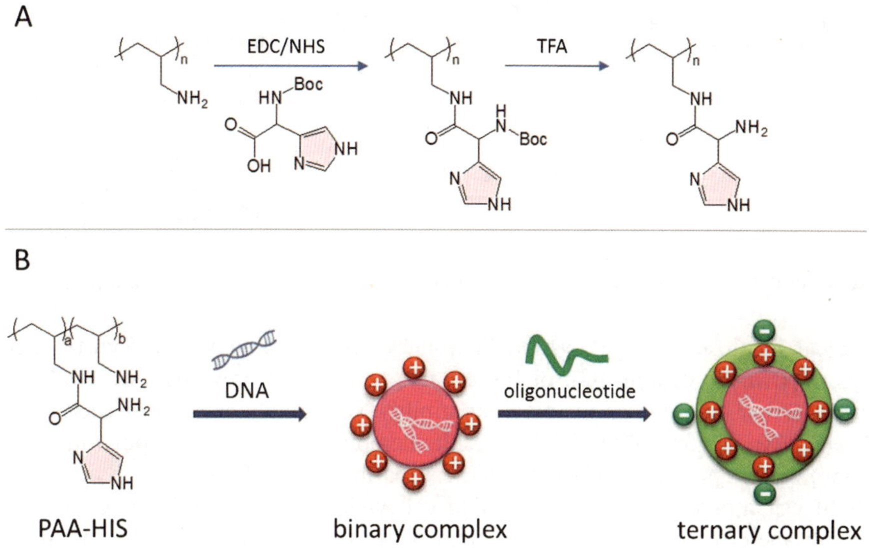

Scheme 4.14   Schematic illustration of (A) PAA-HIS and (B) a ternary polyplex.

cytotoxicity and achieve enhanced transfection efficiency.[108] In addition, Nimesh *et al.* showed polyallylamine complexed with dextran and DNA could simultaneously improve upon transfection efficiency and cell viability.[109] Recently, Chung *et al.* reported the preparation and characterization of ternary nanoparticles with a negative surface charge, which comprises a histidine-conjugated polyallylamine (PAA-HIS)/DNA core complex and a single-stranded oligonucleotide outer layer, to transfect various cell lines (Scheme 4.14).[110]

As a continued effort, the same authors reported their investigations on the endocytotic mechanisms involved in the uptake of the oligonucleotide-coated PAA-HIS/DNA complexes.[111] The results indicated that the oligonucleotide-coated PAA-HIS/DNA complexes could specifically recognize adenosine receptors on the cell surface and were taken up by an adenosine receptor-mediated process. After receptor/ligand binding, oligonucleotide-coated PAA-HIS/DNA/complexes were mainly internalized *via* a caveolae-mediated pathway to result in effective intracellular processing for gene expression.

4.6.3   Linear Poly(amidoamine)s

Recently, linear poly(amidoamine)s, which can be easily synthesized by the polyaddition of primary monoamines or bis(secondary amines) to bisacrylamides, have attracted researchers' attention. Peng and co-workers reported the synthesis, transfection, and intracellular trafficking characteristics of poly(amidoamine)s with a pendant primary amine in the delivery of plasmid DNA to bone marrow stromal cells (BMSCs).[112] It was found that poly(amidoamine)s with a pendant aminobutyl group demonstrated higher transfection efficiency and percentage of

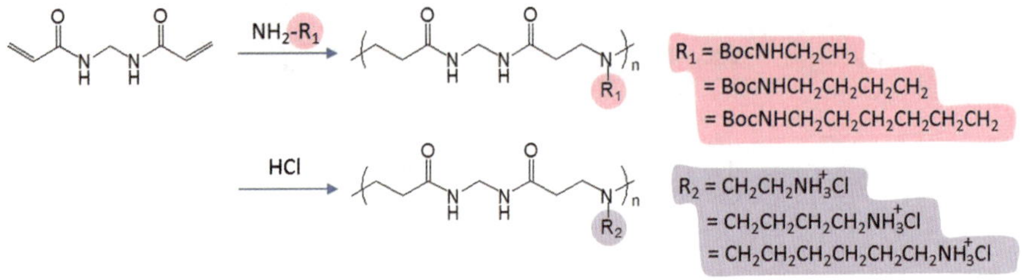

Scheme 4.15   Synthesis of a poly(amidoamine) with pendant primary amines.

nuclear localization than 25 kDa PEI. They also had much lower cytotoxicity against BMSCs with an IC_{50} of 100 mg mL^{-1}. Based on previous research, Peng *et al.* further investigated the biocompatibility, including *in vitro* cytotoxicity and *in vivo* tissue compatibility.[113] The results demonstrated that these poly(amidoamine) vectors possess much better cytocompatibility than 25 kDa PEI, yielding a slight cell morphological change, high cell viability, and a mild effect on cell membrane damage. They also exhibited better tissue compatibility, reflected by no or less inflammatory response in the site of muscle injection.

Liu *et al.* synthesized novel poly(amidoamine)s with pendant primary amines by Michael polyaddition of a diamine to *N,N*-methylenebis(acrylamide) (Scheme 4.15).[114] These polymers all showed high buffering capacity between pH 5–7 and excellent DNA binding ability, which can condense DNA to form nanosized polyelectrolyte complexes with a positive surface charge. These polymers had comparable transfection efficiency and much lower cytotoxicity than that of commercial 25 kDa PEI.

4.6.4   Multi-layer Complexes

Binary complexes of cationic polymers and DNA have been used commonly for DNA delivery, but the excess cationic charge of the binary complexes mainly leads to high toxicity and instability *in vivo*. In attempting to shield the cationic charge and reduce the cytotoxicity, various multi-layer complexes have been designed. Cheng and colleagues prepared a series of self-assembled polyionic complexes (PICs) *via* electrostatic attraction between protamine sulfate (PS) and poly(L-aspartic acid) [P(Asp)] or doxorubicin (DOX)-conjugated P(Asp).[115] An *in vitro* gene transfection investigation revealed that the transfection efficiency of the PICs/DNA complexes was comparable to that of 25 kDa PEI/DNA complex (N/P ratio = 10). Importantly, the gene transfection efficiency of PICs/DNA complexes could be tuned by altering the weight ratio of PS/P(Asp). The suppression of the proliferation activity of HeLa cells could be achieved by replacing P(Asp) with DOX-P(Asp), suggesting a great potential of PICs as effective carriers for combined delivery of drug and gene.

Guo *et al.* developed ternary complexes by coating polyglutamic acid-*graft*-PEG (PGA-g-mPEG) onto binary complexes of polycaprolactone-*graft*-poly(*N,N*-dimethylaminoethyl methacrylate) (PCL-g-PDMAEMA) nanoparticles

(NPs)/DNA for effective and targeted gene delivery (Scheme 4.16A).[116] The ternary complexes of PCL-*g*-PDMAEMA NPs/DNA/PGA-*g*-mPEG demonstrated lower cytotoxicity and higher gene transfection efficiency than the binary complexes *in vitro*. *In vivo* gene transfection experiments indicated that the ternary complexes presented lower toxicity and higher protein expression in HeLa tumor-bearing mice than the binary complexes.

Shen *et al.* have described a combined gene vector composed of PEI and a nuclear protein (HMGB1) containing NLS (Scheme 4.16B).[117] The results of cell viability studies suggested a lower cytotoxicity for the HMGB1/PEI combined carriers. The ternary complexes were formed *via* electrostatic interactions between DNA, HMGB1, and PEI. Transfection efficiencies of the ternary complexes were higher than that of DNA/PEI complexes.

Wang *et al.* have developed a simple method to prepare PGA/PEI/DNA ternary complexes to overcome the serum inhibitory effect of cationic polymers (Scheme 4.16C).[118] Biocompatible PGA containing carboxyl groups could self-assemble with positively charged PEI. The PEI/PGA combined carriers showed lower cytotoxicity than 25 kDa PEI. The transfection efficiency of these terplexes was significantly higher than that of 25 kDa PEI or Lipofectamine 2000 in 10% FBS-containing medium.

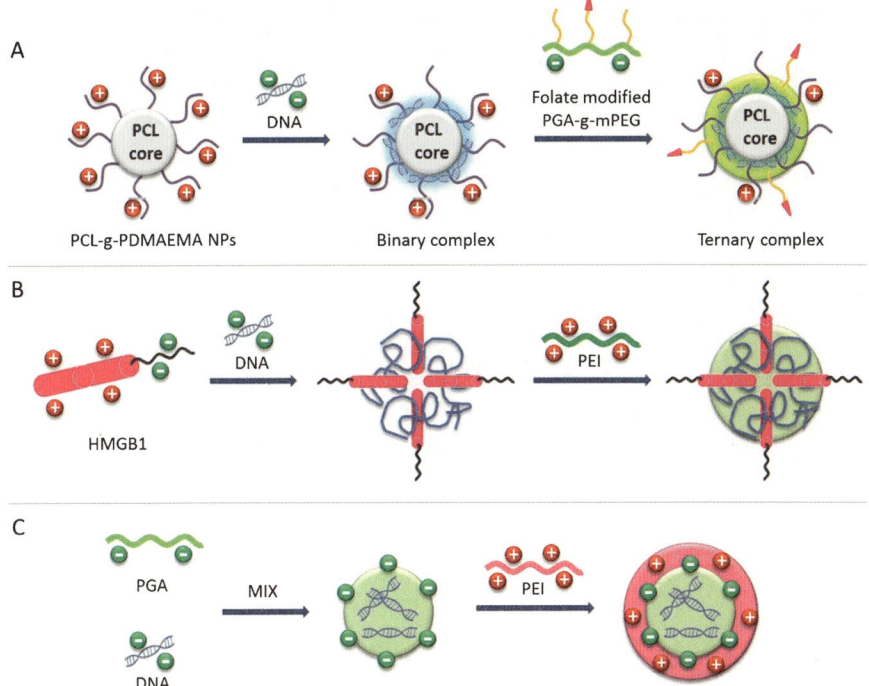

Scheme 4.16 Formation of self-assembled complexes: (A) PCL-*g*-PDMAEMA NPs/ DNA/PGA-*g*-mPEG; (B) HMGB1/DNA/PEI; and (C) PGA/PEI/ DNA ternary polyplex.

4.6.5   Polycarbonates

Polycarbonates provide an attractive option for use as gene delivery vectors owing to their biocompatibility and ease of incorporating functional moieties. In the area of gene delivery, the use of carbonate-based polymers has been less widely reported. Therefore, this versatile class of synthetic polymer presents a very attractive option in the design of novel biodegradable gene delivery vectors. Ong *et al.* described an approach to synthesize cationic biodegradable polycarbonates by an organocatalytic ring-opening polymerization of functional cyclic carbonates containing alkyl halide side chains, followed by a subsequent functionalization step with bis-tertiary amines designed to facilitate gene binding and endosomal escape.[119] The results showed that polycarbonate mediated high luciferase gene expression in HepG2, HEK293, MCF-7, and 4T1 cell lines in the presence of serum, and low cytotoxicity.

Wang *et al.* synthesized PEI-grafted polycarbonates (PMAC-*g*-PEI*x*) as biodegradable polycationic gene vectors (Scheme 4.17A).[120] A backbone polymer, poly(5-methyl-5-allyloxycarbonyl-trimethylene carbonate) (PMAC), was synthesized in bulk, catalyzed by immobilized porcine pancreas lipase

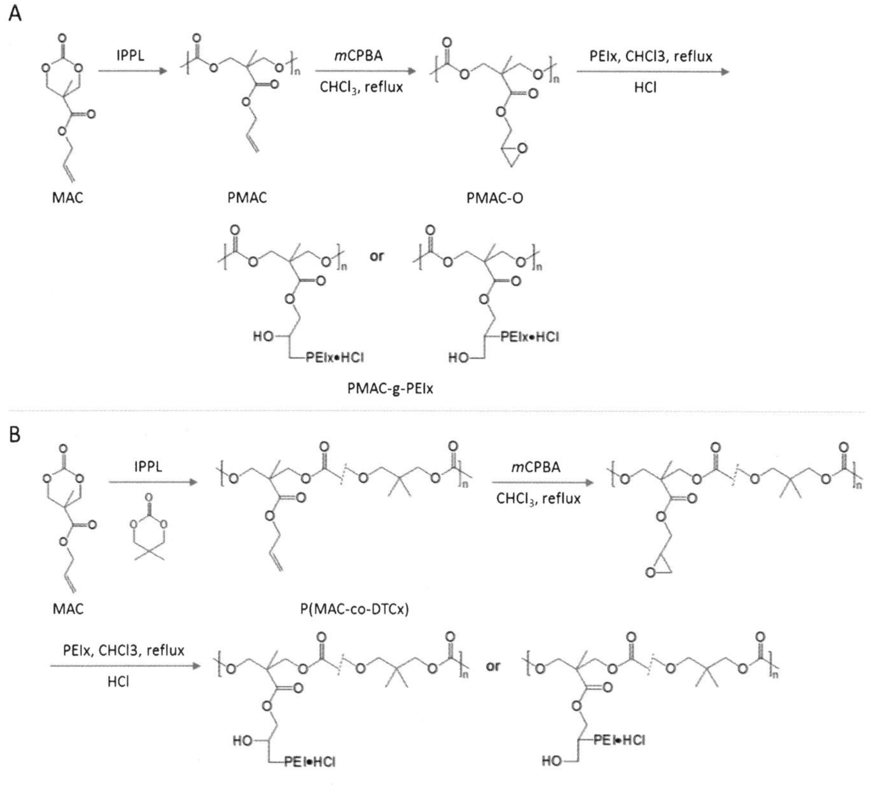

Scheme 4.17   Synthesis of PEI-grafted polycarbonates.

(IPPL). Then, PMAC-O, the allyl epoxidation product of PMAC, was further modified by PEIx with low molecular weight (x = 423,800, or 1,800). *In vitro* experiments demonstrated that the PAMC-g-PEIx showed much lower cytotoxicity and enhanced transfection efficiency in comparison with 25 kDa PEI in 293T cells. The biodegradability of PMAC-g-PEIx can facilitate the efficient release of pDNA from polyplexes and reduce cell cytotoxicity. They further synthesized biodegradable polycations based on polycarbonates with 1800 Da PEI as gene vectors (Scheme 4.17B).[121] The resulting copolymers with different compositions [P(MAC-co-DTCx)] underwent additional allyl epoxidation and were thereby grafted by LMW PEI (1800 Da). Despite a slightly lower DNA binding ability, the PEI-grafted polycarbonates, especially P(MAC-co-DTC45.4)-g-PEI, presented apparently low cytotoxicity and much higher gene transfection efficiency in comparison with 25 kDa PEI in 293T cells. Moreover, preincubation of P(MAC-co-DTC6.7)-g-PEI showed a rapidly weakening DNA binding capacity, while a suitable degradation rate of vectors could facilitate the efficient release of pDNA from polyplexes after cellular uptake and also reduce cell cytotoxicity.

4.6.6 Nanoparticles

Nanoparticles also have been developed and employed as nonviral gene vectors, including gold nanoparticles, silica nanoparticles, carbon nanotubes, lipid-based nanoparticles, quantum dots, and polymeric hydrogels. Compared to cationic carriers, nanoparticles are inert and exhibit less cytotoxicity; compared to liposomes, nanoparticles are more stable with respect to physical stresses. Various surface modifications have been used to improve the transfection efficiency of nanoparticles.

Gold nanoparticles (GNPs) are bioinert and nontoxic, and provide attractive scaffolds for gene delivery vectors. GNPs of small size provide a high surface-to-volume ratio, maximizing the grafting density of target molecules, which allows further tuning of the surface charge and hydrophobicity. Various modifications to the surface of GNPs have been investigated to improve the transfection efficiency. Hu *et al.* synthesized a series of LMW PEI (800 Da)-conjugated gold nanoparticles, which showed 60-fold higher transfection efficiency than that of PEI 25 kDa in 10% serum medium.[122] Recently, Shan *et al.* developed dendrimer-entrapped gold nanoparticles (Au DENPs).[123] The transfection efficiency of Au DENPs was significantly higher than that of G5 PAMAM dendrimers without AuNPs entrapped. The higher gene transfection efficiency of Au DENPs is primarily due to the fact that the entrapment of AuNPs helps preserve the 3D spherical morphology of the dendrimers, allowing for more efficient interaction between dendrimers and DNA.

Silica nanotubes (SNTs) have become a promising material in biomedical applications, owing to their unique properties. Biocompatibility and facile modification through well-known silane chemistry make SNTs even more

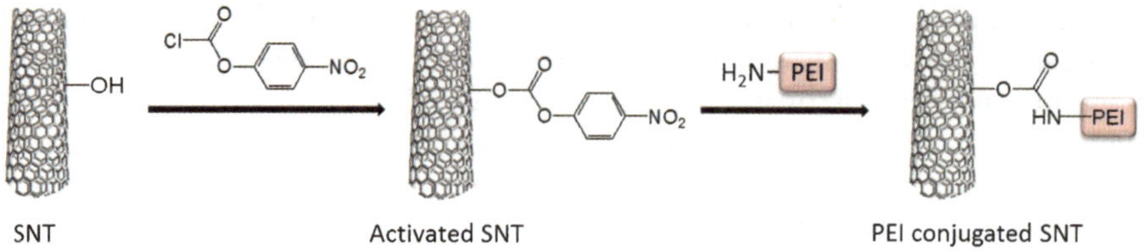

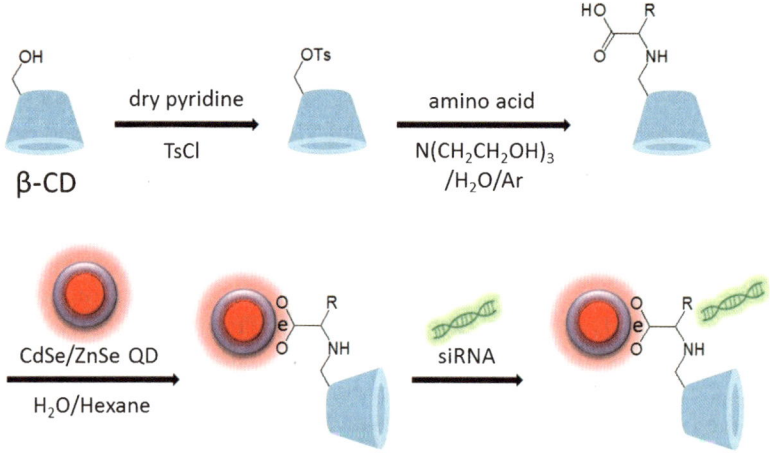

Scheme 4.18   Synthesis of a PEI-conjugated SNT.

attractive tools in various biomedical applications, such as in drug or gene delivery vehicles. To achieve efficient gene delivery, the surface of the SNT must be rendered positive by conjugating cationic materials. Namgung *et al.* functionalized SNTs with magnetic-fluorescence nanocomposites and LMW branched PEI to construct a device (BPEI-SNT) as a gene carrier and a MRI agent (Scheme 4.18).[124] *In vitro* transfection results showed that the transfection efficacy of BPEI-SNT was much higher than BPEI1.8k, and 46 times higher than that of bare SNT in HeLa cells.

Quantum dots (QDs) have the potential to serve as photostable beacons to track siRNA delivery, which is fast becoming an attractive approach to probe gene function in cells. Zhao *et al.* synthesized CdSe/ZnSe QD nanoparticles coated with β-CD coupled to amino acids with different surface charges through direct ligand-exchange reactions and used them to deliver siRNA (Scheme 4.19).[125] Compared with existing transfection agents, the gene-silencing efficiency of the modified QDs was slightly improved for the HPV18 E6 gene in HeLa cells. These findings suggest that the QD

R = L-His, L-Trp, L-Phe or L-Cys residues.

Scheme 4.19   Schematic illustration of the formation of QD nanoparticles coated with β-CD coupled to amino acids and QD-siRNA complexes.

nanoparticles coated with β-CD coupled to amino acids significantly improved the siRNA delivery *in vitro*. They further modified CdSe/ZnSe QDs with L-Arg- or L-His-coupled β-CD to simultaneously deliver DOX and siRNA in HeLa cells.[126] These multifunctional QDs were promising vehicles for the co-delivery of nucleic acids and chemotherapeutics and for real-time tracking of treatment.

Chertok *et al.* developed PEI-modified magnetic nanoparticles (GPEI) as a potential vascular drug/gene carrier to brain tumors. The results showed that GPEI exhibited high cell penetrability and low cell toxicity properties, which are highly desirable for intracellular drug/gene delivery.[127] In addition, GPEI could be magnetically captured in gliomales ions following clinically viable intra-carotid administration. The extent of GPEI accumulation was 5.2-fold higher than that of commercially available G100 magnetic nanoparticles in the tumor lesions, but not in the contra-lateral normal brain, revealing higher target selectivity of cationic nanoparticles. These results warrant further investigation of GPEI as a potential nanocarrier for drug/gene delivery to glioma lesions.

Ribonucleic acid interference (RNAi) is a powerful molecular tool that has potential to revolutionize the treatment of cancer. One major challenge to applying this technology for clinical applications is the lack of site-specific carriers that can effectively deliver short interfering RNA (siRNA) to cancer cells. Veiseh *et al.* developed a magnetic nanoparticle platform consisting of a superparamagnetic iron oxide (Fe_3O_4) core coated with a cationic copolymer of chitosan-grafted-PEG and PEI for nonviral DNA delivery.[128] This unique formulation has shown the ability to safely deliver plasmid DNA *in vivo* and transfect brain tumor cells. They further functionalized these nanovectors with siRNA and a tumor-targeting peptide, chlorotoxin (CTX), to improve tumor specificity and potency.[129] The results showed that cellular internalization and gene knockdown efficiency of the nanovectors were enhanced through targeting with CTX. MRI demonstrated the ability of this nanovector construct to generate specific contrast enhancement of glioblastoma cells. These findings suggested that this CTX-enabled nanoparticle carrier may be well suited for the delivery of RNAi therapeutics to brain cancer cells.

4.6.7 Other Types

Recently, Xu *et al.* reported that ethanolamine (EA)-functionalized poly(-glycidyl methacrylate) (termed PGEA) vectors had excellent transfection efficiency, while exhibiting very low toxicity.[130] As ethylenediamine (ED) has a similar molecular mass to EA but possesses double amino groups, they further investigated the structural effects of EA and ED on transfection. Different EA- and ED-functionalized PGMA (PGEAED) vectors, as well as ED-functionalized PGMA (PGED) vectors, were proposed and compared. The results indicated that the flanking non-ionic hydrophilic hydroxyl groups had a crucial effect on the gene transfection process. PGEAED and PGED showed outstanding transfection efficiency while exhibiting substantially lower toxicity in comparison with 25 kDa PEI. Moreover, the flanking primary amine groups induced by ED could be readily functionalized by glycyrrhetinic acid or cholic acid to improve the biophysical properties of the gene vectors.

Dong *et al.* synthesized a series of biodegradable cationic polymers, poly(ethylene glycol)-*block*-poly(carbonates-*graft*-oligoethylenimine) [mPEG-*b*-P(MCC-*g*-OEI), PPO] copolymers.[131] Different kinds of OEI were grafting onto the mPEG-*b*-PMCC backbone, such as linear OEI423, branched OEI600, and branched OEI1800. The cell toxicity and gene transfection evaluations showed that PPO copolymers, especially PPO1800, exhibited lower cytotoxicity and about three times higher gene transfection efficiency than 25 kDa PEI, both in the absence and presence of serum in the CHO and COS-7 cell lines.

Miyata *et al.* demonstrated that the silica-coated polyplexes could be prepared using a silicic acid condensation reaction for enhanced polyplex stability and transfection activity through shielding of the cationic surface charge (Figure 4.10).[132] The silica-coated polyplex achieved significantly higher transfection efficiency without serious cytotoxicity compared to the polyplex without any silica coating, possibly due to their facilitated endosomal escape.

Zhang and co-workers synthesized folate-PEG modified poly[2-(2-aminoethoxy)ethoxy]phosphazene (folate-PEG-PAEP) as a folate receptor (FR) targeted carrier.[133] Compared with the unmodified nanoparticles, the cytotoxicity of folate-PEG-PAEP decreased significantly at high dose. They also showed much higher transfection efficiency in FR overexpressing HeLa cells, but no significant difference was observed in CHO-k1 cells lacking FR.

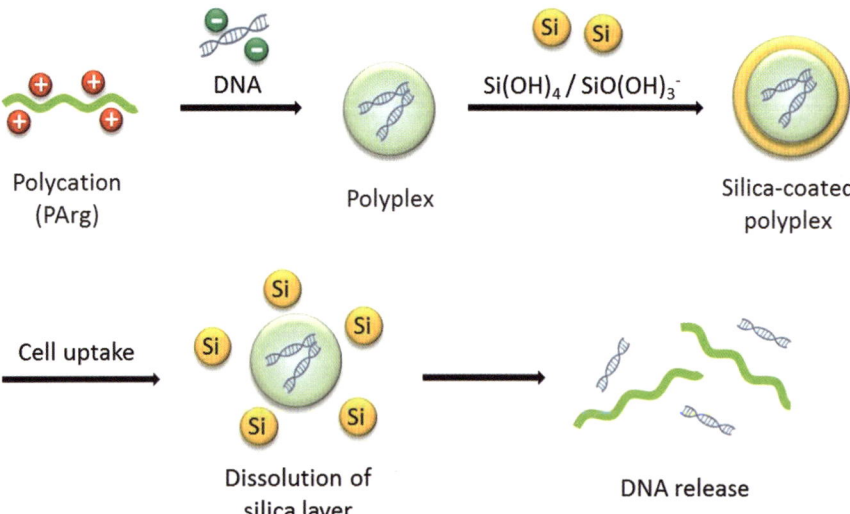

Figure 4.10   Preparation of silica-coated polyplexes.

von Erlach *et al.* investigated different copolymers of poly(L-lysine)-*graft*-poly(2-methyl-2-oxazoline) (PLL-*g*-PMOXA) of variable grafting densities and PMOXA molecular weights for their potential to complex and deliver plasmid DNA.[134] Good transfection efficiency combined with low cytotoxicity of PLL-*g*-PMOXA-DNA condensates was demonstrated in COS-7 cells. This suggested that DNA-PMOXA-*g*-PLL condensate formation for efficient DNA delivery strongly depended on the PMOXA grafting density and molecular weight.

Poly(β-amino esters) (PBAEs) are a new class of polymeric vectors first developed by Lynn *et al.*[135,136] The main advantage offered by PBAEs over PEI is their biodegradability *via* the hydrolytically cleavable ester groups. Using high-throughput synthesis and parallel screening, a large library of PBAEs was created that helped to elucidate the effect of small changes in polymer structure on transfection efficiency.[137,138] Recently, Bhise *et al.* developed novel gene delivery vectors for efficient gene transfer to hard-to-transfect mouse mammary epithelial cells.[139] Ten modified versions of the same base PBAEs, poly(butane-1,4-diol diacrylate-*co*-5-aminopentan-1-ol), were synthesized. Small modifications to the polymer end-capping molecules and tuning of the polymer molecular weight could either significantly enhance the transfection efficacy up to six-fold or instead abolish efficacy completely. These degradable polymers were more effective than FuGENE HD for gene delivery in both 2-D culture and 3-D organotypic culture, and could be used as reagents and therapeutics for breast cancer. Tzeng *et al.* found that PBAEs could transfect glioblastoma 319 astrocytes with comparable or superior efficiency and safety compared to leading commercial reagents (Lipofectamine 2000 and FuGENE HD) (Scheme 4.20).[140]

Recently, Zhang *et al.* developed a new class of cationic transfection agents based on cationic shell-crosslinked knedel-like nanoparticles (cSCKs) that efficiently transfect mammalian cells with both oligonucleotides and plasmid

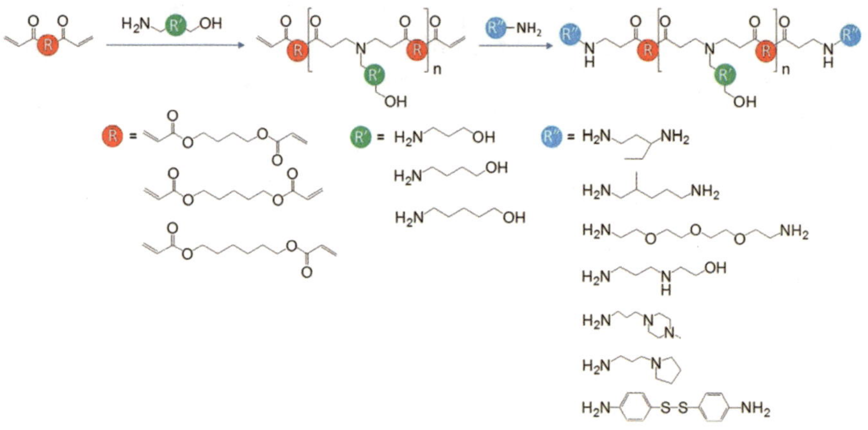

Scheme 4.20   Synthesis of poly(β-amino esters) (PBAEs).

DNA.[141] They further increased the plasmid DNA and phosphorothioate 2'-OMe oliogonucleotides (ps-MeON) transfection efficiency and minimized the cytotoxicity of cSCKs by introducing tertiary amines into the shell by chemical modification of the precursor block copolymer.[142]

The reversible addition–fragmentation chain transfer (RAFT) polymerization technique allows successful and facile synthesis of cationic glycopolymers containing pendant sugar moieties in the absence of protecting group chemistry. Ahmed *et al.* synthesized well-defined cationic glycopolymers of varying molecular weight, cationic chain length, carbohydrate to cationic content ratio, and different architecture (block *versus* random polymers) by RAFT polymerization.[143] It was found that the copolymer architecture largely affected the toxicity, DNA condensation ability, and gene delivery efficacy. These statistical copolymers of 3-aminopropyl methacrylamide (APMA) or 2-aminoethyl methacrylamide (AEMA) and 3-gluconamidopropyl methacrylamide (GAPMA) showed lower toxicity and higher gene expression in the presence and absence of serum, compared to the corresponding diblock copolymers.

Ma *et al.* synthesized poly(aminoethyl methacrylate) (PAEMA), poly(3-amino-2-hydroxypropyl methacrylate) (PAHPMA), poly[2-(2-aminoethylamino)ethyl methacrylate] (PAEAEMA) and poly[3-(2-aminoethylamino)-2-hydroxypropyl methacrylate] (PAEAHPMA) using atom transfer radical polymerization.[144] The results indicated that hydroxyl groups might increase the binding capacity to DNA and decrease the surface charge of the polymer/DNA complexes due to the formation of hydrogen bonds between the polymers and DNA. This effect of hydroxyl groups decreased with increasing amino group density on the polymer.

There has been success in applying a semi-rational approach to nonviral gene delivery vector development using a combinatorial/parallel synthesis approach to construct libraries of materials with unique molecular structures. Gabrielson *et al.* have described a library approach to gene delivery vector development that relies on the supramolecular self-assembly of individual components instead of chemical reactions.[145] Each component in the described system is capable of performing a single and well-defined purpose: DNA binding (dioleylspermine), membrane permeation (oligoarginine), or targeting (folic acid). A combination of electrostatic attraction and the hydrophobic effect is used to bring the individual groups together to form nanoscale complexes with DNA. Because the components responsible for DNA binding, membrane permeation, and targeting are separate, it is possible to alter the balance between hydrophilic and hydrophobic groups by varying the relative amounts in the final formulation. They could readily identify cell-specific formulations that have greater transfection efficiency than the individual components and have superior transfection efficiency to Lipofectamine 2000 under similar conditions.

Takemoto *et al.* reported the synthesis of siRNA-grafted polymers with a disulfide linkage to improve the physicochemical properties and transfection efficacies of the polyion complexes (PICs) as a nanocarrier of siRNA.[146] The disulfide linkage of the siRNA-grafted polymer allowed efficient siRNA release from the PICs under reductive conditions in the cytoplasm. Consequently, the PICs from the siRNA-grafted polymer showed potent gene-silencing effects without cytotoxicity and immunogenicity.

Dai *et al.* synthesized a star cationic polymer (s-PDMAEMA) consisting of a cleavable poly[*N,N*-bis(acryloyl)cystamine] (PBAC) crosslinked core and poly(*N,N*-dimethylethylamine methacrylate) (PDMAEMA) arms by atomic transfer radical polymerization using a one-pot "arm first" method (Figure 4.11).[147] It was shown that s-PDMAEMA achieved higher gene transfection levels relative to the linear precursors, and that s-PDMAEMA200 with longer and more arms exhibited superior transfection efficiencies and lower cytotoxicity compared to 25 kDa PEI. In mild acid milieu the less contracted condensate was formed between DNA and s-PDMAEMA, with longer and more arms due to the stretching of the positively charged arm. They found that star polymers could offer more effective protection of DNA against DNase degradation than the linear counterpart because of the greater buffering capability conferred by the unique molecular architecture.

Xiong *et al.* synthesized a series of cationic fluorine-containing amphiphilic graft copolymers, P(HFMA-St-MOTAC)-*g*-PEG, comprising poly(hexafluorobutyl methacrylate) (PHFMA), poly(methacryloxyethyl trimethylammonium chloride) (PMOTAC), polystyrene (PS) backbones, and PEG side chains.[148] The copolymers showed good binding capacity to DNA, and could be used as a promising nonviral vector.

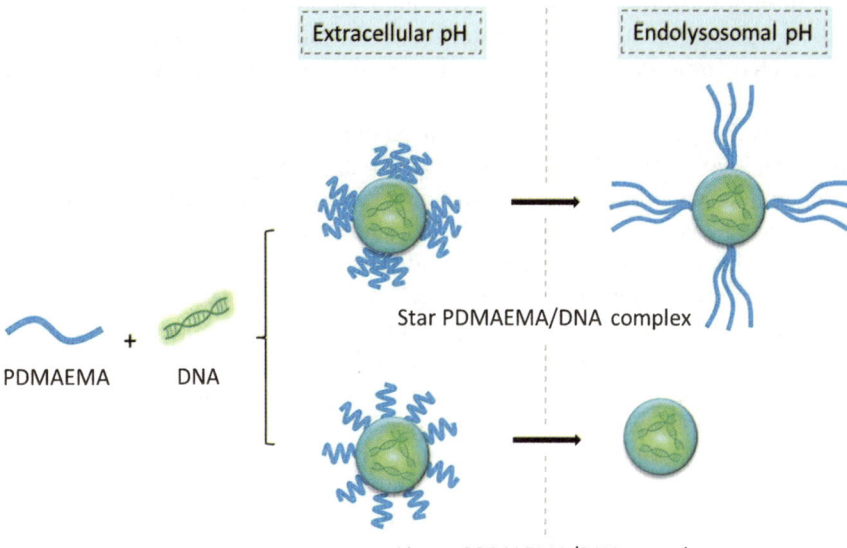

Figure 4.11 pH-dependent morphological change of PDMAEMA/DNA complexes.

4.7   Future Trends

Although human gene therapy has been a rapidly growing field in recent years, several obstacles still exist, among which the lack of effective and safe delivery vectors suitable for clinical use is still the greatest challenge.[149] Based on new understanding of molecular biology and medicine, therapeutic nonviral vectors for cancer gene therapy will be improved for greater safety and efficacy. Currently, we have to confront challenges associated with cell targeting specificity, gene transfer efficiency, gene expression regulation, and vector safety. Hence, the successful nonviral vectors for future gene therapy must be multifunctional systems, and will be more efficient in targeting and transfection. Additionally, they should be biocompatible and biodegradable to prevent vector-induced toxicities and the accumulation of vector components in the host. Accordingly, more efforts should be made in order to facilitate DNA release where LMW polymers are crosslinked or linearly linked together by degradable linkages to form a HMW polymer that can eventually degrade into its LMW components. Various new ideas have been proposed and reported, helping us estimate future trends in development of nonviral gene delivery systems.

4.7.1   Stem Cell Transfection

Stem cells are the precursors for embryonic development as well as adult tissue regeneration. As such, stem cell therapy holds great promise for the field of regenerative medicine. In the past decade, a number of studies have shown that genetic modification of stem cells can improve their therapeutic potential *in vivo* as well as allow for their monitoring *via* the introduction of reporter genes. Research on nonviral vector-mediated stem cell gene transfection has become a hotspot.

Ramasubramanian *et al.* reported a combinatorial nonviral gene delivery platform that targets both inductive and suppressive genes for promoting osteogenic differentiation in human adipose-derived stem cells (hADSCs).[150] They found that co-delivery of poly(β-amino esters)/BMP2 DNA complexes and lipidoid/siGNAS complexes (or lipidoid/siNoggin complexes) significantly accelerated the hADSC differentiation towards osteogenic differentiation, with marked increase in bone marker expression and mineralization.

Roger *et al.* demonstrated that marrow-isolated adult multilineage inducible (MIAMI) cells could serve as cellular carriers for NPs in brain tumors.[151] Two types of NPs loaded with coumarin-6 were investigated: poly(lactic acid) NPs (PLA-NPs) and lipid nanocapsules. The results showed that these NPs could be efficiently internalized into MSCs while cell viability and differentiation were not affected. Furthermore, these NP-loaded cells were able to migrate toward an experimental human glioma model. These data suggested that MSCs could serve as cellular carriers for NPs in brain tumors.

The use of tissue engineering to deliver genes to stem cells has been impeded by low transfection efficiency of the inserted gene and poor retention at the target site. Recently, Jung *et al.* developed a sustained gene delivery system using a fibrous 3D poly(L-lactide) scaffold coated with cell-permeable peptide/DNA complexes (Figure 4.12).[152] The human-derived arginine-rich peptide Hph-1 (YARVRRRGPRR) displayed a higher transfection efficiency and lower toxicity in human adipose-derived stem cells (hADSCs) compared with Lipofectamine. DNA/Hph-1 complexes were released from the scaffolds over 14 days and were successfully transfected into hASCs seeded on the scaffolds.

Target gene transfection for desired cell differentiation has recently become a major issue in stem cell therapy. For the safe and stable delivery of genes into hMSCs, Park *et al.* evaluated the differentiation capability of human mesenchymal stem cells (hMSCs) using the SOX trio genes (master genes for chondrogenic differentiation) as targets after modification with biodegradable PLGA nanoparticles.[153] PEI was polyplexed with a combination of SOX trio fused to reporter genes coated onto PLGA nanoparticles. They found that the SOX trio complexed with PEI-modified PLGA nanoparticles led to increased cell-uptake capacity and subsequent enhanced chondrogenesis of exogenous gene-delivered hMSCs. Park *et al.* designed a nonviral gene delivery system using nanoparticles, with emphasis placed on the ability of the system to

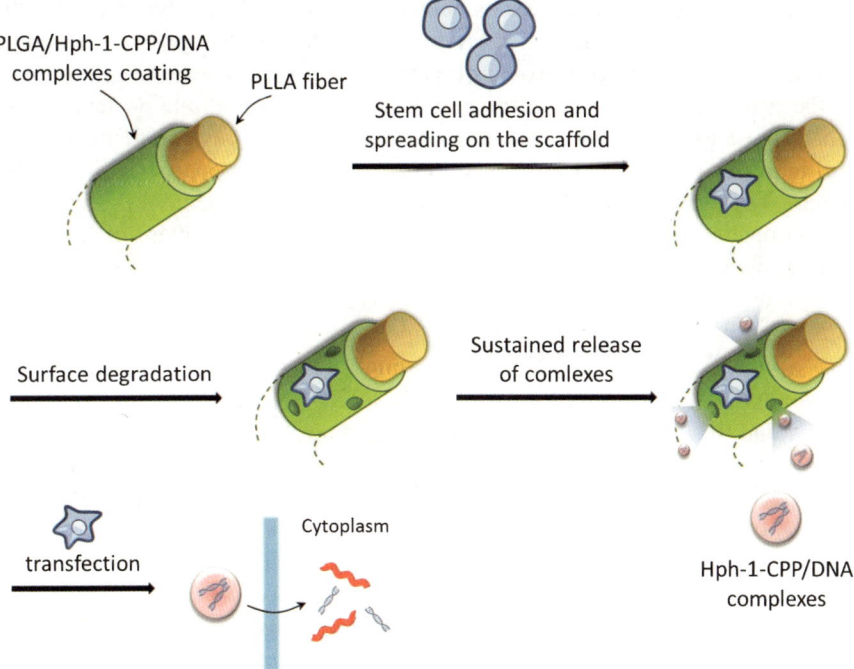

Figure 4.12   Schematic diagram of a scaffold-supported sustained gene delivery system.

mediate high levels of gene expression into hMSCs.[154] By polyplexing with
PEI, the cell-uptake ability of the nanoparticles was enhanced for both *in vitro*
and *in vivo* culture systems, so that transfection efficiency into hMSCs was
enhanced.

Wang *et al.* designed and manufactured a bioactive construct of poly(lac-
tide-*co*-glycolide) (PLGA), fibrin gel, bone marrow mesenchymal stem cells
(BMSCs), and *N,N,N*-trimethylchitosan chloride (TMC)/pDNA encoding
transforming growth factor-β1 (pDNA-TGF-β1) for osteochondral restora-
tion, whose biological performance was evaluated in a rabbit model.[155] The
TMC/DNA complexes had a transfection efficiency of 9% to BMSCs and
showed TGF-β1 expression *in vitro*. *In vivo* culture of the composite constructs
was performed by implantation into full-thickness cartilage defects of rabbit
joints. After implantation for 12 weeks, the cartilage defects were successfully
repaired by the composite constructs of the experimental group, and the neo-
cartilage integrated well with its surrounding tissue and subchondral bone.

To enhance the level and prolong the duration of gene expression for gene-
engineered rat mesenchymal stem cells (MSCs) using a nonviral vector, He
et al. developed a novel transfection system based on the reverse transfection
method and 3D scaffold.[156] Collagen sponge and poly(ethylene terephthalate)
nonwoven fabric were introduced as scaffolds to perform 3D culture with
reverse transfection. pDNA coding transforming growth factor β-1 (TGFβ-1)
was delivered to MSCs to assess its ability in inducing chondrogenesis with the
3D nonviral reverse transfection system. The electric charge of the anionic
gelatin played an important role in this system by affecting the release pattern
of the gene complexes and through the adsorption of serum protein to the
substrate. Also, the 3D scaffold provided a good environment for cell
proliferation and showed enhanced gene expression compared with the 2D
system. Therefore, TGFβ-1 gene-engineered MSCs using a nonviral vector and
the 3D reverse transfection system are promising in the treatment of cartilage-
related disease.

4.7.2   Combinatorial Vectors

Gene vectors with different structures, properties, and transfection mechanisms could be combined together to obtain multifunctional combinatorial
vectors. For example, viral vectors can combine with nonviral vectors, and
polymeric vectors can conjugate with inorganic nanoparticles. Adenoviral
vectors offer many advantages for cancer gene therapy, including high
transduction efficiency, but safety concerns related to severe immunogenicity
and other side effects have led to careful reconsideration of their use in human
clinical trials. To overcome these issues, a strategy of generating hybrid vectors
that combine viral and nonviral elements as more intelligent gene carriers has
been employed. Kim *et al.* coated an adenovirus (Ad) with an arginine-grafted
bioreducible polymer (ABP) *via* electrostatic interaction (Figure 4.13).[157]
Enhanced transduction efficiency was observed in cells treated with the

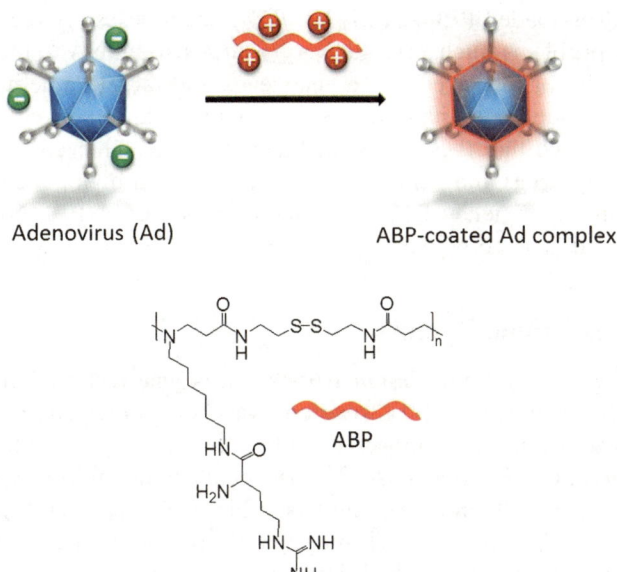

Figure 4.13 Formation of arginine-grafted bioreducible polymer (ABP)-coated adenovirus complex.

cationic ABP polymer-coated Ad complex compared to naked Ad. In both high and low coxsackievirus and adenovirus receptor (CAR)-expressing cells, the ABP-coated Ad complex produced higher levels of transgene expression and minimal toxicity than 25 kDa PEI. Moreover, the ABP-coated Ad complex significantly reduced the innate immune response relative to naked Ad. The combination of Ad with ABP polymer offers the potential to increase the efficiency of vectors for gene therapy by shielding the virus from deactivation by the immune system, and may make systemic administration feasible.

Development of the transfection enhancement of liposomes with attributes of high stability and easy handling in gene therapy is challenging. Li *et al.* reported didodecyldimethylammonium bromide (DDAB, a cationic lipid)-coated gold nanoparticles (DDAB-AuNPs), which enhanced the transfection efficiency generated by two kinds of commercially available cationic liposomes: Lipotap and DOTAP.[158] They showed that DDAB-AuNPs at the optimal concentrations could produce higher gene expression and lower cytotoxicity than lipoplex. The results indicated that DDAB-AuNPs increased the cellular uptake efficiency of DNA molecules, which might account for the enhancement of transfection efficiency.

Wang *et al.* recently developed the PLGA/folate-coated PEGylated polymeric liposome core-shell nanoparticles (PLGA/FPL NPs) for co-delivery of drug and gene.[159] Hydrophobic drugs could be incorporated into the core and the cationic shell of the drug-loaded nanoparticles could be used to bind

DNA. The drug-loaded PLGA/FPL NPs/DNA complexes offered advantages to overcome problems, such as co-delivery of drugs and DNA to improving the chemosensitivity of cancer cells at a gene level, and targeting delivery of drug to the cancer tissue that enhanced the bioavailability and reduced the toxicity. The results showed that these core–shell nanoparticles achieved the possibility of co-delivering drugs and genes to the same cells with high gene transfection and drug delivery efficiency. This suggested that the PLGA/FPL NPs may be a useful drug and gene co-delivery system.

4.7.3 Virus Mimic Vectors

An interesting challenge is to design artificial virus gene vectors that equal their viral counterparts in terms of transfection efficiency, but which are safe to use, target-cell specific, non-immunogenic, and relatively inexpensive to prepare at scales that permit their clinical use.[160] It is a challenging multidisciplinary task that requires a broad and deep understanding of chemistry, biology, and physiology, as well as medicine. The design of artificial viruses is also a good example of nanotechnology, which involves the creation of a fully functional gene-delivery system through the manipulation and assembly of macromolecular structures. There are three main structural components of artificial gene delivery systems (Figure 4.14): (1) the plasmid vector, engineered for optimal expression; (2) the artificial virus core, consisting of pDNA, condensing agents, and functional peptides; and (3) the hydrophilic shell, exposing targeting ligands for cell-type-specific gene delivery. However, at present, the concept of an artificial virus as a gene delivery system is still immature, and more efforts are needed to meet the objective of efficient, targeted gene delivery and long-lasting gene expression *in vivo*.

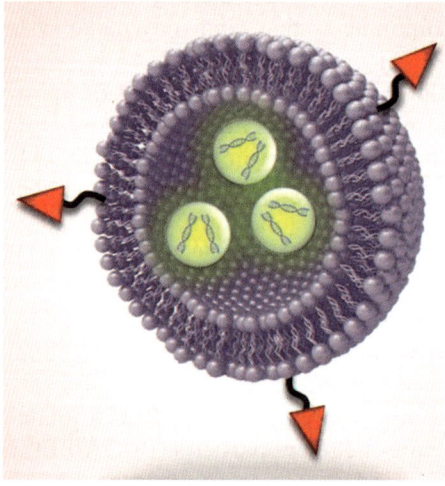

Figure 4.14 Artificial virus gene delivery system.

Xiong *et al.* have developed a series of virus-like siRNA carriers, based on poly(ethylene oxide)-*block*-poly(ε-caprolactone) (PEO-*b*-PCL) micelles decorated with integrin $\alpha_v\beta_3$ targeting peptide (RGD4C) and/or cell penetrating peptide (TAT) on the PEO shell, and modified with a polycation (spermine) in the PCL core for siRNA binding and protection.[161] They demonstrated that RGD and TAT modification increased cellular uptake of siRNA formulated micelles. These peptide functionalized micelles, especially the RGD/TAT micelles containing mdr1 siRNA, effectively silenced P-gp expression, increased DOX intracellular uptake, improved DOX penetration into nuclei, and finally enhanced DOX cytotoxicity in MDA435/LCC6 DOX-resistant cells. They further synthesized the degradable PEO-*b*-PCL block copolymers for traceable co-delivery of siRNA and DOX, and demonstrated that incorporation of fluorescent probes in the micellar core allows for *in vitro* and *in vivo* tracking of micelles.[162]

4.7.4　Therapeutic Genes

The efficient delivery of therapeutic genes into cells of interest is a critical challenge to a broad application of nonviral vectors. In cancer gene therapy, therapeutic agents including functional normal tumor suppressor genes, inflammatory immune cytokine genes, and microRNAs are delivered to the tumor cell using a carrier.

Recently, Dai *et al.* synthesized a ternary copolymer (PEG-*b*-PLL-*g*-lPEI; PPI) by grafting LMW linear PEI onto PEGylated PLL.[163] These compounds demonstrated higher cell viability and enhanced gene transfer efficiency compared to PEG-*b*-PLL. Moreover, remarkable tumor cell apoptosis and growth inhibition were achieved when the folate-encoded PPI was used to transfer the TNF-related apoptosis-inducing ligand gene (TRAIL).

Transforming growth factor-β (TGF-β) is a well-known cytokine related to tissue fibrosis. Recent studies demonstrated that the regulation of TGF-β levels resulted in a significant therapeutic effect on liver fibrosis. Park *et al.* developed a liver-targeted vector for TGF-β siRNA using reducible PEI-grafted hyaluronic acid (HA) conjugates.[164] This system showed low toxicity and obvious therapeutic effects on liver cirrhosis due to HA receptor-mediated endocytosis. The TGF-β siRNA/reducible polymer complex showed a feasible therapeutic effect on liver cirrhosis. This system can be exploited for the target-specific systemic treatment of various liver diseases.

Arote *et al.* developed a biodegradable folate conjugated poly(ester amine) (FP-PEA) vector for dominant-negative c-Jun (TAM67) delivery.[165] FP-PEA showed marked antitumor activity against FR-positive human KB tumors in nude mice, with no evidence of toxicity after therapy using the TAM67 gene. Furthermore, the therapeutic effect occurred in the apparent absence of weight loss or noticeable tumor apoptosis.

4.8 Conclusion

In summary, although intensive studies have been made on nonviral gene vectors and numerous research papers have been published in recent years, their commercial development and clinical applications are still limited. Therefore, great emphasis should be placed on *in vivo* trials and development of new commercial transfection agents during the next decade. In order to obtain the permission for clinical application of nonviral gene vector-mediated gene therapy, sustained efforts should be made to combine those efficient synthetic vectors with therapeutic genes, make clinical safety assessments, and provide indications how animal models correlate with clinical experience. It is very important for us to remember the mechanisms, applications, and limitations of current gene delivery systems, and obtain more novel information from chemical, medical, biological, and various other fields so as to design innovational successful gene vectors. Gene therapy has rapidly developed, and it will play an important role in future medical research and clinical applications. It is possible to predict that every genetic disease will have gene therapy as its treatment in the future.

Acknowledgements

This work was financially supported by the Ministry of Science and Technology of China (2009CB930300, 2011CB606202) and the Natural Science Foundation of Hubei Province, China (2009CDA024).

References

1. M. A. Mintzer and E. E. Simanek, *Chem. Rev.*, 2009, **302**, 259–302.
2. D. W. Pack, A. S. Hoffman, S. Pun and P. S. Stayton, *Nat. Rev. Drug Discovery*, 2005, **4**, 581–593.
3. S. Y. Wonga, J. M. Peletb and D. Putnam, *Prog. Polym. Sci.*, 2007, **32**, 799–837.
4. X. Z. Zhang, X. Zeng, Y. X. Sun and R. X. Zhuo, in *Bioactive Materials in Medicine: Design and Applications*, ed. X. Zhao, J. M. Courtney and H. Qian, Woodhead, Cambridge, 2011, pp. 179–219.
5. H. Petersen, P. M. Fechner, D. Fischer and T. Kissel, *Macromolecules*, 2002, **35**, 6867–6874.
6. H. Debus, P. Baumhof, J. Probst and T. Kissel, *J. Controlled Release*, 2010, **148**, 334–343.
7. O. M. Merkel, R. Urbanics, P. Bedocs, Z. Rozsnyay, L. Rosivall, M. Toth, T. Kissel and J. Szebeni, *Biomaterials*, 2011, **32**, 4936–4942.
8. X. A. Chen, L. J. Zhang, Z. J. He, W. W. Wang, B. Xu, Q. Zhong, X. T. Shuai, L. Q. Yang and Y. B. Deng, *Int. J. Nanomed.*, 2011, **6**, 843–853.

9. A. Beyerle, A. Braun, A. Banerjee, N. Ercal, O. Eickelberg, T. H. Kissel and T. Stoeger, *Biomaterials*, 2011, **32**, 8694–8701.

10. N. D. Weber, O. M. Merkel, T. Kissel and M. A. Muñoz-Fernández, *J. Controlled Release*, 2012, **157**, 55–63.

11. L. R. Tsai, M. H. Chen, C. T. Chien, M. K. Chen, F. S. Lin, K. M. Lin, Y. K. Hwu, C. S. Yang and S. Y. Lin, *Biomaterials*, 2011, **32**, 3647–3653.

12. H. Huang, H. Yu, G. Tang, Q. Wang and J. Li, *Biomaterials*, 2010, **31**, 1830–1838.

13. M. Liu, Z. H. Li, F. J. Xu, L. H. Lai, Q. Q. Wang, G. P. Tang, W. T. Yang, *Biomaterials*, 2012, **33**, 2240–2250.

14. H. Yao, S. S. Ng, W. O. Tucker, Y. K. Tsang, K. Man, X. M. Wang, B. K. Chow, H. F. Kung, G. P. Tang and M. C. Lin, *Biomaterials*, 2009, **30**, 5793–5803.

15. Q. Y. Jiang, L. H. Lai, J. Shen, Q. Q. Wang, F. J. Xu and G. P. Tang, *Biomaterials*, 2011, **32**, 7253–7262.

16. Q. Hu, J. Wang, J. Shen, M. Liu, X. Jin, G. Tang and P. K. Chu, *Biomaterials*, 2012, **33**, 1135–1145.

17. W. Xiao, Y. X. Sun, H. Cheng, X. Zeng, X. Z. Zhang and R. X. Zhuo, *J. Microencapsul.*, 2010, **27**, 447–452.

18. X. H. Luo, F. W. Huang, S. Y. Qin, H. F. Wang, J. Feng, X. Z. Zhang and R. X. Zhuo, *Biomaterials*, 2011, **32**, 9925–9939.

19. D. Dey, M. Inayathullah, A. S. Lee, M. C. LeMieux, X. Zhang, Y. Wu, D. Nag, P. E. De Almeida, L. Han, J. Rajadas and J. C. Wu, *Biomaterials*, 2011, **32**, 4647–4658.

20. Q. Peng, Z. L. Zhong and R. X. Zhuo, *Bioconjugate Chem.*, 2008, **19**, 499–506.

21. Q. Peng, C. Hu, J. Cheng, Z. L. Zhong and R. X. Zhuo, *Bioconjugate Chem.*, 2009, **20**, 340–346.

22. W. Xia, P. Wang, C. Lin, Z. Li, X. Gao, G. Wang and X. Zhao, *J. Controlled Release*, 2012, **157**, 427–436.

23. Y. X. Sun, X. Zeng, Q. F. Meng, X. Z. Zhang, S. X. Cheng and R. X. Zhuo, *Biomaterials*, 2008, **29**, 4356–4365.

24. J. Z. Deng, Y. X. Sun, H. Y. Wang, C. Li, F. W. Huang, S. X. Cheng, R. X. Zhuo and X. Z. Zhang, *Acta Biomater.*, 2011, **7**, 2200–2208.

25. H. Koo, G. W. Jin, H. Kang, Y. Lee, K. Nam, Z. C. Bai and J. S. Park, *Biomaterials*, 2010, **31**, 988–997.

26. T. K. Endres, M. Beck-Broichsitter, O. Samsonova, T. Renette and T. H. Kissel, *Biomaterials*, 2011, **32**, 7721–7731.

27. F. W. Huang, J. Feng, J. Nie, S. X. Cheng, X. Z. Zhang and R. X. Zhuo, *Macromol. Biosci.*, 2009, **9**, 1176–1184.

28. F. W. Huang, H. Y. Wang, C. Li, H. F. Wang, Y. X. Sun, J. Feng, X. Z. Zhang and R. X. Zhuo, *Acta Biomater.*, 2010, **6**, 4285–4295.

29. F. He, C. F. Wang, T. Jiang, B. Han and R. X. Zhuo, *Biomacromolecules*, 2010, **11**, 3028–3035.

30. N. Zhao, S. Roesler and T. Kissel, *Int. J. Pharm.*, 2011, **411**, 197–205.

31. H. Y. Tian, X. S. Chen, H. Lin, C. Deng, P. B. Zhang, Y. Wei and X. B. Jing, *Chem.–Eur. J.*, 2006, **12**, 4305–4312.
32. H. Y. Tian, W. Xiong, J. Z. Wei, Y. Wang, X. S. Chen, X. B. Jing and Q. Y. Zhu, *Biomaterials*, 2007, **28**, 2899–2907.
33. H. Y. Tian, L. Lin, J. Chen, X. S. Chen, T. G. Park and A. Maruyama. *J. Controlled Release*, 2011, **155**, 47–53.
34. J. Yao, Y. Fan, R. Du, J. Zhou, Y. Lu, W. Wang, J. Ren and X. Sun, *Biomaterials*, 2010, **31**, 9357–9365.
35. B. Liang, M. L. He, C. Y. Chan, Y. C. Chen, X. P. Li, Y. Li, D. Zheng, M. C. Lin, H.F. Kung, X. T. Shuai and Y. Peng, *Biomaterials*, 2009, **30**, 4014–4020.
36. C. Zhang, S. Gao, W. Jiang, S. Lin, F. Du, Z. Li and W. Huang, *Biomaterials*, 2010, **31**, 6075–6086.
37. H. Oliveira, R. Fernandez, L. R. Pires, M. C. Martins, S. Simões, M. A. Barbosa and A. P. Pêgo, *J. Controlled Release*, 2010, **143**, 350–358.
38. D. W. Hwang, S. Son, J. Jang, H. Youn, S. Lee, D. Lee, Y. S. Lee, J. M. Jeong, W. J. Kim and D.S. Lee, *Biomaterials*, 2011, **32**, 4968–4975.
39. Y. Pi, X. Zhang, J. Shi, J. Zhu, W. Chen, C. Zhang, W. Gao, C. Zhou and Y. Ao, *Biomaterials*, 2011, **32**, 6324–6332.
40. X. Zeng, Y. X. Sun, X. Z. Zhang, S. X. Cheng and R. X. Zhuo, *Pharm. Res.*, 2009, **26**, 1931–1941.
41. X. Zeng, Y. X. Sun, X. Z. Zhang and R. X. Zhuo, *Org. Biomol. Chem.*, 2009, **7**, 4201–4210.
42. X. Zeng, Y. X. Sun, X. Z. Zhang and R. X. Zhuo, *Mol. Biosyst.*, 2010, **6**, 1933–1940.
43. X. Zeng, Y. X. Sun, X. Z. Zhang and R. X. Zhuo, *Biomaterials*, 2010, **31**, 4771–4780.
44. S. J. Kim, H. Ise, M. Goto, K. Komura, C. S. Cho and T. Akaike, *Biomaterials*, 2011, **32**, 3471–3480.
45. Q. Peng, F. Chen, Z. L. Zhong and R. X. Zhuo, *Chem. Commun.*, 2010, **46**, 5888–5890.
46. S. Li, Y. Wang, J. Zhang, W. H. Yang, Z. H. Dai, W. Zhu and X. Q. Yu, *Mol. Biosyst.*, 2011, **7**, 1254–1262.
47. A. Swami, R. K. Kurupati, A. Pathak, Y. Singh, P. Kumar and K. C. Gupta, *Biochem. Biophys. Res. Commun.*, 2007, **362**, 835–841.
48. R. Goyal, R. Bansal, S. Tyagi, Y. Shukla, P. Kumar and K. C. Gupta, *Mol. Biosyst.*, 2011, **7**, 2055–2065.
49. R. Goyal, S. K. Tripathi, S. Tyagi, A. Sharma, K. R. Ram, D. K. Chowdhuri, Y. K. Shukla, P. Kumar and K. C. Gupta, *Nanomedicine*, 2012, **8**, 167–175.
50. V. B. Morris and C. P. Sharma, *Biomaterials*, 2010, **31**, 8759–8769.
51. C. Pfeifer, G. Hasenpusch, S. Uezguen, M. K. Aneja, D. Reinhardt, J. Kirch, M. Schneider, S. Claus, W. Friess and C. Rudolph, *J. Controlled Release*, 2011, **154**, 69–76.
52. J. Yang, P. Zhang, L. Tang, P. Sun, W. Liu, P. Sun, A. Zuo and D. Liang, *Biomaterials*, 2010, **31**, 144–155.

53. J. Q. Gao, Q. Q. Zhao, T. F. Lv, W. P. Shuai, J. Zhou, G. P. Tang, W. Q. Liang, Y. Tabata and Y. L. Hu, *Int. J. Pharm.*, 2010, **387**, 286–294.
54. Y. Ping, C. Liu, Z. Zhang, K. L. Liu, J. Chen and J. Li, *Biomaterials*, 2011, **32**, 8328–8341.
55. B. Lu, D. Q. Wu, H. Zheng, C. Y. Quan, X. Z. Zhang and R. X. Zhuo, *Mol. Biosyst.*, 2010, **6**, 2529–2538.
56. Z. Mohammadi, M. Abolhassani, F. A. Dorkoosh, S. Hosseinkhani, K. Gilani, T. Amini, A. R. Najafabadi and M. R. Tehrani, *Int. J. Pharm.*, 2011, **409**, 307–313.
57. H. D. Han, L. S. Mangala, J. W. Lee, M. M. Shahzad, H. S. Kim, D. Shen, E. J. Nam, E. M. Mora, R. L. Stone, C. Lu, S. J. Lee, J. W. Roh, A. M. Nick, G. Lopez-Berestein and A. K. Sood, *Clin. Cancer Res.*, 2010, **16**, 3910–3922.
58. Y. Z. Du, L. L. Cai, J. Li, M. D. Zhao, F. Y. Chen, H. Yuan and F. Q. Hu, *Int. J. Nanomed.*, 2011, **6**, 1559–1568.
59. D. Lee, W. Zhang, S. A. Shirley, X. Kong, G. R. Hellermann, R. F. Lockey and S. S. Mohapatra, *Pharm. Res.*, 2007, **24**, 157–167.
60. C. X. Li, T. Y. Guo, D. Z. Zhou, Y. L. Hu, H. Zhou, S. F. Wang, J. T. Chen and Z. P. Zhang, *J. Controlled Release*, 2011, **154**, 177–188.
61. A. K. Varkouhi, R. J. Verheul, R. M. Schiffelers, T. Lammers, G. Storm and W. E. Hennink, *Bioconjugate Chem.*, 2010, **21**, 2339–2346.
62. X. Zhao, L. Yin, J. Ding, C. Tang, S. Gu, C. Yin and Y. Mao, *J. Controlled Release*, 2010, **144**, 46–54.
63. Z. X. Liao, Y. C. Ho, H. L. Chen, S. F. Peng, C. W. Hsiao and H. W. Sung, *Biomaterials*, 2010, **31**, 8780–8788.
64. P. Opanasopit, J. Tragulpakseerojn, A. Apirakaramwong, T. Ngawhirunpat and T. Rojanarata, *Int. J. Pharm.*, 2011, **410**, 161–168.
65. E. A. Klausner, Z. Zhang, R. L. Chapman, R. F. Multack and M. V. Volin, *Biomaterials*, 2010, **31**, 1814–1820.
66. Y. N. Xue, M. Liu, L. Peng, S. W. Huang and R. X. Zhuo, *Macromol. Biosci.*, 2010, **10**, 404–414.
67. H. M. Wu, S. R. Pan, M. W. Chen, Y. Wu, C. Wang, Y. T. Wen, X. Zeng and C. B. Wu, *Biomaterials*, 2011, **32**, 1619–1634.
68. H. Yu, Y. Nie, C. Dohmen, Y. Li and E. Wagner, *Biomacromolecules*, 2011, **12**, 2039–2047.
69. Q. Zhang, S. Chen, R. X. Zhuo, X. Z. Zhang and S. X. Cheng, *Bioconjugate Chem.*, 2010, **21**, 2086–2092.
70. Y. Liu, R. Q. Huang, L. Han, W. L. Ke, K. Shao, L. Y. Ye, J. N. Lou and C. Jiang, *Biomaterials*, 2009, **30**, 4195–4202.
71. R. Q. Huang, W. L. Ke, L. Han, J. F. Li, S. H. Liu and C. Jiang, *Biomaterials*, 2011, **32**, 2399–2406.
72. S. X. Huang, J. F. Li, L. Han, S. H. Liu, H. J. Ma, R. Q. Huang and C. Jiang, *Biomaterials*, 2011, **32**, 6832–6838.
73. W. L. Ke, K. Shao, R. Q. Huang, L. Han, Y. Liu, J. F. Li, Y. Y. Kuang, L. Y. Ye, J. N. Lou and C. Jiang, *Biomaterials*, 2009, **30**, 6976–6985.

74. S. Koppu, Y. J. Oh, R. Edrada-Ebel, D. R. Blatchford, L. Tetley, R.J. Tate and C. Dufès, *J. Controlled Release*, 2010, **143**, 215–221.

75. F. Lemarié, D. R. Croft, R. J. Tate, K. M. Ryan and C. Dufès, *Biomaterials*, 2011, **33**, 2701–2709.

76. H. Aldawsari, R. Edrada-Ebel, D. R. Blatchford, R. J. Tate, L. Tetley and C. Dufès, *Biomaterials*, 2011, **32**, 5889–5899.

77. O. Taratula, O. Garbuzenko, R. Savla, Y. A. Wang, H. He and T. Minko, *Curr. Drug Delivery*, 2011, **8**, 59–69.

78. O. Taratula, O. Garbuzenko, P. Kirkpatrick, I. Pya, R. Savla, V. P. Pozharov, H. He and T. Minko, *J. Controlled Release*, 2009, **140**, 284–293.

79. J. Hao, X. Sha, Y. Tang, Y. Jiang, Z. Zhang, W. Zhang, Y. Li and X. Fang, *Arch. Pharm. Res.*, 2009, **32**, 1045–1054.

80. K. Luo, C. Li, G. Wang, Y. Nie, B. He, Y. Wu and Z. Gu, *J. Controlled Release*, **155**, 77–87.

81. K. T. Al-Jamal, W. T. Al-Jamal, S. Akerman, J. E. Podesta, A. Yilmazer, J. A. Turton, A. Bianco, N. Vargesson, C. Kanthou, A. T. Florence, G. M. Tozer and K. Kostarelos, *Proc. Natl. Acad. Sci. U. S. A.*, 2010, **107**, 3966–3971.

82. C. Liu, W. Yu, Z. Chen, J. Zhang and N. Zhang, *J. Controlled Release*, 2011, **151**, 162–175.

83. K. Numata, J. Hamasaki, B. Subramanian and D. L. Kaplan, *J. Controlled Release*, 2010, **146**, 136–143.

84. J. H. Kang, D. Asai, S. Yamada, R. Toita, J. Oishi, T. Mori, T. Niidome and Y. Katayama, *Proteomics*, 2008, **8**, 2006–2011.

85. J. H. Kang, D. Asai, J. H. Kim, T. Mori, R. Toita, T. Tomiyama, Y. Asami, J. Oishi, Y. T. Sato, T. Niidome, B. Jun, H. Nakashima and Y. Katayama, *J. Am. Chem. Soc.*, 2008, **130**, 14906–14907.

86. J. H. Kang, J. Oishi, J. H. Kim, M. Ijuin, R. Toita, B. Jun, D. Asai, T. Mori, T. Niidome, K. Tanizawa, S. Kuroda and Y. Katayama, *Nanomedicine*, 2010, **6**, 583–589.

87. J. X. Chen, H. Y. Wang, C. Y. Quan, X. D. Xu, X. Z. Zhang and R. X. Zhuo, *Org. Biomol. Chem.*, 2010, **8**, 3142–3148.

88. H. Hyun, Y. W. Won, K. M. Kim, J. Lee, M. Lee and Y. H. Kim, *Biomaterials*, 2010, **31**, 9128–9134.

89. H. Y. Nam, A. McGinn, P. H. Kim, S. W. Kim and D. A. Bull, *Biomaterials*, 2010, **31**, 8081–8087.

90. T. J. Harris, J. J. Green, P. W. Fung, R. Langer, D. G. Anderson and S. N. Bhatia, *Biomaterials*, 2010, **31**, 998–1006.

91. H. Y. Wang, J. X. Chen, Y. X. Sun, J. Z. Deng, C. Li, X. Z. Zhang and R. X. Zhuo, *J. Controlled Release*, 2010, **155**, 26–33.

92. S. Yamano, J. Dai, C. Yuvienco, S. Khapli, A. M. Moursi and J. K. Montclare, *J. Controlled Release*, 2011, **152**, 278–285.

93. S. Govindarajan, J. Sivakumar, P. Garimidi, N. Rangaraj, J. M. Kumar, N. M. Rao and V. Gopal, *Biomaterials*, 2012, **33**, 2570–2582.

94. S. Chen, R. X. Zhuo and S. X. Cheng, *J. Gene Med.*, 2010, **12**, 705–713.

95. S. H. Min, D. M. Kim, M. N. Kim, J. Ge, D. C. Lee, I. Y. Park, K. C. Park, J. S. Hwang, C. W. Cho and Y. I. Yeom, *Biomaterials*, 2010, **31**, 1858–1864.

96. M. R. Capecchi, *Cell*, 1980, **22**, 479–488.

97. D. A. Dean, *Exp. Cell Res.*, 1997, **230**, 293–302.

98. H. Y. Wang, C. Li, W. J. Yi, Y. X. Sun, S. X. Cheng, R. X. Zhuo and X. Z. Zhang, *Bioconjugate Chem.*, 2011, **22**, 1567–1575.

99. N. Kanayama, S. Fukushima, N. Nishiyama, K. Itaka, W. D. Jang, K. Miyata, Y. Yamasaki, U. I. Chung and K. Kataoka, *ChemMedChem*, 2006, **1**, 439–444.

100. M. Oba, K. Miyata, K. Osada, R. J. Christie, M. Sanjoh, W. Li, S. Fukushima, T. Ishii, M. R. Kano, N. Nishiyama, H. Koyama and K. Kataoka, *Biomaterials*, 2011, **32**, 652–663.

101. K. Itaka, S. Ohba, K. Miyata, H. Kawaguchi, K. Nakamura, T. Takato, U. I. Chung and K. Kataoka, *Mol. Ther.*, 2007, **15**, 1655–1662.

102. S. Uchida, K. Itaka, Q. Chen, K. Osada, K. Miyata, T. Ishii, M. Harada-Shiba and K. Kataoka, *J. Controlled Release*, 2011, **155**, 296–302.

103. T. C. Lai, K. Kataoka and G. S. Kwon, *Biomaterials*, 2011, **32**, 4594–4603.

104. P. Li, D. Liu, X. Sun, C. Liu, Y. Liu and N. Zhang, *Nanotechnology*, 2011, **22**, 245104.

105. Y. Hu, K. Li, L. Wang, S. Yin, Z. Zhang and Y. Zhang, *J. Controlled Release*, 2010, **144**, 75–81.

106. R. S. Shirazi, K. K. Ewert, C. Leal, R. N. Majzoub, N. F. Bouxsein and C. R. Safinya, *Biochim. Biophys. Acta*, 2011, **1808**, 2156–2166.

107. Q. D. Huang, W. J. Ou, H. Chen, Z. H. Feng, J. Y. Wang, J. Zhang, W. Zhu and X. Q. Yu, *Eur. J. Pharm. Biopharm.*, 2011, **78**, 326–335.

108. A. Pathak, A. Aggarwal, R. K. Kurupati, S. Patnaik, A. Swami, Y. Singh, P. Kumar, S. P. Vyas and K. C. Gupta, *Pharm. Res.*, 2007, **24**, 1427–1440.

109. S. Nimesh, R. Kumar and R. Chandra, *Int. J. Pharm.*, 2006, **320**, 143–149.

110. Y. C. Chung, W. Y. Hsieh and T. H. Young, *Biomaterials*, 2010, **31**, 4194–4203.

111. Y. C. Chung, T. Y. Cheng and T. H. Young, *Biomaterials*, 2011, **32**, 4471–4480.

112. L. Peng, M. Liu, Y. N. Xue, S. W. Huang and R. X. Zhuo, *Biomaterials*, 2009, **30**, 5825–5833.

113. L. Peng, Y. Gao, Y. N. Xue, S. W. Huang and R. X. Zhuo, *Biomaterials*, 2010, **31**, 4467–4476.

114. M. Liu, J. Chen, Y. P. Cheng, Y. N. Xue, R. X. Zhuo and S. W. Huang, *Macromol. Biosci.*, 2010, **10**, 384–392.

115. H. Cheng, Y. Y. Li, X. Zeng, Y. X. Sun, X. Z. Zhang and R. X. Zhuo, *Biomaterials*, 2009, **30**, 1246–1253.

116. S. Guo, Y. Huang, W. Zhang, W. Wang, T. Wei, D. Lin, J. Xing, L. Deng, Q. Du, Z. Liang, X. J. Liang and A. Dong, *Biomaterials*, 2011, **32**, 4283–4292.

117. Y. Shen, H. Peng, J. Deng, Y. Wen, X. Luo, S. Pan, C. Wu and M. Feng, *Int. J. Pharm.*, 2009, **375**, 140–147.

118. C. Wang, M. Feng, J. Deng, Y. Zhao, X. Zeng, L. Han, S. Pan and C. Wu, *Int. J. Pharm.*, 2010, **398**, 237–245.

119. Z. Y. Ong, K. Fukushima, D. J. Coady, Y. Y. Yang, P. L. Ee and J. L. Hedrick, *J. Controlled Release*, 2011, **152**, 120–126.

120. C. F. Wang, Y. X. Lin, T. Jiang, F. He and R. X. Zhuo, *Biomaterials*, 2009, **30**, 4824–4832.

121. F. He, C. F. Wang, T. Jiang, B. Han and R. X. Zhuo, *Biomacromolecules*, 2010, **11**, 3028–3035.

122. C. Hu, Q. Peng, F. Chen, Z. L. Zhong and R. X. Zhuo, *Bioconjugate Chem.*, 2010, **21**, 836–843.

123. Y. Shan, T. Luo, C. Peng, R. Sheng, A. Cao, X. Cao, M. Shen, R. Guo, H. Tomás and X. Shi, *Biomaterials*, 2012, **33**, 3025–3035.

124. R. Namgung, Y. Zhang, Q. L. Fang, K. Singha, H. J. Lee, I. K. Kwon, Y. Y. Jeong, I. K. Park, S. J. Son and W. J. Kim, *Biomaterials*, 2011, **32**, 3042–3052.

125. M. X. Zhao, J. M. Li, L. Du, C. P. Tan, Q. Xia, Z. W. Mao and L. N. Ji, *Chem.–Eur. J.*, 2011, **17**, 5171–5179.

126. J. M. Li, Y. Y. Wang, M. X. Zhao, C. P. Tan, Y. Q. Li, X. Y. Le, L. N. Ji and Z. W. Mao, *Biomaterials*, 2012, **33**, 2780–2790.

127. B. Chertok, A. E. David and V. C. Yang, *Biomaterials*, 2010, **31**, 6317–6324.

128. F. M. Kievit, O. Veiseh, N. Bhattarai, C. Fang, J. W. Gunn, D. Lee, R. G. Ellenbogen, J. M. Olson and M. Zhang, *Adv. Funct. Mater.*, 2009, **19**, 2244–2251.

129. O. Veiseh, F. M. Kievit, C. Fang, N. Mu, S. Jana, M. C. Leung, H. Mok, R. G. Ellenbogen, J. O. Park and M. Zhang, *Biomaterials*, 2010, **31**, 8032–8042.

130. F. J. Xu, Y. Zhu, M. Y. Chai and F. S. Liu, *Acta Biomater.*, 2011, **7**, 3131–3140.

131. X. Dong, H. Tian, L. Chen, J. Chen and X. Chen, *J. Controlled Release*, 2011, **152**, 135–142.

132. K. Miyata, N. Gouda, H. Takemoto, M. Oba, Y. Lee, H. Koyama, Y. Yamasaki, K. Itaka, N. Nishiyama and K. Kataoka, *Biomaterials*, 2010, **31**, 4764–4770.

133. P. Zhang, Z. Zhang, Y. Yang and Y. Li, *Int. J. Pharm.*, 2010, **392**, 241–248.

134. T. von Erlach, S. Zwicker, B. Pidhatika, R. Konradi, M. Textor, H. Hall and T. Lühmann, *Biomaterials*, 2011, **32**, 5291–5303.

135. D. M. Lynn and R. Langer, *J. Am. Chem. Soc.*, 2000, **122**, 10761–10768.

136. D. M. Lynn, D. G. Anderson, D. Putnam and R. Langer, *J. Am. Chem. Soc.*, 2001, **123**, 8155–8156.
137. D. G. Anderson, D. M. Lynn and R. Langer, *Angew. Chem. Int. Ed.*, 2003, **42**, 3153–3158.
138. J. J. Green, R. Langer and D. G. Anderson, *Acc. Chem. Res.*, 2008, **41**, 749–759.
139. N. S. Bhise, R. S. Gray, J. C. Sunshine, S. Htet, A. J. Ewald and J. J. Green, *Biomaterials*, 2010, **31**, 8088–8096.
140. S. Y. Tzeng, H. Guerrero-Cázares, E. E. Martinez, J. C. Sunshine, A. Quiñones-Hinojosa and J. J. Green, *Biomaterials*, 2011, **32**, 5402–5410.
141. K. Zhang, H. Fang, Z. Wang, J. S. Taylor and K. L. Wooley, *Biomaterials*, 2009, **30**, 968–977.
142. K. Zhang, H. Fang, Z. Wang, J. S. Taylor and K. L. Wooley, *Biomaterials*, 2010, **31**, 1805–1813.
143. M. Ahmed and R. Narain, *Biomaterials*, 2011, **32**, 5279–5290.
144. M. Ma, F. Li, Z. F. Yuan and R. X. Zhuo, *Acta. Biomater.*, 2010, **6**, 2658–2665.
145. N. P. Gabrielson and J. Cheng, *Biomaterials*, 2010, **31**, 9117–9127.
146. H. Takemoto, A. Ishii, K. Miyata, M. Nakanishi, M. Oba, T. Ishii, Y. Yamasaki, N. Nishiyama and K. Kataoka, *Biomaterials*, 2010, **31**, 8097–8105.
147. F. Dai, P. Sun, Y. Liu and W. Liu, *Biomaterials*, 2010, **31**, 559–569.
148. S. D. Xiong, L. Li, J. Jiang, L. P. Tong, S. Wu, Z. S. Xu and P. K. Chu, *Biomaterials*, 2010, **31**, 2673–2685.
149. G. McLachlan, H. Davidson, E. Holder, L. A. Davies, I. A. Pringle, S. G. Sumner-Jones, A. Baker, P. Tennant, C. Gordon, C. Vrettou, R. Blundell, L. Hyndman, B. Stevenson, A. Wilson, A. Doherty, D. J. Shaw, R. L. Coles, H. Painter, S. H. Cheng, R. K. Scheule, J. C. Davies, J. A. Innes, S. C. Hyde, U. Griesenbach, E. W. Alton, A. C. Boyd, D. J. Porteous, D. R. Gill and D. D. Collie, *Gene Ther.*, 2011, **18**, 996–1005.
150. A. Ramasubramanian, S. Shiigi, G. K. Lee and F. Yang, *Pharm. Res.*, 2011, **28**, 1328–1337.
151. M. Roger, A. Clavreul, M. C. Venier-Julienne, C. Passirani, L. Sindji, P. Schiller, C. Montero-Menei and P. Menei, *Biomaterials*, 2010, **31**, 8393–8401.
152. M. R. Jung, I. K. Shim, E. S. Kim, Y. J. Park, Y. I. Yang, S. K. Lee and S. J. Lee, *J. Controlled Release*, 2011, **152**, 294–302.
153. J. S. Park, H. N. Yang, D. G. Woo, S. Y. Jeon, H. J. Do, H. Y. Lim, J. H. Kim and K. H. Park, *Biomaterials*, 2011, **32**, 3679–3688.
154. J. S. Park, K. Na, D. G. Woo, H. N. Yang, J. M. Kim, J. H. Kim, H. M Chung and K. H. Park, *Biomaterials*, 2010, **31**, 124–132.
155. W. Wang, B. Li, Y. Li, Y. Jiang, H. Ouyang and C. Gao, *Biomaterials*, 2010, **31**, 5953–5965.

156. C. X. He, N. Li, Y. L. Hu, X. M. Zhu, H. J. Li, M. Han, P. H. Miao, Z. J. Hu, G. Wang, W. Q. Liang, Y. Tabata and J. Q. Gao, *Pharm. Res.*, 2011, **28**, 1577–1590.

157. P. H. Kim, T. I. Kim, J. W. Yockman, S. W. Kim and C. O. Yun, *Biomaterials*, 2010, **31**, 1865–1874.

158. D. Li, G. Li, P. Li, L. Zhang, Z. Liu, J. Wang and E. Wang, *Biomaterials*, 2010, **31**, 1850–1857.

159. H. Wang, P. Zhao, W. Su, S. Wang, Z. Liao, R. Niu and J. Chang, *Biomaterials*, 2010, **31**, 8741–8748.

160. K. Itaka and K. Kataoka, *Eur. J. Pharm. Biopharm.*, 2009, **71**, 475–483.

161. X. B. Xiong, H. Uludağ and A. Lavasanifar, *Biomaterials*, 2010, **31**, 5886–5893.

162. X. B. Xiong and A. Lavasanifar, *ACS Nano*, 2011, **5**, 5202–5213.

163. J. Dai, S. Zou, Y. Pei, D. Cheng, H. Ai and X. Shuai, *Biomaterials*, 2011, **32**, 1694–1705.

164. K. Park, S. W. Hong, W. Hur, M. Y. Lee, J. A. Yang, S. W. Kim, S. K. Yoon and S. K. Hahn, *Biomaterials*, 2011, **32**, 4951–4958.

165. R. B. Arote, S. K. Hwang, H. T. Lim, T. H. Kim, D. Jere, H. L. Jiang, Y. K. Kim, M. H. Cho and C. S. Cho, *Biomaterials*, 2010, **31**, 2435–2445.

CHAPTER 5

Functional Hyperbranched Polymers for Drug and Gene Delivery

YUE JIN AND XINYUAN ZHU*

Shanghai Jiao Tong University, School of Chemistry and Chemical Engineering, State Key Laboratory of Metal Matrix Composites, 800 Dongchuan Road, Shanghai 200240, P. R. China
*E-mail: xyzhu@sjtu.edu.cn

5.1 Introduction

Hyperbranched polymers (HBPs) with an intrinsic globular structure are one important subclass of dendritic polymers.[1] Compared to the conventional linear, branched, and crosslinked polymers, HBPs show unique features which mainly derive from their dendritic architecture, adequate spatial cavities, and numerous terminal groups. Correspondingly, HBPs exhibit the advantages of non/low-entanglement, low melt viscosity, good solubility, and a large number of end-groups that are prone to be chemically modified.[1–5] Besides, the one-pot synthetic procedure of HBPs is facile by comparison with that of dendrimers, which is conducive to their large-scale scientific and industrial applications. Given these merits, the development of HBPs has drawn considerable attention during the past decades, especially in the applications of drug and gene delivery.[6] Nowadays, the great demands of *in vivo* programmable and controlled delivery and release of drug and gene call for smart and functional carriers or vectors. Taking the outstanding architectural advantages of HBPs

RSC Polymer Chemistry Series No. 3
Functional Polymers for Nanomedicine
Edited by Youqing Shen
© The Royal Society of Chemistry 2013
Published by the Royal Society of Chemistry, www.rsc.org

into account, HBPs are one of the most promising materials to be designed as functional vehicles for drug and gene delivery. By means of various synthetic methodologies, HBPs can be blessed with numerous specific functions, such as responsiveness (pH, temperature, magnetic field, redox sensitivity, *etc.*),[7–12] recognition (saccharide, enzyme, folic acid, antibody, *etc.*),[13–15] and imaging (quantum dots, conjugated polymers, bioluminescent proteins, *etc.*).[16–19] These excellent performances of HBPs can, to a considerable degree, improve *in vivo* delivery behavior and increase the related auxiliary functions of drug and gene vehicles. Moreover, these functions for drug delivery and gene transfection can be controlled and optimized through modulating the branched structures of HBPs.[2,6] In a nutshell, the design and preparation of functional HBPs are of great significance for the exploitation of novel drug and gene delivery systems. This field has been extending into its own niche and more valuable works are underway. The combination of recent progress, the synthetic approaches, the functionalization of HBPs, and their applications in drug and gene delivery are summarized in this chapter.

5.2   Preparation of Functional HBPs

A variety of synthetic strategies for HBPs have been proposed and achieved during the past decades.[3–5] As previously outlined, several different methodologies have been developed. Firstly, the main methods of synthesizing the highly branched architecture will be briefly discussed. Subsequently, the functionalization strategies of HBPs will be presented.

5.2.1   Preparation of HBPs

HBPs have versatile chemical structures (linear units, dendritic units, and terminal units) and exhibit unique properties. The majority of HBPs are prepared through two pivotal methods, namely the step growth approach and the chain growth approach. Step growth was initially developed according to Flory's theory. With the discovery of self-condensing vinyl polymerization by Frechét, more and more attention was focused on the chain growth method. Occasionally, copolymerization of different monomers requires a hybrid of two synthetic methods. Furthermore, supramolecular interaction has also been introduced to construct and control the highly branched architecture as well.

5.2.1.1   Step Growth

There have emerged two kinds of monomers for the step growth method, namely AB_x and $A_2 + B_y$. A broad portfolio of suitable A and B structures can result in different HBPs. On the one hand, polycondensation of AB_x monomers is the classical way to prepare HBPs with and without core moieties. However, there only exist a limited number of AB_x monomers, which are not easy to obtain. On the other hand, HBPs can be prepared by the $A_2 +$

B_y approach from easily available monomers as well. Among them, the method of $A_2 + B_3$ polymerization provides a facile approach to prepare HBPs. Step growth has been widely used to synthesize hyperbranched poly(sulfone-amine)s,[20–22] hyperbranched poly(amide-amine)s,[16,23] and hyperbranched poly(ester-amine)s[24,25] with commercially available monomers.

5.2.1.2 Chain Growth

Self-condensing vinyl polymerization (SCVP) was first proposed by Frechét *et al.*[26,27] The synthetic method is based on the design of an inimer bearing both a vinyl monomer and an initiating group. These monomers can accomplish chain growth and step growth through propagation of double bonds and also initiation of double bonds, thus leading to HBPs in a one-pot reaction.

Self-condensing ring-opening polymerization (SCROP) can be called ring-opening multibranching polymerization (ROMBP) as well. As a matter of fact, the mechanism of SCROP is similar to that of SCVP. A vinyl group is replaced by a heterocyclic group as the monomer part of the inimer. Besides, irreversible reactions are considered for SCVP, whereas reversible preconditions usually occur for SCROP.[28,29] The SCROP approaches stem from the classical ring-opening reaction mechanism (ROP), for instance polyesters.

Proton-transfer polymerization (PTP) was also proposed by Frechét.[30] PTP is featured with a proton transfer for the activation of the nucleophile used in epoxide opening in each propagation step. The addition of a catalytic amount of initiator such as OH^- can initiate polymerization of the H-AB$_2$ monomer, thus affording the reactive nucleophile, which is able to further react with another monomer unit to afford a dimer. The latent reactivity now passes on to one of the original epoxy groups.

Hybrids of different polymerization methods can give rise to a brand-new HBP and meanwhile render itself unique structures and characteristics.[8]

5.2.2 Functionalization of HBPs

Different synthetic approaches are suitable for different monomers, but these methods basically play a role in constructing the backbones of HBPs, most of which do not possess specific functions. With the motivation of biomedical applications, the resulting HBPs require further functionalization by tethering external functional components or alteration of the backbone structures from the intrinsic chemical components. Benefiting from three-dimensional topological architecture and numerous terminal groups, HBPs are inclined to be further modified compared with other polymeric materials. Functionalization of HBPs generally involves terminal modification, hybrid modification, and backbone modification, which is of great importance to their further applications, as shown in Figure 5.1.

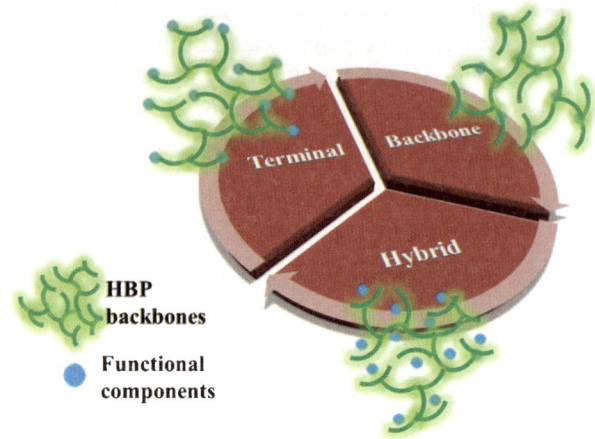

Figure 5.1   Three strategies for functionalization of HBPs.

5.2.2.1   Terminal Modification

The terminal groups of HBPs occupy a great proportion in the macro-molecules compared with linear polymers, so the alteration of the terminals is capable, to a considerable degree, of transforming the characteristics of HBPs. Nowadays, the numerous terminal groups of HBPs have been used to realize their functionalization, thus revealing the great influence of the terminal groups on the attributes of HBPs. Generally speaking, the terminal modification is mainly based on the functional groups at the periphery of HBPs, such as hydroxyl, carboxyl, amine, thiol, and halide terminals. Through these various end-groups, many specific functional components can be grafted onto the periphery of HBPs, including different functional end-groups, small molecules (drug, organic dye, folic acid, *etc.*),[31,32] oligomers or polymers (so-called "graft to" method),[10] or directly initiating the grafting polymerization of functional monomers (so-called "graft from" method) to prepare block HBPs and further obtain self-assemblies of HBPs. The interactions between terminal groups and modified components vary from covalent bonds to noncovalent bonds, thus affording many modified methods. Such a synthetic methodology of functionalizing HBPs is termed as "terminal groups replacing" (TGR).[9] Figure 5.2 summarizes the primary approaches to TGR; how to obtain the expected functions through TGR will be discussed in the section on functionality of delivery (Section 5.3).

5.2.2.2   Hybrid Modification

The driving force of hybrid modification mainly stems from host–guest interaction, complexation interaction, hydrogen bonding interaction, and electrostatic attraction. By means of these weak interactions, the external components such as metal ions can be settled into the cavities of HBPs, thus

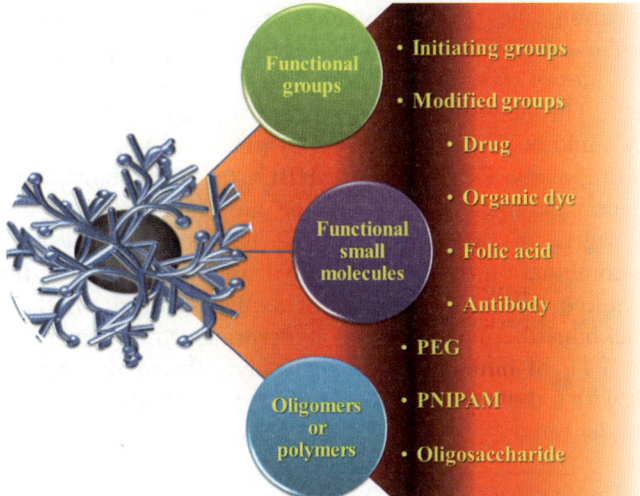

Figure 5.2 Primary approaches to terminal group replacement.

achieving the goal of functionalization. Inclusion in the cavities endows HBPs unique functions, making this hybrid modification of HBPs more reliable for the merits of exterior components and the interactions between them, which are not exclusively polymeric backbones. The backbones of HBPs can act as a template, or rather a nanoreactor to prepare these functional particles, and *vice versa*. At present, functional inorganic ions or nanoparticles, luminescent organic molecules, and new functional polymers are continually being developed, and meanwhile the hybrid approaches of HBPs and other components are expanding their own niche, enriching the family of exterior functional matters and topological complexity. Therefore, hybrid modification has been highlighted in many recent reports on biodetection and imaging. For instance, quantum dots and magnetic particles have already been added to the cavities of HBPs, achieving the functions of biodetection[16] and magnetofection *in vitro* as magnetic nonviral gene vectors,[12] respectively.

5.2.2.3 Backbone Modification

Backbone modification differs from the previously mentioned modification approaches, because this method focuses on altering the intrinsic characteristics of HBPs instead of introducing the exterior components into the original systems. In order to implement the backbone modification of HBPs, the elaborate selection of both suitable monomers and effective polymerization methods are required to attain the orchestrated molecular architectures. Both functional monomers and nonfunctional monomers are capable of achieving functional HBPs by means of certain polymerization methods. In general, not only the functional groups of monomers but also their arrangement have a great impact on the functional behavior of HBPs. The functionality resulting from the location, the special distribution of different monomers or the macromolecular topological structure of HBPs, plays a significant role in the inherent attributes of the macromolecular architecture except for the selection of monomers. Through selection of an appropriate synthetic method, the functionality could be further controlled by means of adjusting the molecular structure and topology. Particularly for HBPs, the functionalization hinges on the fundamental parameters involving the molecular weight, polydispersity (PDI), and degree of branching (DB). It has been well realized that an appropriate combination of hydrophilic/hydrophobic balance in the polymer chains is required for the occurrence of a phase transition. Based on this concept, backbone-thermoresponsive hyperbranched polymers have been synthesized:[7] PTP of butane-1,4-diol diglycidyl ether and various triols can afford HBPs with a thermoresponsive backbone, and both drugs and genes can be efficiently delivered.

5.3 Functionality of Delivery

Currently, we have witnessed a rapidly increasing amount of publications related to applications of HBPs and their assemblies in drug and gene delivery because of their advantages of morphological varieties, excellent template ability, facile functionalization, and smart responsiveness. These advantages are paving the way for biomedical applications of HBPs and their self-assemblies towards the direction of smart, programmable, and controlled delivery. Modification of pharmaceutical nanocarriers is normally used to control their biological properties in a desirable fashion and make them simultaneously perform various diagnostic or therapeutic functions in the delivery process. Therefore, functional materials are of great significance for drug and gene delivery. The applications of HBPs and their self-assemblies in this field are emerging just in the nick of time. Herein, we will classify the following work on HBP nanocarriers into five parts based on the functionality of delivery.

5.3.1 Responsiveness

The effect of drug therapy or gene transfection is closely related to the release behavior of the delivery system in the body. The responsiveness is expected to control the release behavior, making the delivery more efficient. Driven by a need to control drug or gene release in response to an appropriate stimulus, various stimuli-responsive HBPs and their assemblies have been developed as smart vehicles and studied extensively, comprising temperature-, pH-, redox-, light-, magnetic-, and enzyme-responsive ones, *etc.* The typical external stimuli can be logically categorized into three groups: physical (light, electric, magnetic, ultrasound, mechanical, and temperature), chemical (pH, ionic strength, solvent, electrochemical, and specific substances), and biological (enzymes and receptors). Stimuli-responsive HBPs have the capacity to respond to a tiny variation in environmental conditions, with subsequent obvious alteration in some aspect like shape, volume, phase state, or electrical, optical, mechanical, or surface properties. Therefore, HBPs with responsiveness are widely exploited to improve the functionality of delivery. In this section, we focus on stimuli-responsive HBPs which are sensitive to pH, temperature, and reduction, respectively.

5.3.1.1 pH Responsiveness

As we know, most tumor cells have an elevated acid level. In tumor tissue, extracellular pH ($pH_e \approx 6.8$) is slightly more acidic than the pH in normal tissue ($pH \approx 7.4$), mainly due to the anaerobic respiration and subsequent glycolysis together with the tumor phenotype. In the organelles, such as the endosome or lysosome ($pH_{endo} \approx 5.0$), the acid level is further increased, promoting drug or gene release from the endocytosed nanocarriers.[33] Therefore, the pH-sensitive polymeric self-assemblies which can rapidly respond to the mild acidic pH trigger provide an opportunity for the achievement of programmable and controlled drug delivery. Up to now, a series of acid-cleavable covalent linkages have been designed and applied in polymeric carriers to obtain pH-sensitive materials. Introducing pH-sensitive moieties into HBPs endows them with responsive functions. For example, the ionizable moieties (carbonyl and amine groups, *etc.*) and acid-cleavable covalent bonds (acetal, orthoester, and *cis*-acotinyl, *etc.*) render HBPs pH responsive. On the one hand, driven by the electrostatic and hydrogen bonding interactions, the protonation of HBPs directly leads to a change of the electric charge, ionic polarity, solubility, and even the whole chemical environment of HBPs, which reflects the pH-dependent manner of the smart materials. On the other hand, the acid-sensitive covalent bonds, including dynamic covalent bonds (hydrazone, imine, and oxime bonds), can reversibly react with hydrogen ions, resulting in changes of the chemical properties of HBPs.

According to our work published previously, hyperbranched poly[3-methyl-3- (hydroxymethyl)oxetane] (HPMHO) was prepared. Subsequently, succinic anhydride (SA) was tethered onto HPMHO by Xia *et al.*, thus obtaining carboxyl-modified HPMHOs with different carboxyl terminal contents (Suc-HPMHOs) through esterification of HPMHO and SA.[34] Because of the existence of the hydrophobic core and hydrophilic ionization of the carboxylic acid at the periphery, the HBP could homogenously disperse in water. The pH-responsive range of these Suc-HPMHOs can be easily adjusted by controlling the degree of carboxylation, involving the extracellular pH ($pH_e \approx 6.8$). Terminal modification of HPMHO allows the carboxyl groups to bind cisplatin through electrostatic interaction, constructing a HBP–cisplatin complex for antitumor investigation.

The pH-labile linkages endow HBPs with the function of pH responsiveness, and realize the programmable release of drug or gene according to the environmental pH alteration. For instance, Zhu *et al.* prepared a pH-sensitive hyperbranched polyacylhydrazone (HPAH) through the $A_2 + B_3$ polycondensation of butane-2,3-dione and 1-(2-aminoethyl)piperazine tripropionylhydrazine.[35] This hyperbranched dynamer HPAH was regarded as an excellent carrier for drug delivery, since it featured good aqueous solubility and low cytotoxicity. On the one hand, since the acylhydrazone bonds are stable in physiological conditions (pH $\approx$ 7.4), the HPAH–doxorubicin (DOX) nanosized micelles can avoid removal by the reticuloendothelial system (RES), kidneys, and intestines, perform the enhanced permeability and retention (EPR) effect, and reduce systemic side effects. On the other hand, after the endocytosis of HPAH–DOX micelles, the acidic lysosomes (pH $\approx$ 5–6) trigger cleavage of the acylhydrazone bonds, resulting in the release of DOX.

From the aforementioned examples, the pH sensitivity endows delivery systems with controlled release behavior.

5.3.1.2   *Thermo-responsiveness*

In response to the environmental temperature alteration, thermo-responsive HBPs exhibiting unique changes of chemical and physical properties are promising materials for various biomedical applications, especially in smart drug/gene delivery systems and tissue engineering. HBPs with a lower critical solution temperature (LCST) are readily soluble in water below their LCST, and their aqueous solutions undergo a phase transition to an insoluble state with the increasing temperature. In particular, polymers with a LCST around 37 °C have received much attention due to their phase transition close to body temperature. The most well-known class of thermo-responsive polymer, poly(*N*-isopropylacrylamide) (PNIPAM), exhibits a rapid coil-to-globule conversion at its LCST around 32 °C in aqueous solution.

Through PTP of triethanolamine, trimethylolpropane, and glycidyl methacrylate with potassium hydride as a catalyst, novel thermo-responsive hyperbranched poly(amine-ester)s were successfully designed and synthesized in one pot by Pang *et al.*[36] As shown in Figure 5.3, *in vitro* evaluation suggested that hyperbranched poly(amine-ester)s exhibited low cell cytotoxicity and efficient cell internalization against COS-7 cells. Moreover, DOX as a model drug was encapsulated into hyperbranched poly(amine-ester)s in aqueous solution, availing of a phase transition at the LCST. Thermo-responsive hyperbranched poly(amine-ester)s can be utilized to construct promising drug delivery systems for cancer therapy.

The combination of temperature-sensitive polymer nanoparticles (nanogel) and magnetically activated superparamagnetic nanoparticles has realized reversible, on–off drug or gene release by means of exertion and removal of an oscillating magnetic field.[37,38] The external oscillating magnetic field activates the superparamagnetic nanoparticles, which behave as local heat sources. When arriving at the phase transition temperature, the temperature-sensitive vehicles collapse and subsequently release their cargoes (drug or gene)

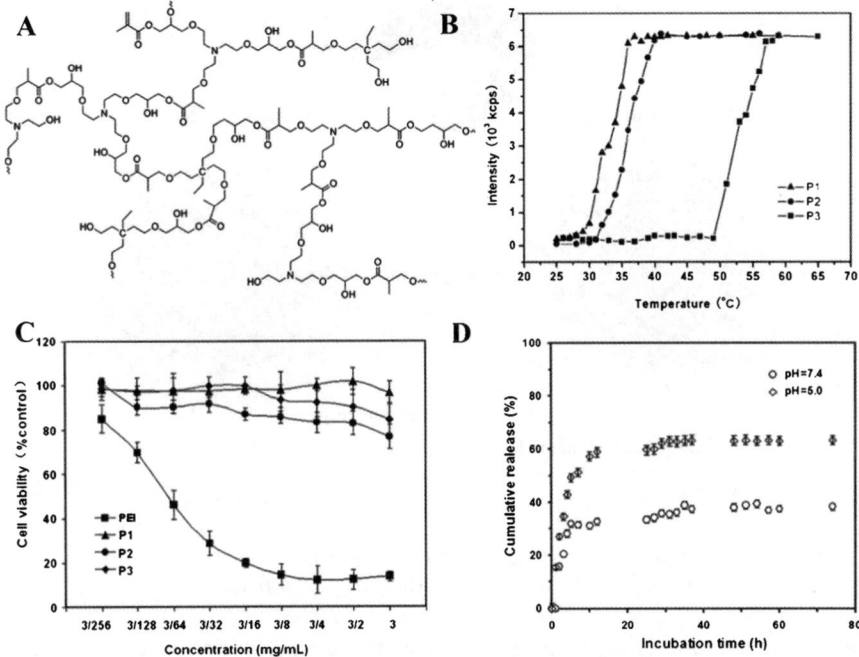

Figure 5.3   (A) A novel thermo-responsive hyperbranched poly(amine-ester). (B) Temperature dependence of the light scattering intensity. (C) Cell viability against COS-7 cells at different polymer concentrations compared to the control (PEI). (D) Cumulative release of drug from a DOX-loaded delivery system.[36]

into the matrix. Therefore, sustained release technology comes true through controlling the external oscillating magnetic field, as shown in Figure 5.4.

5.3.1.3   Redox Responsiveness

As a thiol-containing tripeptide, glutathione (GSH) is found in human blood plasma in micromolar concentrations, whereas it is around 10 mM in the cytosol. Especially, the concentration of cytosolic GSH in some tumor cells has been found to be a couple of times higher than that in normal cells.[39] Owing to the existence of a different order of magnitude in the redox potential between the mildly oxidizing extracellular milieu and the reducing intracellular fluids, redox-sensitive micelles have drawn great attention for intracellular drug and gene delivery among those various smart vehicles. Materials incorporating disulfide bonds have been explored for this purpose, which can be cleaved in the presence of reducing agents such as GSH. Therefore, disulfide-containing polymeric micelles can facilitate intracellular release of the encapsulated drug and gene by the cleavage of this bond.[40] Until now, redox-responsive amphiphilic HBPs and their nanosized assemblies have become one of the

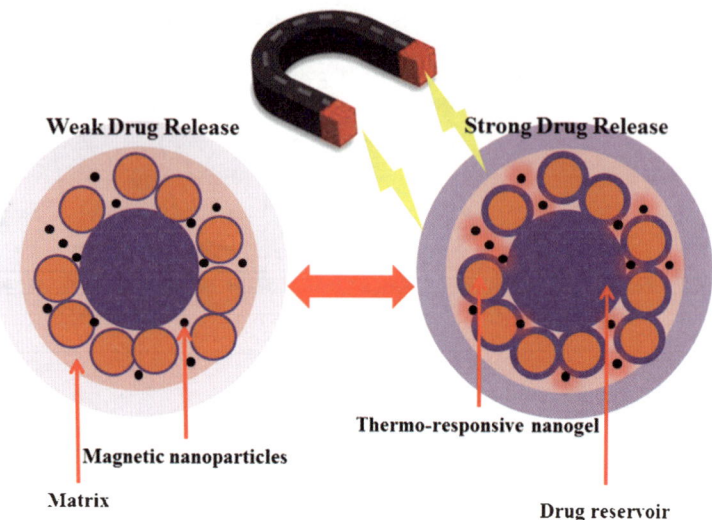

Figure 5.4   Sustained release technology of thermo-responsive HBPs through controlling the external oscillating magnetic field.

most attractive materials to construct drug and gene delivery systems, because of the existence of a redox potential gradient between the extra- and intracellular space. However, most of them fail in preclinical studies because of their high toxicity. The design of biocompatible materials is of significant need.

Liu *et al.* designed and prepared a series of redox-responsive HBPs.[11,41] For instance, synthesis of an amphiphilic hyperbranched homopolymer with alternating hydrophobic disulfide and hydrophilic polyphosphate segments along the highly branched structure by SCROP of the monomer, consisting of both hydrophobic and hydrophilic moieties, as shown in Figure 5.5. This hyperbranched polyphosphate (HPHDP) has excellent bioavailability and is further utilized in the construction of hydrophobic anticancer drug delivery system, which can be rapidly and efficiently transported into the nuclei of tumor cells and shows enhanced inhibition of cell proliferation.

Such redox-stimulus biodegradable micelles self-assembled from functional HBPs are promising biomaterials to improve both drug and gene delivery efficacy.

5.3.2   Targeting

Chemotherapy is detrimental to both cancer cells and normal cells. Despite ceaseless and intense efforts to discover highly effective oncology drugs, conventional chemotherapeutic agents still perform with poor specificity in arriving at tumor tissues and are often restricted by dose-limiting cytotoxicity. The functional delivery systems are capable of making drugs accumulate

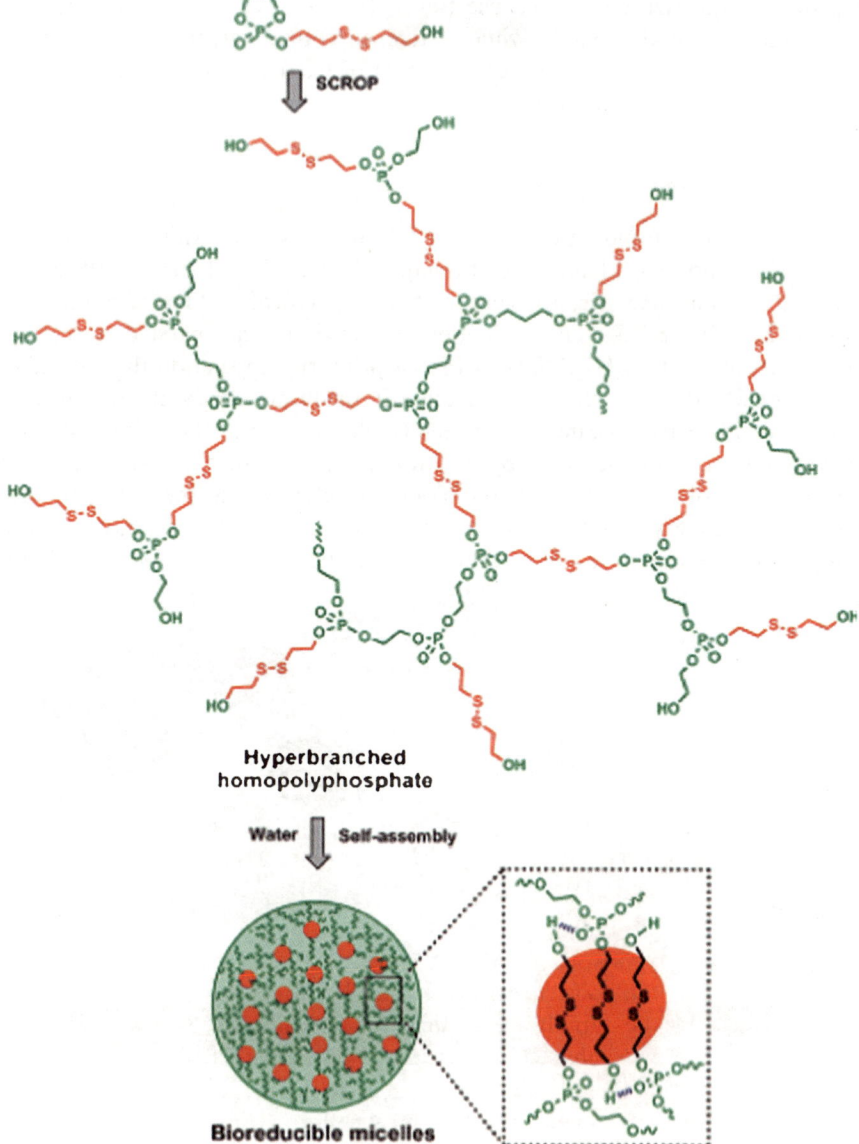

Figure 5.5 Synthesis of HPHDP and schematic representation of the self-assembled micelles. (Reproduced from Liu *et al.*[41] with permission from Wiley-VCH.)

specifically or nonspecifically in the required pathological zone. By smart vehicles, the active or passive targeting to the tumor tissue can, to a considerable degree, decrease the lethality of normal cells and improve the efficacy of transportation. Generally speaking, passive targeting mainly depends on the size effect, because tumor tissues have a larger intercellular space than normal tissues, while active targeting refers to the specific recognition between functional components of carriers and cancer cells. The details are discussed as follows.

5.3.2.1   Passive Targeting

Owing to their rapid growth, many tumor tissues exhibit fenestrated vasculature and poor lymphatic drainage, thus leading to the EPR effect, which allows nanoparticles to accumulate nonspecifically at the tumor site, as shown in Figure 5.6. This is often referred to as passive targeting. Nanoparticles such as **HBP** self-assemblies primarily depend on the size effect to accumulate at the specific tumor tissues. In the human body, the equilibrium is involved in pharmacokinetics between (i) elimination of the polymeric drug delivery system from the blood by the kidneys, liver, and other organs and (ii) extravasation of the drug out of the blood vasculature into the tumor tissues. Polymeric delivery systems whose hydrodynamic diameters have a size comparable to the pores of the kidney are inclined to be eliminated; by

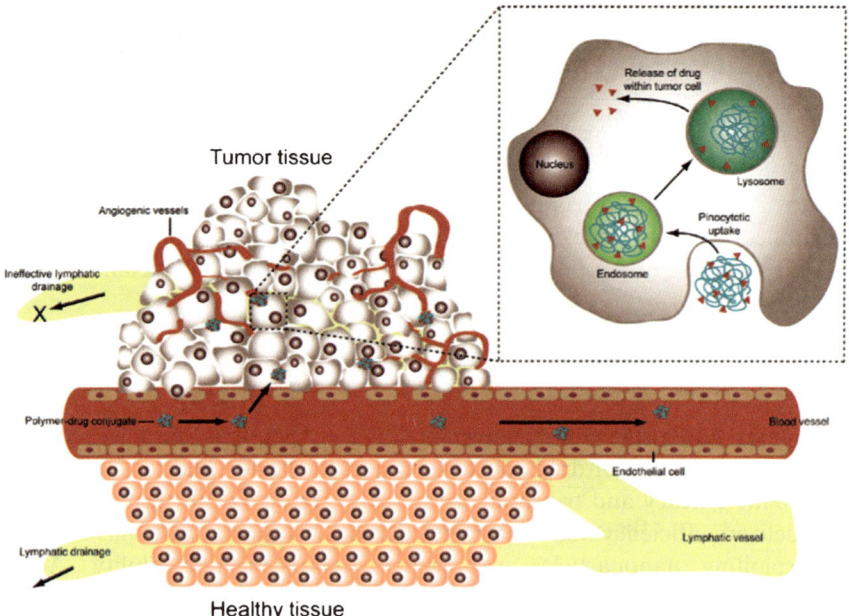

Figure 5.6   Schematic representation of the EPR effect: passive targeting to tumor tissue is achieved by extravasation of polymers through the increased permeability of the tumor vasculature and ineffective lymphatic drainage. Passively targeted polymer drug conjugates are taken up by cancer cells through pinocytosis and processed by endosomes and lysosomes (*inset*). (Reproduced from Fox *et al.*[42] with permission from the American Chemical Society.)

contrast, if the vehicles in the blood stream are able to traverse tumors, they are assumed to possess an order of magnitude smaller hydrodynamic diameter than the openings of the vessels. Therefore, a broad group of features in the macromolecular architecture, such as its molecular weight, molecular conformation, chain flexibility, branching, and location of the attached drug, have a great impact on the elimination of polymeric carriers through the kidney, while the architectural factors have a much smaller effect on the extravasation of the drug delivery systems into the tumor tissues.[42] From the structural perspective, the molecular architecture of HBPs is prone to be adjusted, benefiting from the various terminal groups and three-dimensional topological architecture. Therefore, passive targeting of HBP self-assemblies can be independently controlled in the aspects of both elimination and extravasation. Design of HBP-based drug carriers with an effective delivery behavior requires that the polymers must eventually be eliminated from the body.

Hyperbranched polyglycerols (HPGs) combine several remarkable features together, including a highly flexible aliphatic polyether backbone, multiple hydrophilic groups, and excellent biocompatibility. Moreover, the controlled synthesis of well-defined HPGs *via* SCROP of glycidol has realized the adjustment of fundamental parameters involving molecular weight, PDI, DB, and end-group functionality. A variety of linear–hyperbranched block copolymers of HPBs and their assemblies with a diameter less than 200 nm have been reported. These inherent merits of HPGs and their assemblies are useful for a number of biomedical applications, such as drug encapsulation or surface attachment.[43,44] The backbone of HPGs often appears in the drug delivery systems with a passive targeting ability. Similarly, if other HPB self-assemblies have a suitable diameter, they can take advantage of the EPR effect to transport drug or gene cargo and manifest the preferential accumulation to tumor tissues.

5.3.2.2 Active Targeting

One major limitation with passive targeting is that it is hard to realize a sufficiently high level of drug concentration at the tumor site, resulting in low therapeutic efficacy and triggering severe side effects. To further improve the drug delivery efficiency and cancer specificity, a strong emphasis has been put on exploiting nanoparticles with active tumor-targeting ability. Active targeting can be achieved by functionalizing nanocarriers with targeting ligands such as small molecules (*e.g.* folic acid and saccharide), antibodies, lectin, and peptides. These functional components can recognize and bind to specific receptors that are unique to cancer cells. Non-immunogenic folic acid (FA) possesses a high affinity for FA-binding proteins that are selectively overexpressed on the surface of human tumor cells, including breast, lung, brain, endometrial, renal, and colon cancer cells. Since the FA receptor is frequently overexpressed in tumor cells, the experimental results from flow cytometry and confocal laser scanning microscopy (CLSM) analysis generally show that FA-conjugated nanoparticles exhibit higher cellular uptake than the ones without FA conjugation. Meanwhile, other recognition agents have been developed to improve the targeting efficiency, such as Arg-Gly-Asp oligopeptides (RGD) and boronic acid. RGD with biological activity has been widely applied in drug delivery systems, which selectively targets at the receptors of $\alpha_v\beta_3$ integrin overexpressed in rapidly growing tumor capillaries during angiogenesis. Antibodies are also widely used in targeting the specific tumor tissues, such as the anti-EGFR antibody and the anti-RON antibody.[45–55] By comparison with common antibody–antigen recognition, multivalent sugar–protein binding has received much attention for the fabrication of targeted delivery systems, and has a lower cost and immunogenicity. It should also be mentioned that the effective combination of boronic acid with saccharide can be applied when recognizing saccharide on the surface of glycoproteins in the tumor tissue. The utility of peptide and boronic acids may accelerate the access to vast diagnostic methods economically.[56,57] Another obvious phenomenon occurring in tumor tissues is that multivalent saccharide binding by lectins is important for the process of cancer metastasis. The sequence transformation of carbohydrate arrays of glycoprotein on the cancer cells has altered the tethering ability of lectin and saccharide during carcinogenesis, which is a hint for targeting. The mutual interaction of lection and glycoprotein is similar to that of antibody and antigen. Interest in lectin specifically recognizing saccharide moieties involved in the glycoprotein is increasing due to their potential as therapeutics/diagnostics for a variety of diseases such as cancers.[58] The efficient modification of these functional targeting groups on the delivery systems is the key point for realizing the functional transportation to the tumor tissues. Therefore, HBPs with numerous terminal groups and cavities have burgeoned rapidly over the past decades.

Prabaharan *et al.* prepared a folate-conjugated amphiphilic hyperbranched block copolymer [H40-*star*-(PLA-*b*-PEG/PEG-FA)] as a carrier for tumor-targeted drug delivery.[59] The synthetic drug delivery system comprised a dendritic Boltorn H40 core, a hydrophobic poly(L-lactide) (PLA) inner shell, and a hydrophilic methoxy-poly(ethylene glycol) (PEG) and folate-conjugated poly(ethylene glycol) (PEG-FA) outer shell. The hydrophobic anticancer drug DOX was encapsulated into the core of the micelles. The release pattern of the DOX-loaded micelles manifested an initial burst release within 4 h and an ensuing sustained release of the entrapped DOX over a period of about 40 h. Cellular uptake of the DOX-loaded HBP micelles was found to be higher than that of the DOX-loaded H40-*star*-(PLA-*b*-PEG) micelles because of the folate-receptor-mediated endocytosis, thereby providing higher cytotoxicity against the 4T1 mouse mammary carcinoma cell line. The result indicates that the H40-*star*-(PLA-*b*-PEG/PEG-FA) micelles prepared from HBP are promising as tumor-targeted drug delivery nanocarriers. With many terminal groups on H40, the effect of targeting is better than that of linear polymers. The folate ligands at the periphery of H40 play a significant role in active targeting at the dangerous cancer cells.

5.3.3 Imaging

During the clinical drug research and development phase, there are narrow choices for evaluating pharmaceutical and pharmacological properties of new drugs. Among them, molecular imaging, which aims at direct/indirect visualization of biomarkers and biological processes in living organisms, becomes a robust tool to survey *in vivo* drug behavior during both the preclinical and clinical phases of discovery and development. Hence, the combination of anticancer therapy along with real time *in vivo* imaging of the targeted cancer cells is very promising for delivery systems in the clinical arena. Through this noninvasive functionalization, the determination of therapeutic responses of the HBP-based delivery systems can be detected and tracked *in vivo* anytime, which is conducive to pharmacokinetic research. The imaging functionalization renders the HBP-based delivery system broader applications for observing and further modulating the delivery system *in vivo*. In fact, the panorama of the physiological environment and the concrete process of metabolism in tumor tissues can be viewed by means of molecular imaging.[60] Likewise, proper labeling of these HBP delivery systems allows noninvasive and continuous visualization and tracking of their delivery kinetics and biodistribution *in vivo*. Generally speaking, optical probes or contrast agents of imaging includes quantum dots (*e.g.* CdS and CdTe), organic dyes (*e.g.* near-infrared fluorescent molecules), bioluminescent proteins (green fluorescent protein), and conjugated polymers. How to design these components into HBP-based delivery systems requires reasonable methods. Blue photoluminescence was monitored in an aqueous solution of biodegradable hyperbranched poly(amino ester)s by Liu *et al.*, which is an inherent merit of these HBPs.[61] Chen *et al.* also synthesized the luminescent hyperbranched poly(amido amine)s (HPAMAMs).[62] Different amounts of β-cyclodextrin (β-CD) were fastened onto the terminal of HPAMAMs by Michael addition, thus obtaining HPAMAM-CDs. Owing to the existence of numerous amino groups, HPAMAM-CDs could be exploited as nonviral gene delivery vectors, and the corresponding gene transfection could be visualized by utilization of the HPAMAM-CDs' photoluminescence. As expected, the cellular uptake and gene transfection processes could be tracked by flow cytometry and CLSM without any fluorescent labeling. HPAMAM-CDs with numerous amine groups had high DNA condensation ability, realizing effective gene transfection. Moreover, the inner cavities of β-CDs in HPAMAM-CDs could encapsulate drugs through host–guest interaction. Therefore, the HPAMAM-CDs may have potential application in the combination of gene therapy and chemotherapy, and also provide the imaging function. There are other approaches to offer visualization of cancer cells, such as introducing QDs into the cavities of HBPs[16] or constructing amphiphilic hyperbranched conjugated polymers.[63] Although the field of imaging functionalization of HBP-based delivery systems is still in its infancy, interdisciplinary research will facilitate its maturation in this area and assist the exploitation of drugs.

5.3.4 Biodegradability and Biocompatibility

The long-term accumulation of these external vehicles, which cannot be degraded or metabolized, is detrimental to the human body. Therefore, excellent biodegradability and biocompatibility are of particular importance for the design of new biomedical polymeric vehicles. From this perspective, hyperbranched polyesters have become increasingly important in the biomedical field due to the general ease of metabolization of the degradation products. In addition, the introduction of functional components is able to facilitate the property alteration of the polymers, *e.g.* post-modification with hydrophilic blocks or bio-ligands. Thurecht *et al.* constructed water-soluble, degradable core–shell nanoparticles.[64] The nanoparticles were composed of the biodegradable core–shell HBPs *via* reversible addition–fragmentation chain transfer polymerization (RAFT) of N,N-dimethylamino-2-ethyl methacrylate and ring opening polymerization (ROP) of ε-caprolactone. The accelerating degradation of the hyperbranched materials with different crosslinking densities suggested that they could be used in the manufacture of controlled drug delivery systems. In addition, Liu *et al.* prepared a series of biodegradable hyperbranched polyphosphates which exhibited potential value as drug carriers.[65–67] The promising delivery systems of the newly emerging hyperbranched polyphosphates show high biodegradability and biocompatibility. With the readily available feedstocks and newly developed chemistries, more and more biodegradable and biocompatible materials will be discovered for biomedical applications.

5.3.5 Multifunctionality

The single function of the delivery system is not able to satisfy the needs of both smart and programmable delivery systems, so multifunctional delivery systems need to be designed, such as the combination of imaging and responsiveness. On the one hand, preparing multifunctional HBP-based nanocarriers with controlled properties requires the conjugation of recognition agents, imaging agents, reporter groups, or other functional ligands to the carrier surface. On the other hand, functional components may be loaded inside the nanocarriers or dispersed within the HBP structures by means of noncovalent attachments, such as hydrogen bonding interactions, host–guest interactions, and hydrophobic adsorption of certain intrinsic or specially inserted hydrophobic groups. Multifunctional pharmaceutical HBP-based nanocarriers could enlarge the horizon of producing highly efficient and specialized delivery systems for drugs and genes. Such multifunctional systems should allow pharmaceutical agents to release in a required temporal and spatial controllability and programmable pattern. The incorporation of multifunctional components requires taking advantage of versatile linking tools, such as the click reaction. Dong *et al.* prepared two-photon sensitive and sugar-targeted nanocarriers from biodegradable amphiphiles.[68] Both of the linkages between the polymeric backbone and functional ligands depended on click chemistry. More and more outstanding methodologies of integration are emerging, such as degradable covalent bonds, reversible dynamic covalent bonds, and noncovalent interactions.

5.4 Applications in Drug and Gene Delivery

Various drugs and genes have been exploited to treat different malignancies. Given that the side effects of these chemotherapeutic drugs are severe and the exterior genes are rapidly degraded by various enzymes in the blood, delivery vehicles such as polymeric micelles, vesicles, and conjugates should be adopted.

The common drugs contain doxorubicin, cisplatin, paclitaxel, methotrexate, or 5-fluorouracil. Although the discovery of various antitumor drugs for cancer therapy has made a giant leap, clinical outcomes are disappointing because most of them exhibit severe side effects. Utilization of nanocarrier systems like self-assembled polymeric micelles might overcome this problem. The polymeric micelle carriers have several unique advantages, such as enhancing the aqueous solubility and bioavailability of drugs, prolonging the circulation time, improving the preferential accumulation at the tumor site by the EPR effect, and reducing systemic side effects.

In the meantime, gene therapy has drawn significant attention over the past decades as a potential method for treating genetic disorders and an alternative approach to traditional chemotherapy applied in treating cancer. However, free oligonucleotides and DNA are rapidly degraded by serum nucleases in the blood when injected intravenously, so the design of effective vectors that compact and protect oligonucleotides becomes indispensible for gene therapy. The gene carriers mainly include two kinds, namely viral vectors and nonviral vectors. For viral vectors, they are very efficient at transfection, but meanwhile plagued by some downsides, such as immunogenicity, limitations in scale-up procedures, and safety concerns. By comparison, nonviral vectors are endowed with many advantages, like low/absent immunogenicity, easy manufacture, high compound stability, and facile chemical modification, so nonviral gene vectors are receiving more and more attention.[69,70]

5.4.1 Application as Drug Carriers

Owing to their outstanding pharmacokinetics and biodistribution profiles *via* the EPR effect, drug delivery systems in the form of nanocarriers have exhibited exciting efficacy for cancer treatments. Five drug-loaded patterns of HBPs have been summarized.[6] As shown in Figure 5.7, the drugs are tied to HBPs through noncovalent interactions for the drug complexes (type 1) or

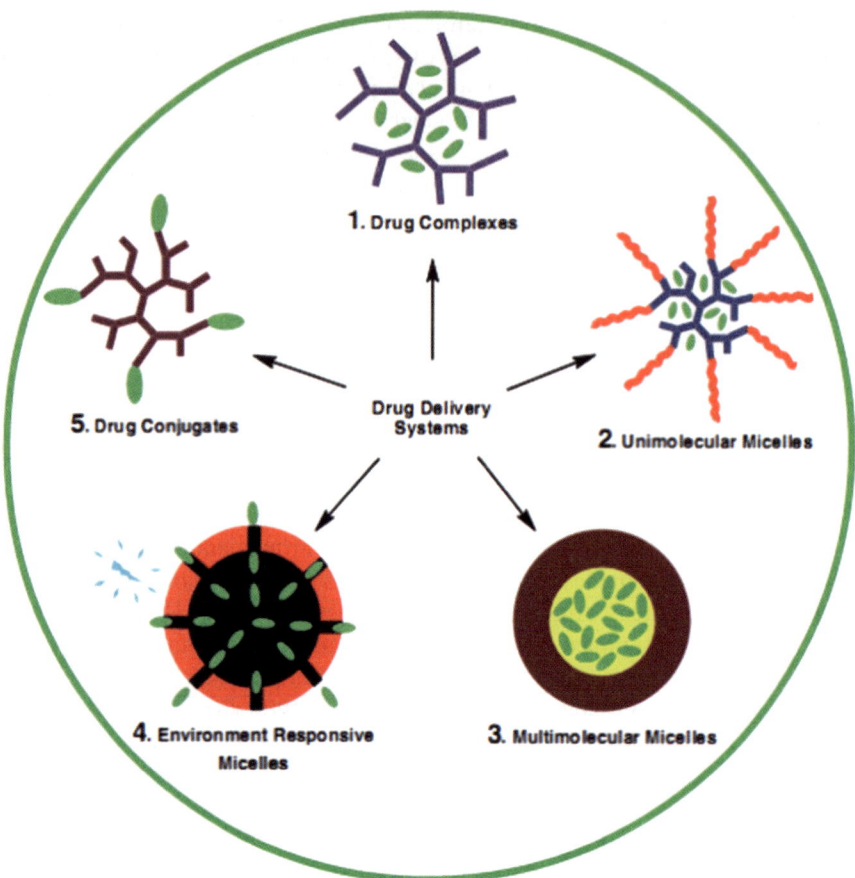

Figure 5.7   HBP carriers for drug delivery. (Reproduced from Zhou *et al.*[6] with permission from Wiley-VCH.)

entrapped inside the nanocavities of dendritic structures for the unimolecular micelles (type 2). However, the unimolecular micelles of HBPs can only load a small amount of guest molecules due to the limited volume of the interior cavities in one molecule. By contrast, multimolecular micelles possess much larger hydrophobic cores, thus enhancing the amount of encapsulating drug and the controllability of drug delivery. For the multimolecular micelles like type 3, the drugs are loaded in the micelle cores or shells according to their hydrophilicity. For the responsive HBP-based micelles like type 4, the exterior stimuli can trigger the transformation of the hydrophilic/hydrophobic balance and further alter the HBP self-assembling morphologies, thus controlling the release rate of the drugs. Type 5 represents drug conjugates in which drugs are covalently conjugated to the terminal groups of HBPs. Apart from the common advantages of the micelles, these HBP-based drug delivery systems have unique merits, such as better temporal and spatial controllability due to their hierarchy architecture and facile modification owing to their numerous terminal groups.[6]

5.4.2 Application as Gene Vectors

The nonviral gene vectors can be categorized into three kinds: cationic lipids, peptides, and cationic polymers. The method of liposome-mediated gene transfer is one of the earliest strategies used to introduce exogenous genetic material into cancer cells. The headgroup of the lipid can be adjusted to cationic states, thus increasing the cellular uptake. As the second vector, peptide oligonucleotide conjugates offer a unique strategy of delivering genes into cells with high efficiency and cell specificity. To deliver oligonucleotides into cells, these peptide vectors rely on short sequences of basic amino acid residues which readily penetrate the plasma membrane.[71] Although the development of liposomes and peptides persists, researchers have given more attention to the exceptional gene vehicles of cationic polymers and their dendritic derivatives.[72] The typical polymeric materials include poly(L-lysine) (PLL), polyethylenimine (PEI), polymethacrylate (*e.g.* PDMAEMA), carbohydrate-based polymers, and poly(amido-amine) (PAMAM). Most of these structures comprise amine groups, which is conducive to compacting DNA and penetrating the cell membrane. By introduction of a dendritic structure into the polymer backbone, the cationic polymers exhibit compact and globular structures in combination with a great number of various amine groups, which facilitates DNA condensation, gene delivery, and transfection improvement. Our group has developed several methods to prepare nonviral gene vectors.[12,23,62,73-75] Recently, Dong *et al.* developed a facile supramolecular approach for the preparation of charge-tunable dendritic polycations *via* the host–guest interaction between two different cationic β-CD derivative hosts and an adamantane-modified hyperbranched polyglycerol (HPG-AD) guest as gene vectors.[75] The multifunctionality was given to the dendritic polymers by the dynamic-tunable ability of supramolecular polymers *via* noncovalent interactions, thus making the optimization of structural parameters much easier than that of covalent polymerizations or modifications. For example, the parameters of charge density and charge distribution on the surface of vectors can be modulated, which fully embodies the advantages of HBPs and supramolecular chemistry in gene transfection. The functional HBPs provide a potential strategy to promote cationic polymers as safe and efficient nonviral gene carriers with satisfied gene delivery performance.

5.5 Summary

A series of synthetic strategies for HBPs have been developed, such as SCVP and SCROP, and meanwhile other novel preparation methodologies are continuously being proposed. Subsequently, additional functions enhance the inherent advantages of HBPs. All the above efforts are a prerequisite for the wide applications of these polymeric materials. It should be particularly mentioned that biomedical application in drug or gene delivery is one of the most promising research fields. The functionalization of HBPs paves the effective path for smart delivery systems by granting them responsiveness, targeting, biodegradability, imaging, *etc.* With the assistance of the delivery systems of the functional HBP self-assemblies, the cargoes can be successfully transported with auxiliary functionalities like long circulation times, specific accumulation at the targeted tissues, stimuli-responsive release, reduction of cytotoxicity, and *in vivo* diagnosis. The excellent results shed light on the feasibility of functional HBPs as a robust tool for smart delivery systems. However, there still exist certain controversial issues to be solved in the HBP-based vehicles. For example, the precise architectural controllability, the exploitation of new smart components, and the compatibility of versatile modified functional components with HBPs all need improving. Therefore, it is imperative that controlled synthetic approaches and better ligating methods of HBPs should be discovered. Despite the existing problems of functional HBP-based delivery systems for drugs and genes, they are certainly one of the most promising smart materials in the biomedical applications due to their intrinsic attributions.

Acknowledgements

This work was financially supported by the National Basic Research Program (2009CB930400, 2012CB821500) and National Natural Science Foundation of China (20974062), and the China National Funds for Distinguished Young Scientists (21025417).

References

1. *Hyperbranched Polymers: Synthesis, Properties, and Applications*, ed. D. Yan, C. Gao and H. Frey, Wiley, Hoboken, NJ, 2011.
2. X. Zhu, Y. Zhou and D. Yan, *J. Polym. Sci., Part B: Polym. Phys.*, 2011, **49**, 1277–1286.
3. C. Gao and D. Yan, *Prog. Polym. Sci.*, 2004, **29**, 183–275.
4. B. Voit, *J. Polym. Sci., Part A: Polym. Chem.*, 2000, **38**, 2505–2525.
5. B. Voit and A. Lederer, *Chem. Rev.*, 2009, **109**, 5924–5973.
6. Y. Zhou, W. Huang, J. Liu, X. Zhu and D. Yan, *Adv. Mater.*, 2010, **22**, 4567–4590.
7. Z. Jia, H. Chen, X. Zhu and D. Yan, *J. Am. Chem. Soc.*, 2006, **128**, 8144–8145.
8. Z. Jia, G. Li, Q. Zhu, D. Yan, X. Zhu, H. Chen, J. Wu, C. Tu and J. Sun, *Chem.–Eur. J.*, 2009, **15**, 7593–7600.
9. W. Dong, Y. Zhou, D. Yan, H. Li and Y. Liu, *Phys. Chem. Chem. Phys.*, 2007, **9**, 1255–1262.

10. C. Kojima, K. Yoshimura, A. Harada, Y. Sakanishi and K. Kono, *J. Polym. Sci., Part A: Polym. Chem.*, 2010, **48**, 4047–4054.
11. J. Liu, Y. Pang, W. Huang, Z. Zhu, X. Zhu, Y. Zhou and D. Yan, *Biomacromolecules*, 2011, **12**, 2407–2415.
12. Y. Shi, L. Zhou, R. Wang, Y. Pang, W. Xiao, H. Li, Y. Su, X. Wang, B. Zhu, X. Zhu, D. Yan and H. Gu, *Nanotechnology*, 2010, **21**, 115103.
13. C. Grandjean, G. Angyalosi, E. Loing, E. Adriaenssens, O. Melnyk, V. Pancré, C. Auriault and H. Gras-Masse, *ChemBioChem*, 2001, **2**, 747–757.
14. K.-Y. Pu, J. Shi, L. Cai, K. Li and B. Liu, *Biomacromolecules*, 2011, **12**, 2966–2974.
15. S. Yu, W. Zhang, J. Zhu, Y. Yin, H. Jin, L. Zhou, Q. Luo, J. Xu and J. Liu, *Macromol. Biosci.*, 2011, **11**, 821–827.
16. L. Zhu, Y. Shi, C. Tu, R. Wang, Y. Pang, F. Qiu, X. Zhu, D. Yan, L. He, C. Jin and B. Zhu, *Langmuir*, 2010, **26**, 8875–8881.
17. D. Konkolewicz, S. Gaillard, A. G. West, Y. Y. Cheng, A. Gray-Weale, T. W. Schmidt, S. P. Nolan and S. B. Perrier, *Organometallics*, 2011, **30**, 1315–1318.
18. C. Chen, G. Liu, X. Liu, S. Pang, C. Zhu, L. Lv and J. Ji, *Polym. Chem.*, 2011, **2**, 1389–1397.
19. K. Y. Pu, K. Li, J. Shi and B. Liu, *Chem. Mater.*, 2009, **21**, 3816–3822.
20. D. Yan and C. Gao, *Macromolecules*, 2000, **33**, 7693–7699.
21. C. Gao, D. Yan, X. Zhu and W. Huang, *Polymer*, 2001, **42**, 7603–7610.
22. H. Wan, Y. Chen, L. Chen, X. Zhu, D. Yan, B. Li, T. Liu, L. Zhao, X. Jiang and G. Zhang, *Macromolecules*, 2008, **41**, 465–470.
23. R. Wang, L. Zhou, Y. Zhou, G. Li, X. Zhu, H. Gu, X. Jiang, H. Li, J. Wu, L. He, X. Guo, B. Zhu and D. Yan, *Biomacromolecules* 2010, **11**, 489–495.
24. Y. Liu, D. Wu, Y. Ma, G. Tang, S. Wang, C. He, T. S. Chung and S. H. Goh, *Chem. Commun.*, 2003, 2630–2631.
25. D. Wu, Y. Liu, L. Chen, C. He, T. S. Chung and S. H. Goh, *Macromolecules*, 2005, **38**, 5519–5525.
26. J. M. J. Fréchet, M. Henmi, I. Gitsov, S. Aoshima, M. R. Leduc and R. B. Grubbs, *Science*, 1995, **269**, 1080–1083.
27. C. J. Hawker, J. M. J. Fréchet, R. B. Grubbs and J. Dao, *J. Am. Chem. Soc.*, 1995, **117**, 10763–10764.
28. X. L. Lou, F. Cheng, P. F. Cao, Q. Tang, H. J. Liu and Y. Chen, *Chem.-Eur. J.*, 2009, **15**, 11566–11572.
29. A. Sunder, R. Hanselmann, H. Frey and R. Mülhaupt, *Macromolecules*, 1999, **32**, 4240–4246.
30. H.-T. Chang and J. M. J. Fréchet, *J. Am. Chem. Soc.*, 1999, **121**, 2313–2314.
31. Z. Guo, Y. Zhang, W. Huang, Y. Zhou and D. Yan, *Macromol. Rapid Commun.*, 2008, **29**, 1746–1751.
32. X. Meng, Q. G. He, H. M. Cao and J. G. Cheng, *Chin. Chem. Lett.*, 2011, **22**, 725–728.

33. J.-Z. Du, X.-J. Du, C.-Q. Mao and J. Wang, *J. Am. Chem. Soc.*, 2011, **133**, 17560–17563.
34. Y. Xia, Y. Wang, Y. Wang, C. Tu, F. Qiu, L. Zhu, Y. Su, D. Yan, B. Zhu and X. Zhu, *Colloids Surf., B*, 2011, **88**, 674–681.
35. L. Zhu, C. Tu, B. Zhu, Y. Su, Y. Pang, D. Yan, J. Wu and X. Zhu, *Polym.Chem.*, 2011, **2**, 1761–1768.
36. Y. Pang, J. Liu, Y. Su, J. Wu, L. Zhu, X. Zhu, D. Yan and B. Zhu, *Polym. Chem.*, 2011, **2**, 1661–1670.
37. T. Hoare, B. P. Timko, J. Santamaria, G. F. Goya, S. Irusta, S. Lau, C. F. Stefanescu, D. Lin, R. Langer and D. S. Kohane, *Nano Lett.*, 2011, **11**, 1395–1400.
38. E. Amstad, J. Kohlbrecher, E. Müller, T. Schweizer, M. Textor and E. Reimhult, *Nano Lett.*, 2011, **11**, 1664–1670.
39. G. Saito, J. A. Swanson and K.-D. Lee, *Adv. Drug Delivery Rev.*, 2003, **55**, 199–215.
40. J. Chen, C. Wu and D. Oupický, *Biomacromolecules*, 2009, **10**, 2921–2927.
41. J. Liu, W. Huang, Y. Pang, P. Huang, X. Zhu, Y. Zhou and D. Yan, *Angew. Chem. Int. Ed.*, 2011, **50**, 9162–9166.
42. M. E. Fox, F. C. Szoka and J. M. J. Fréchet, *Acc. Chem. Res.*, 2009, **42**, 1141–1151.
43. M. Calderón, M. A. Quadir, S. K. Sharma and R. Haag, *Adv. Mater.*, 2010, **22**, 190–218.
44. D. Wilms, S.-E. Stiriba and H. Frey, *Acc. Chem. Res.*, 2010, **43**, 129–141.
45. T. Noh, Y. H. Kook, C. Park, H. Youn, H. Kim, E. T. Oh, E. K. Choi, H. J. Park and C. Kim, *J. Polym. Sci., Part A: Polym. Chem.*, 2008, **46**, 7321–7331.
46. C. A. Boswell, D. B. Tesar, K. Mukhyala, F.-P. Theil, P. J. Fielder and L. A. Khawli, *Bioconjugate Chem.*, 2010, **21**, 2153–2163.
47. S. Guin, H.-P. Yao and M.-H. Wang, *Mol. Pharmaceutics*, 2010, **7**, 386–397.
48. H. He, C. Xie and J. Ren, *Anal. Chem.*, 2008, **80**, 5951–5957.
49. X. Huang, I. H. El-Sayed, W. Qian and M. A. El-Sayed, *J. Am. Chem. Soc.*, 2006, **128**, 2115–2120.
50. X. Huang, I. H. El-Sayed, W. Qian and M. A. El-Sayed, *Nano Lett.*, 2007, **7**, 1591–1597.
51. A. F. Hussain, F. Kampmeier, V. von Felbert, H.-F. Merk, M. K. Tur and S. Barth, *Bioconjugate Chem.*, 2011, **22**, 2487–2495.
52. J. Lee, Y. Choi, K. Kim, S. Hong, H.-Y. Park, T. Lee, G. J. Cheon and R. Song, *Bioconjugate Chem.*, 2010, **21**, 940–946.
53. S. Quiles, K. P. Raisch, L. L. Sanford, J. A. Bonner and A. Safavy, *J. Med. Chem.*, 2009, **53**, 586–594.
54. C. Wängler, G. Moldenhauer, M. Eisenhut, U. Haberkorn and W. Mier, *Bioconjugate Chem.*, 2008, **19**, 813–820.
55. J. Yu, D. Javier, M. A. Yaseen, N. Nitin, R. Richards-Kortum, B. Anvari and M. S. Wong, *J. Am. Chem. Soc.*, 2010, **132**, 1929–1938.

56. P. R. Westmark, S. J. Gardiner and B. D. Smith, *J. Am. Chem. Soc.*, 1996, **118**, 11093–11100.
57. J. B. Crumpton, W. Zhang and W. L. Santos, *Anal. Chem.*, 2011, **83**, 3548–3554.
58. K. H. Schlick, R. A. Udelhoven, G. C. Strohmeyer and M. J. Cloninger, *Mol. Pharmaceutics*, 2005, **2**, 295–301.
59. M. Prabaharan, J. J. Grailer, S. Pilla, D. A. Steeber and S. Gong, *Biomaterials*, 2009, **30**, 3009–3019.
60. Z.-R. Lu, *Mol. Pharmaceutics*, 2006, **3**, 471–471.
61. D. Wu, Y. Liu, He and S. H. Goh, *Macromolecules*, 2005, **38**, 9906–9909.
62. Y. Chen, L. Zhou, Y. Pang, W. Huang, F. Qiu, X. Jiang, X. Zhu, D. Yan and Q. Chen, *Bioconjugate Chem.*, 2011, **22**, 1162–1170.
63. F. Qiu, C. Tu, R. Wang, L. Zhu, Y. Chen, G. Tong, B. Zhu, L. He, D. Yan and X. Zhu, *Chem. Commun.*, 2011, **47**, 9678–9680.
64. Y. Zheng, W. Turner, M. Zong, D. J. Irvine, S. M. Howdle and K. J. Thurecht, *Macromolecules*, 2011, **44**, 1347–1354.
65. J. Liu, W. Huang, Y. Pang, X. Zhu, Y. Zhou and D. Yan, *Langmuir*, 2010, **26**, 10585–10592.
66. J. Liu, Y. Pang, W. Huang, X. Huang, L. Meng, X. Zhu, Y. Zhou and D. Yan, *Biomacromolecules*, 2011, **12**, 1567–1577.
67. J. Liu, Y. Pang, W. Huang, X. Zhu, Y. Zhou and D. Yan, *Biomaterials*, 2010, **31**, 1334–1341.
68. L. Sun, Y. Yang, C. M. Dong and Y. Wei, *Small*, 2011, **7**, 401–406.
69. M. A. Mintzer and E. E. Simanek, *Chem. Rev.*, 2008, **109**, 259–302.
70. B. Walch, T. Breinig, M. J. Schmitt and F. Breinig, *Gene Ther.*, 2012, **19**, 237–245.
71. W.-J. Yi, J. Yang, C. Li, H.-Y. Wang, C.-W. Liu, L. Tao, S.-X. Cheng, R.-X. Zhuo and X.-Z. Zhang, *Bioconjugate Chem.*, 2011, **23**, 125–134.
72. J. H. Jeong, S. H. Kim, L. V. Christensen, J. Feijen and S. W. Kim, *Bioconjugate Chem.*, 2010, **21**, 296–301.
73. Y. Shi, J. Du, L. Zhou, X. Li, Y. Zhou, L. Li, X. Zang, X. Zhang, F. Pan, H. Zhang, Z. Wang and X. Zhu, *J. Mater. Chem.*, 2012, **22**, 355–360.
74. M. Chen, J. Wu, L. Zhou, C. Jin, C. Tu, B. Zhu, F. Wu, Q. Zhu, X. Zhu and D. Yan, *Polym. Chem.*, 2011, **2**, 2674–2682.
75. R. Dong, L. Zhou, J. Wu, C. Tu, Y. Su, B. Zhu, H. Gu, D. Yan and X. Zhu, *Chem. Commun.*, 2011, **47**, 5473.

CHAPTER 6

Functional Polymersomes for Controlled Drug Delivery

FENGHUA MENG, RU CHENG, CHAO DENG AND
ZHIYUAN ZHONG*

Biomedical Polymers Laboratory, and Jiangsu Key Laboratory of Advanced
Functional Polymer Design and Application, Department of Polymer Science
and Engineering, College of Chemistry, Chemical Engineering and Materials
Science, Soochow University, Suzhou 215123, P. R. China
*E-mail: zyzhong@suda.edu.cn

6.1 Introduction

Polymersomes (also referred to as polymeric vesicles) have attracted rapidly
growing interest since the early work of Eisenberg, Discher, Hammer, and co-
workers,[1,2] due to their intriguing aggregation phenomena and cell- and virus-
mimicking functions, as well as their tremendous potential applications in
medicine, pharmacy, and biotechnology. Polymersomes have fluid-filled cores
with membranes that consist of entangled chains separating the core from the
outside medium.[3] Together with micelles, they are the most common
morphological structures of amphiphiles in water. However, unlike micelles,
which mostly encapsulate hydrophobic compounds, polymersomes can not
only encapsulate hydrophobic molecules within the membrane but also load
hydrophilic molecules within the aqueous interior. While liposomes are made
from small phospholipid molecules, polymersomes are microscopic assemblies
of macromolecular amphiphiles of vastly different architectures (diblock,

RSC Polymer Chemistry Series No. 3
Functional Polymers for Nanomedicine
Edited by Youqing Shen
© The Royal Society of Chemistry 2013
Published by the Royal Society of Chemistry, www.rsc.org

triblock, graft, hyperbranched, *etc.*) and molecular weights (ranging from hundreds to tens of thousands dalton).

The formation of polymersomes is mainly dictated by the fraction of the hydrophilic block (f_{phil}), the molecular weight, and the effective interaction parameter of the hydrophobic block with water (χ). For block copolymers with a high χ, vesicles are favored when $f_{phil} = 20\text{--}40\%$, which is the same for natural lipids. At higher f_{phil} (>45%), cylindrical or spherical micelles are predominantly formed. It should be noted that the block copolymer vesicles show extremely slow chain exchange dynamics and thus exhibit a much lower critical aggregation concentration (CAC) compared to liposomes. Eisenberg and co-workers found that block copolymer vesicles are thermodynamically stabilized with short hydrophilic chains located preferentially at the watery core and long hydrophilic chains at the outer surface.[4] The sizes of polymersomes span from nano to micro scales, depending on their macromolecular parameters, *e.g.* structures, compositions, and molecular weights (thermodynamics), as well as fabrication methods, *e.g.* direct dissolution, phase inversion, film rehydration (kinetics).

The membrane thickness (d) of polymersomes was reported to scale up with the molecular weights of the hydrophobic blocks (M_{phobe}) as $d \approx M_{phobe}^{1/2}$ by Discher and co-workers.[5,6] The bending rigidity of polymersomes increases with increasing membrane thickness as a result of enhanced interdigitation. The membrane fluidity, on the other hand, usually decreases with increasing molecular weight and becomes more pronounced when the hydrophobic chains are long enough to entangle. The permeability of polymersomes is governed by the membrane thickness according to the Fick's first law. The diffusion coefficients of polymersome membranes are one order lower compared with those of liposomes ($D \approx 0.1 \text{ mm}^2 \text{ s}^{-1}$ or less *versus* $1 \text{ mm}^2 \text{ s}^{-1}$). The permeability of polymersomes can be fine tuned by adjusting M_{phobe} and the nature of polymersome membrane, or by incorporating degradable or stimuli-sensitive hydrophobic blocks into the vesicle membrane.[7,8]

The polymersome surface chemistry and topology can be tailored using block copolymers with complementary hydrophilic chains. For example, Battaglia and co-workers reported that binary mixtures of poly[2-(methacryloyloxy)ethyl-phosphorylcholine]-*b*-poly[2-(diisopropylamino)ethyl methacrylate] (PMPC-*b*-PDPA) and PEG-*b*-PDPA copolymers produced hybrid polymersomes with specific PEG or PMPC domains within their exterior envelopes, which along with polymersome size play a significant role in cellular uptake.[9] Discher and co-workers reported the formation of spotted polymersomes from a mixture of the charged block copolymer PAA-*b*-PBD and the neutral block copolymer PEO-*b*-PBD in the presence of divalent cations such as calcium and copper that induced domain formation by selective binding with polyanionic amphiphiles (Figure 6.1).[10]

As a result of the significantly higher molecular weights, polymersomes usually have thicker membranes (in the range of 5–30 nm *versus* 3–5 nm for liposomes), superb colloidal stability, enhanced mechanical strength, and

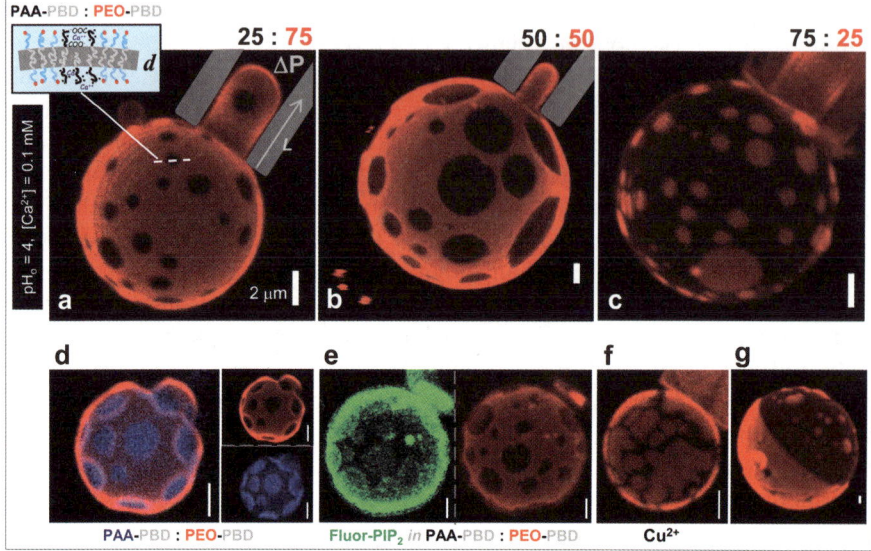

Figure 6.1   Spotted vesicles imaged by *z*-sectioning confocal microscopy while aspirated in micropipettes. Cation-induced, lateral phase segregation of charged PAA-PBD and neutral, fluorescently-labeled PEO-PBD diblock copolymers formed at pH 4, 0.1 mM calcium at PAA-PBD content of (a) 25%, (b) 50%, and (c) 75%. (Reproduced from Christian *et al.*[10] with permission from Macmillan.)

reduced chemical permeability compared to liposomes. To achieve prolonged circulation time *in vivo*, liposomes have to be modified with poly(ethylene glycol) (PEG), which are then known as "stealth liposomes". In contrast, most polymersomes are intrinsically stealthed in that they are typically made from amphiphilic copolymers consisting of nonfouling polymers such as PEG, dextran, and poly(acrylic acid) (PAA).

The presence of large watery interiors and robust hydrophobic walls makes polymersomes suitable for encapsulation and controlled delivery of both hydrophilic (*e.g.* proteins, siRNA, DNA, chelated Gd) and hydrophobic species (*e.g.* paclitaxel, doxorubicin , quantum dots). For instance, Li *et al.* have reported loading the hydrophobic anticancer drug paclitaxel (PTX) into PEO-PBD polymersome membranes.[11] The Ji and Qiu groups investigated the encapsulation of the hydrophilic anticancer drug doxorubicin hydrochloride (DOX·HCl) into the polymersome lumen.[12,13] Kataoka *et al.* reported that myoglobin encapsulated inside polyion complex vesicles could perform reversible oxygenation/deoxygenation reactions.[14] Palmer *et al.* reported that hemoglobin-loaded PEG-poly(ε-caprolactone) (PEG-PCL) polymersomes had similar oxygen affinities to human red blood cells.[15] Hammer *et al.* showed that near-infrared (NIR)-emissive polymersomes prepared by stably incorporating large multimeric porphyrin-based NIR fluorophores into the thick polymersome membrane were able to penetrate through the dense tumor tissue

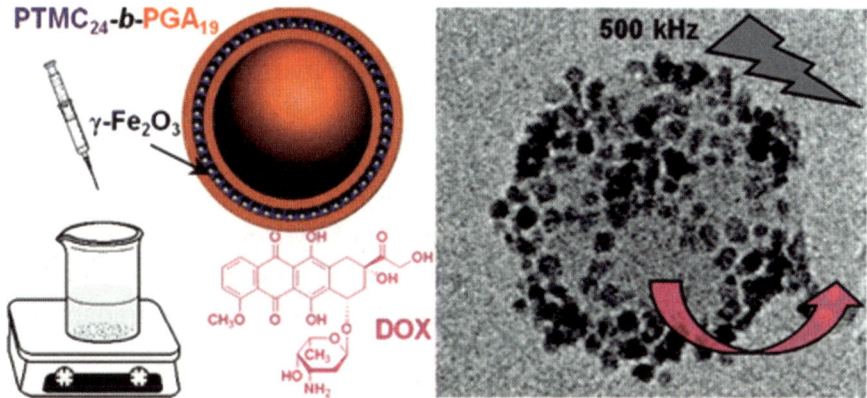

Figure 6.2 PTMC-*b*-PGA polymersomes simultaneously loaded with DOX and Fe$_2$O$_3$ nanoparticles. (Reproduced from Sanson *et al.*[18] with permission from the American Chemical Society.)

of a 9L glioma-bearing rat.[16] Polymersomes also offer good carriers for both hydrophobic and hydrophilic substances simultaneously. Discher *et al.* reported that intravenous injection of DOX·HCl- and PTX-loaded PEG-PCL polymersomes into nude mice bearing human breast carcinoma brought about growth arrest and shrinkage of rapidly growing tumors.[17] Lecommandoux *et al.* developed theranostic PTMC-*b*-PGA polymersomes simultaneously loaded with DOX and Fe$_2$O$_3$ nanoparticles, which displayed enhanced contrast properties and triggered DOX release by radiofrequency magnetic hyperthermia (Figure 6.2).[18]

Drug release profiles from polymersomes can be elegantly controlled by manipulating polymersome degradability, membrane permeability, and stimuli-responsive properties.[19,20] It should further be noted that targeting polymersomes that selectively deliver payloads to the sites of action can be prepared by decorating polymersomes with a specific ligand such as an antibody, peptide, or folate (Figure 6.3).[21] In this chapter we will highlight the recent development of functional polymersomes, including stimuli-sensitive polymersomes, chimaeric polymersomes, biomimetic polymersomes, and tumor-targeting polymersomes.

6.2 Stimuli-Responsive Polymersomes

Stimuli-sensitive polymersomes have emerged as novel programmable delivery systems in which the release of the encapsulated content is modulated by a stimulus.[20,22] In particular, pH and reduction are the two most appealing stimuli as they exist naturally in certain pathological sites and/or intracellular compartments.[23,24] For example, taking advantage of acidic endo/lysosomal compartments, Forster *et al.* developed pH-sensitive polymersomes from PEO-poly(2-vinylpyridine) (PEO-P2VP),[25] Armes *et al.* reported pH-sensitive

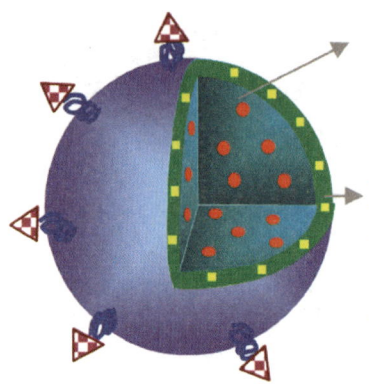

Watery Core
− Loading of hydrophilic species *e.g.*
  drugs, proteins, peptides, DNA,
  siRNA.

Hydrophobic Wall
− Loading of hydrophobic species *e.g.*
  anticancer drugs, imaging probes.
− Control over stability, degradability
  and drug permeability.

Targeting Ligand
− Specific cellular uptake

Figure 6.3 Schematic polymersome structure, function, and applications. (Reproduced from Meng and Zhong[21] with permission from the American Chemical Society.)

PMPC-PDPA polymersomes for intracellular DOX and DNA release,[26,27] and Lecommandoux *et al.* designed pH and temperature dual-responsive polymersomes based on poly[2-(diethylamino)ethyl methacrylate]-PGA (PDEA-PGA).[28] We designed pH-sensitive degradable polymersomes demonstrating pH-dependant release of PTX (hydrophobic) and DOX·HCl (hydrophilic) from block copolymers of PEG and an acid-labile polycarbonate, poly(2,4,6-trimethoxybenzylidenepentaerythritol carbonate) (PTMBPEC).[29] Battaglia *et al.* reported that pH-responsive PMPC-PDPA polymersomes encapsulated with GFP-encoding DNA plasmid could deliver GFP to primary human dermal fibroblast cells and Chinese hamster ovary cells.[27] Hubbell *et al.* reported oxidation-responsive polymersomes based on PEG-*b*-poly(propylene sulfide) (PEG-*b*-PPS) that underwent rapid destabilization in the presence of H_2O_2 (oxidative agent) due to conversion of the PPS hydrophobe into a hydrophile, poly(propylene sulfoxide) and ultimately poly(propylene sulfone).[30] They have also designed reduction-responsive polymersomes based on a PEG-SS-PPS block copolymer containing an intervening disulfide bond, wherein cellular uptake and disruption of polymersomes leading to efficient cytoplasmic release of encapsulated substances were observed in cells following 10 min incubation.[31] There is a high concentration of glutathione tripeptides (reducing agent) in the cytosol and cell nucleus, which makes reduction-responsive nano-vehicles highly promising for intracellular drug delivery.[32] Lately, we have designed pH and reduction dual-responsive PEG-SS-PDEA polymersomes that could encapsulate therapeutic proteins with high efficiencies by adjusting the pH to 7.4 and then release the proteins intracellularly upon cell entry and thus induce greatly enhanced apoptosis of MCF-7 cells compared to free protein and reduction-insensitive controls (Figure 6.4).[33]

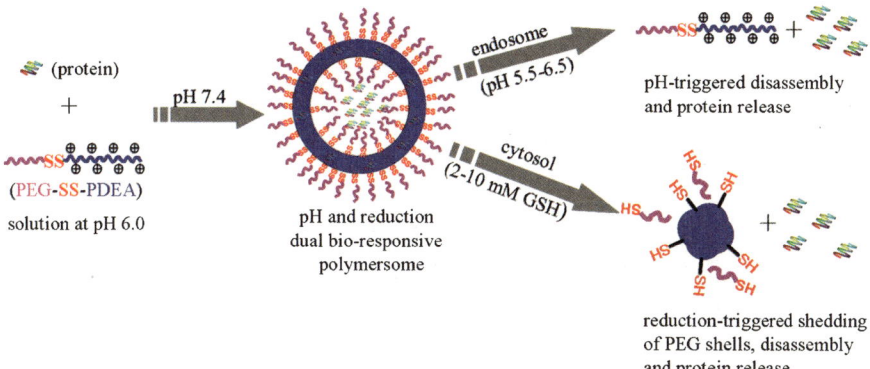

Figure 6.4 Illustration of pH and reduction dual-bioresponsive polymersomes based on PEG-SS-PDEA diblock copolymer for facile loading and triggered intracellular release of proteins. (Reproduced from Zhang *et al.*[33] with permission from the American Chemical Society.)

We have devised robust dual-responsive polymersomes based on a temperature-sensitive PEO-*b*-poly(acrylic acid)-*b*-poly(*N*-isopropylacrylamide) (PEO-PAA-PNIPAM) triblock copolymer and crosslinked with cystamine *via* carbodiimide chemistry.[34] The crosslinked polymersomes, while showing remarkable stability against dilution and decrease of temperature, were otherwise rapidly dissociated into monomers under reductive conditions mimicking an intracellular environment. These polymersomes provided efficient protein loading and triggered intracellular protein release, resulting in appreciable apoptosis of cancer cells.[35] Grubbs *et al.* reported that ABC triblock copolymers with PEO as the A block, temperature-sensitive PNIPAM or poly(ethylene oxide-*stat*-butylene oxide) as the B block, and polyisoprene as the C block underwent reversible transitions from spherical micelles to polymersomes upon increasing the temperature to above their lower critical solution temperature (LCST).[36,37] Kim *et al.* recently reported monosaccharide-responsive polymersomes for controlled release of insulin by sugars under physiologically relevant pH conditions.[38]

Dmochowski *et al.* prepared photoactive composite polymersomes by incorporating horse spleen ferritin (HSF) in the aqueous interior and a bis[(porphinato)zinc] (PZn$_2$) chromophore in the membrane of PEO-*b*-PBD block copolymer vesicles.[39] The polymersomes were deformed and destructed on the minute timescale upon exposure to near-UV to near-IR light. Meier *et al.* developed photosensitive polymersomes based on the block copolymer PMCL$_{76}$-ONB-PAA$_{16}$ with the light-sensitive nitrobenzene (ONB) group in between the two blocks. Under UV irradiation the ONB groups could be cleaved, resulting in polymersome disruption and the release of the encapsulated fluorescein and proteins (Figure 6.5).[40]

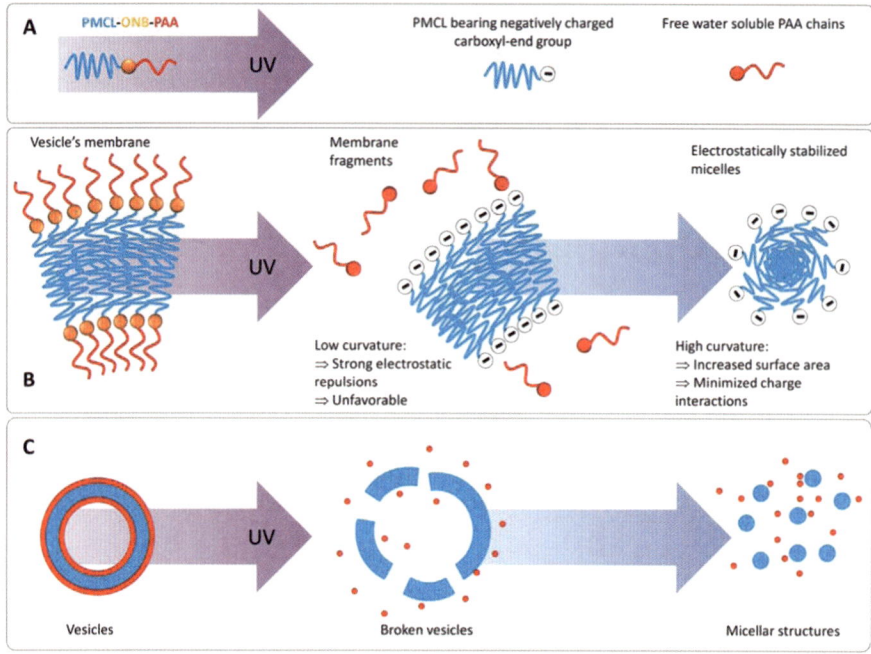

Figure 6.5 Proposed photodegradation mechanism of PMCL$_{76}$-ONB-PAA$_{16}$ poly-mersomes: (A) the diblock copolymer is rapidly cleaved upon UV irradiation; (B) as a result of chain scission, packing of the PMCL chains forming the membrane is progressively destabilized and evolves into a more favorable arrangement; (C) from vesicles to broken vesicles to stabilized micellar structures.[40]

6.3 Chimaeric Polymersomes

Chimaeric polymersomes that have distinct interior environments separated from the outside by an asymmetric membrane (analogous to the cellular membrane) have received significant recent interest. We have shown that biodegradable chimaeric polymersomes, formed from asymmetric PEG-PCL-PDEA ($M_{n,PEG} > M_{n,PDEA}$) triblock copolymers by a film rehydration method, efficiently delivered and released exogenous proteins into cancer cells.[41] The longer PEG block is preferentially oriented at the polymersome outer layer, thereby offering excellent biocompatibility and stability in the circulation; the shorter cationic PDEA block is preferentially located inside the polymersomes, which on the one hand facilitates efficient encapsulation and stabilization of proteins and on the other hand may assist polymersomes escaping from endosomes, resulting in the efficient cytoplasmic delivery of proteins. Very recently, we developed pH-sensitive degradable chimaeric polymersomes based on PEG-*b*-poly[2,4,6-trimethoxybenzylidene-1,1,1-tris(hydroxymethyl)ethane methacrylate]-*b*-PAA (PEG-PTTMA-PAA) triblock copolymers ($M_{n,PEG} > M_{n,PAA}$). The shorter PAA block inside the polymersomes promoted active and

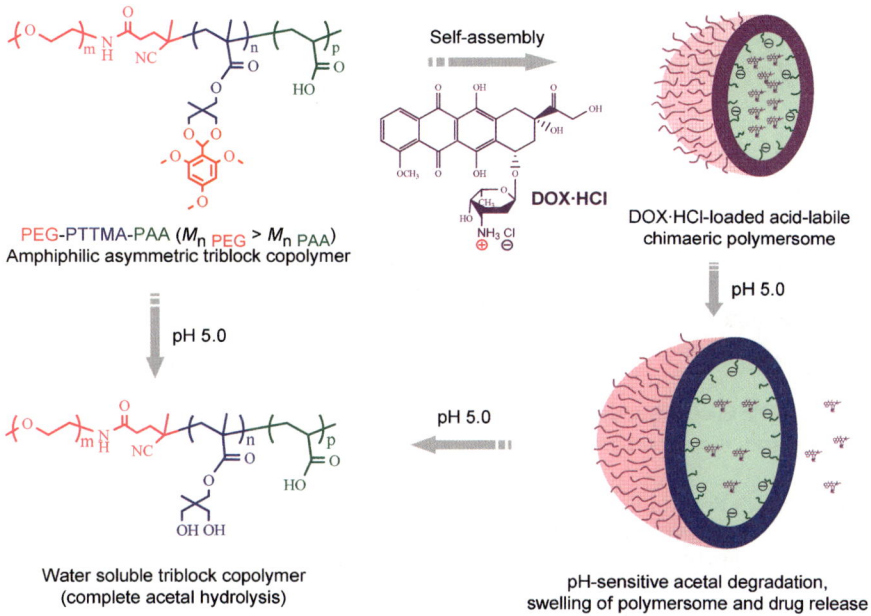

Figure 6.6 Schematic presentation of pH-sensitive biodegradable chimaeric polymersomes based on an asymmetric PEG-PTTMA-PAA triblock copolymer. (Reproduced from Du *et al.*[42] with permission from Elsevier.)

efficient loading of the hydrophilic anticancer drug DOX·HCl. The PTTMA block triggered efficient intracellular release and high cytotoxicity of DOX·HCl (IC_{50} close to that of free DOX·HCl) due to rapid hydrolysis of the acetal groups of PTTMA at mildly acidic pH (Figure 6.6).[42] Armes, Ryan and co-workers have reported the formation of asymmetric polymersomes from PEO-PDPA-poly[2-(dimethylamino)ethyl methacrylate] (PEO-PDPA-PDMA) triblock copolymers in aqueous solution.[43]

Jin *et al.* have reported fabrication of asymmetric polymersomes from PEG-*b*-PCL and dextran-*b*-PCL (DEX-*b*-PCL) by "phase-guided assembly" in which DEX-*b*-PCL formed the inner leaflet around the dispersed dextran phase and PEG-*b*-PCL formed the outer leaflet with the PEG block facing the PEG continuous phase (Figure 6.7).[44] These asymmetric polymersomes could encapsulate erythropoietin with high efficiency. Li *et al.* prepared asymmetric polymersomes from a mixture of PEG-*b*-PBD and the UV-sensitive liquid-crystalline copolymer PEG-*b*-PMAzo, which upon UV illumination resulted in instantaneous bursting of the polymersomes.[45]

6.4 Biomimetic Polymersomes

Polymersomes spanning from nano to micro scales provide an ideal technology platform for mimicking viruses, cell organelles, and cells. Choi and

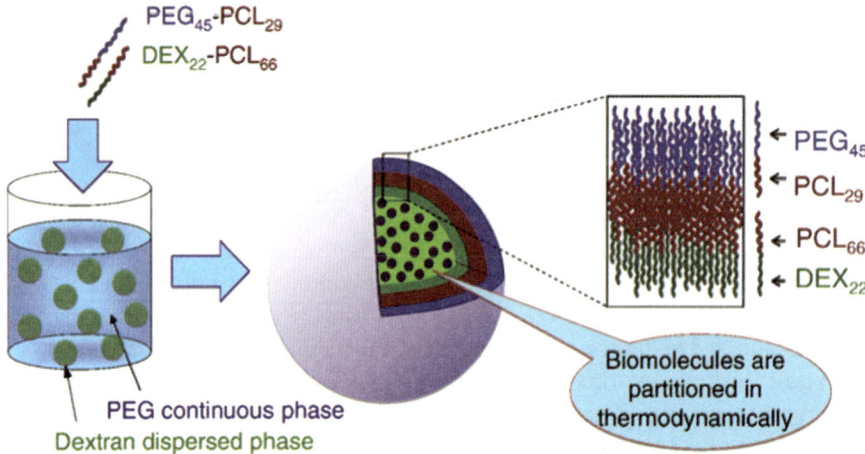

Figure 6.7 Preparation of polymersomes from PEG-*b*-PCL and DEX-*b*-PCL by "phase-guided assembly". (Reproduced from Zhang *et al.*[44] with permission from Elsevier.)

Montemagno reported ATP-producing cellular mimetic proteopolymersomes in which ATP had been generated by coupling reactions between bacteriorhodopsin, a light-driven transmembrane proton pump, and F_0F_1-ATP synthase motor protein, reconstituted in polymersomes.[46] These hybrid proteopolymersomes are interesting as artificial organelles for *in vitro* investigation of cellular metabolism as well as for synthesis of functional smart materials. Nolte *et al.* reported PS-PIAT polymersome nanoreactors containing tandem CALB, GOX, and HRP enzymes whose activities were coupled.[47] Meier *et al.* reported the design of artificial cell organelles based on polymersomes encapsulated with a set of enzymes, which entered the cells in a target-specific fashion and showed intact biochemical functionality in the cellular environment.[48]

Deming and co-workers prepared biomimetic pH-responsive polymersomes based on an amphiphilic ethylene glycol-modified polyleucine-*b*-polylysine copolypeptide, in which vesicle size and structure were dictated in a manner similar to viral capsids, essentially by the ordered conformations of polypeptides.[49] In a following work, polyleucine-*b*-polyarginine copolypeptide was designed to form biomimetic polymersomes, in which polyarginine not only directed vesicle formation but also facilitated intracellular trafficking of polymersomes due to its mimicking of protein transduction domains (PTDs).[50] In an effort to mimic viral capsids, Lecommandoux and co-workers designed polysaccharide-*b*-polypeptide (simple glycoprotein analog) copolymer vesicles.[51] The same authors reported glycopeptide-based biologically active polymersomes, with bioactive galactose units in the shell showing selective lectin binding.[52] Interestingly, biomimetic polymersomes using hyaluronan as a hydrophilic stabilizing block were demonstrated to target cancer cells overexpressing CD44 glycoprotein receptors.[53]

6.5 Tumor-Targeting Polymersomes

Targeting polymersomes were developed to enhance the drug outcome of the polymersome formulations by decorating the polymersome surface with targeting ligands like antibodies, RGD, and folic acid. For instance, Jiang and co-workers reported that lactoferrin-conjugated polymersomes with loaded tetrandrine and DOX could pass the blood–brain barrier (BBB), accumulate at the tumor site, and shrink the tumor in glioma-bearing rats.[54] Kokkoli and co-workers reported that PR_b-functionalized PEO-PBD polymersomes (PR_b is an effective $\alpha5\beta1$ targeting peptide) efficiently delivered tumor necrosis factor-α (TNFα) to LNCaP human prostate cancer cells, resulting in a dramatic enhancement of the cytotoxic potential of TNFα.[55] The targeting effect significantly outperforms GRGDSP-functionalized polymersomes in terms of promoting cell binding/internalization and DOX cytotoxicity.[56] Feijen and co-workers have developed anti-EGFR modified polymersomes based on a PEG-pep-PDLLA copolymer containing a lysosomal enzyme cathepsin B cleavable peptide for systemic cancer chemotherapy (Figure 6.8).[57] Recently, Lecommandoux and co-workers

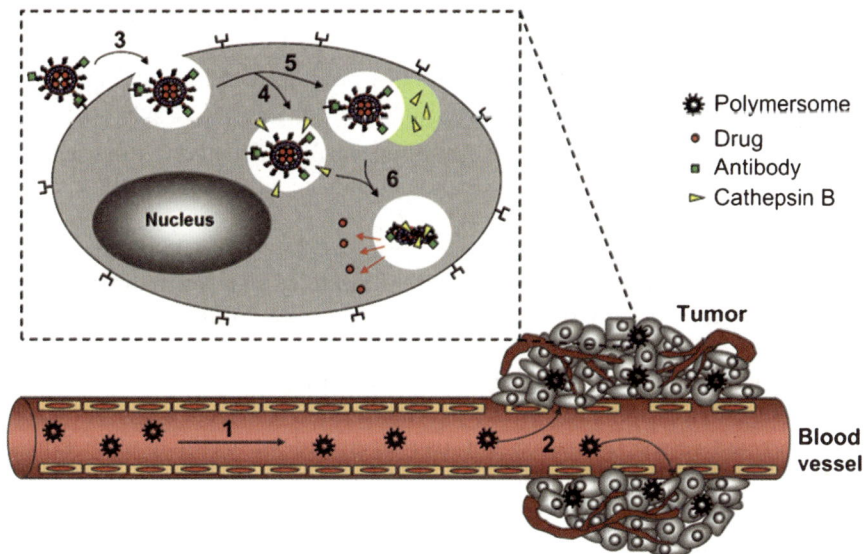

Figure 6.8 Schematic illustration of systemic targeting drug delivery by using antibody conjugated and peptide-containing PEG-pep-PDLLA polymersomes in which therapeutic drugs or proteins are present. (Reproduced from Lee *et al.*[57] with permission from Elsevier.)

reported that self-targeting polymersomes based on poly(γ-benzyl L-glutamate)-*b*-hyaluronan could deliver DOX to CD44 overexpressing cells and suppress tumor growth *in vivo*.[58,59]

6.6 Conclusion and Perspectives

The past decade has witnessed rapid progress in the design, preparation, and controlled drug delivery applications of polymersomes. With enhanced stability and remarkable chemical versatility, polymersomes have emerged as an advanced alternative to widely applied liposomes. It remains a challenge, however, to produce polymersomes with precisely controlled physicochemical properties (including vesicle size, surface, shape, membrane thickness, permeability, *etc.*), which are of utmost importance to their practical applications. Moreover, polymersomes will in particular play a pivotal role in targeted intracellular delivery of hydrophilic pharmaceutics and biopharmaceutics, including anticancer drugs, proteins, and nucleic acids, as well as in combination cancer therapy in which targeted delivery of two or more reagents using one single vehicle brings about synergistic treatment effects (*e.g.* for recurrent or resistant tumors). The current polymersomes for drug delivery applications, however, encounter several challenges, including cumbersome polymersome preparation procedures (*e.g.* often involving organic solvent), low drug loading content and loading efficiency, inferior *in vivo* specificity, and/or an uncontrolled drug release profile. In the future, more effort should be directed to the development of novel multifunctional biocompatible polymersomes that are preferably formed directly in water, have high drug loading levels, are stable in circulation, have programmed biodegradability *in vivo*, and preferentially accumulate and release drugs at the targeted sites in a controlled manner.

Acknowledgements

This work is financially supported by research grants from the National Natural Science Foundation of China (NSFC 51173126, 20974073, and 50973078), and a Project Funded by the Priority Academic Program Development of Jiangsu Higher Education Institutions.

References

1. B. M. Discher, Y. Y. Won, D. S. Ege, J. C. M. Lee, F. S. Bates, D. E. Discher and D. A. Hammer, *Science*, 1999, **284**, 1143–1146.
2. D. E. Discher and A. Eisenberg, *Science*, 2002, **297**, 967–973.
3. G. Battaglia and A. J. Ryan, *J. Am. Chem. Soc.*, 2005, **127**, 8757–8764.
4. L. B. Luo and A. Eisenberg, *J. Am. Chem. Soc.*, 2001, **123**, 1012–1013.

5. H. Bermudez, A. K. Brannan, D. A. Hammer, F. S. Bates and D. E. Discher, *Macromolecules*, 2002, **35**, 8203–8208.
6. G. Srinivas, D. E. Discher and M. L. Klein, *Nat. Mater.*, 2004, **3**, 638–644.
7. Z. L. Cheng, D. L. J. Thorek and A. Tsourkas, *Adv. Funct. Mater.*, 2009, **19**, 3753–3759.
8. K. T. Kim, J. Cornelissen, R. J. M. Nolte and J. C. M. van Hest, *Adv. Mater.*, 2009, **21**, 2787–2791.
9. M. Massignani, C. LoPresti, A. Blanazs, J. Madsen, S. P. Armes, A. L. Lewis and G. Battaglia, *Small*, 2009, **5**, 2424–2432.
10. D. A. Christian, A. W. Tian, W. G. Ellenbroek, I. Levental, K. Rajagopal, P. A. Janmey, A. J. Liu, T. Baumgart and D. E. Discher, *Nat. Mater.*, 2009, **8**, 843–849.
11. S. L. Li, B. Byrne, J. Welsh and A. F. Palmer, *Biotechnol. Prog.*, 2007, **23**, 278–285.
12. G. Y. Liu, Q. Jin, X. S. Liu, L. P. Lv, C. J. Chen and J. Ji, *Soft Matter*, 2011, **7**, 662–669.
13. C. Zheng, L. Y. Qiu and K. J. Zhu, *Polymer*, 2009, **50**, 1173–1177.
14. A. Kishimura, A. Koide, K. Osada, Y. Yamasaki and K. Kataoka, *Angew. Chem. Int. Ed.*, 2007, **46**, 6085–6088.
15. S. Rameez, H. Alosta and A. F. Palmer, *Bioconjugate Chem.*, 2008, **19**, 1025–1032.
16. P. P. Ghoroghchian, P. R. Frail, K. Susumu, D. Blessington, A. K. Brannan, F. S. Bates, B. Chance, D. A. Hammer and M. J. Therien, *Proc. Natl. Acad. Sci. U. S. A.*, 2005, **102**, 2922–2927.
17. F. Ahmed, R. I. Pakunlu, G. Srinivas, A. Brannan, F. Bates, M. L. Klein, T. Minko and D. E. Discher, *Mol. Pharmaceutics*, 2006, **3**, 340–350.
18. C. Sanson, O. Diou, J. Thevenot, E. Ibarboure, A. Soum, A. Brulet, S. Miraux, E. Thiaudiere, S. Tan, A. Brisson, V. Dupuis, O. Sandre and S. Lecommandoux, *ACS Nano*, 2011, **5**, 1122–1140.
19. C. LoPresti, H. Lomas, M. Massignani, T. Smart and G. Battaglia, *J. Mater. Chem.*, 2009, **19**, 3576–3590.
20. F. H. Meng, Z. Y. Zhong and J. Feijen, *Biomacromolecules*, 2009, **10**, 197–209.
21. F. H. Meng and Z. Y. Zhong, *J. Phys. Chem. Lett.*, 2011, **2**, 1533–1539.
22. M. H. Li and P. Keller, *Soft Matter*, 2009, **5**, 927–937.
23. N. Rapoport, *Prog. Polym. Sci.*, 2007, **32**, 962–990.
24. F. H. Meng, W. E. Hennink and Z. Y. Zhong, *Biomaterials*, 2009, **30**, 2180–2198.
25. U. Borchert, U. Lipprandt, M. Bilang, A. Kimpfler, A. Rank, R. Peschka-Suss, R. Schubert, P. Lindner and S. Forster, *Langmuir*, 2006, **22**, 5843–5847.
26. J. Z. Du, Y. P. Tang, A. L. Lewis and S. P. Armes, *J. Am. Chem. Soc.*, 2005, **127**, 17982–17983.
27. H. Lomas, I. Canton, S. MacNeil, J. Du, S. P. Armes, A. J. Ryan, A. L. Lewis and G. Battaglia, *Adv. Mater.*, 2007, **19**, 4238–4243.

28. W. Agut, A. Brulet, C. Schatz, D. Taton and S. Lecommandoux, *Langmuir*, 2010, **26**, 10546–10554.
29. W. Chen, F. H. Meng, R. Cheng and Z. Y. Zhong, *J. Controlled Release*, 2010, **142**, 40–46.
30. A. Napoli, M. Valentini, N. Tirelli, M. Muller and J. A. Hubbell, *Nat. Mater.*, 2004, **3**, 183–189.
31. S. Cerritelli, D. Velluto and J. A. Hubbell, *Biomacromolecules*, 2007, **8**, 1966–1972.
32. F. H. Meng, W. E. Hennink and Z. Y. Zhong, *Biomaterials*, 2009, **30**, 2180–2198.
33. J. C. Zhang, L. L. Wu, F. H. Meng, Z. J. Wang, C. Deng, H. Y. Liu and Z. Y. Zhong, *Langmuir*, 2012, **28**, 2056–2065.
34. H. F. Xu, F. H. Meng and Z. Y. Zhong, *J. Mater. Chem.*, 2009, **19**, 4183–4190.
35. R. Cheng, F. H. Meng, S. B. Ma, H. F. Xu, H. Y. Liu, X. B. Jing and Z. Y. Zhong, *J. Mater. Chem.*, 2011, **21**, 19013–19020.
36. A. Sundararaman, T. Stephan and R. B. Grubbs, *J. Am. Chem. Soc.*, 2008, **130**, 12264–12265.
37. Y. Cai, K. B. Aubrecht and R. B. Grubbs, *J. Am. Chem. Soc.*, 2011, **133**, 1058–1065.
38. H. Kim, Y. J. Kang, S. Kang and K. T. Kim, *J. Am. Chem. Soc.*, 2012, **134**, 4030–4033.
39. G. P. Robbins, M. Jimbo, J. Swift, M. J. Therien, D. A. Hammer and I. J. Dmochowski, *J. Am. Chem. Soc.*, 2009, **131**, 3872–3874.
40. E. Cabane, V. Malinova, S. Menon, C. G. Palivan and W. Meier, *Soft Matter*, 2011, **7**, 9167–9176.
41. G. J. Liu, S. B. Ma, S. K. Li, R. Cheng, F. H. Meng, H. Y. Liu and Z. Y. Zhong, *Biomaterials*, 2010, **31**, 7575–7585.
42. Y. F. Du, W. Chen, M. Zheng, F. H. Meng and Z. Y. Zhong, *Biomaterials*, 2012, **33**, 7291–7299.
43. A. Blanazs, M. Massignani, G. Battaglia, S. P. Armes and A. J. Ryan, *Adv. Funct. Mater.*, 2009, **19**, 2906–2914.
44. Y. L. Zhang, F. Wu, W. E. Yuan and T. Jin, *J. Controlled Release*, 2010, **147**, 413–419.
45. E. Mabrouk, D. Cuvelier, F. Brochard-Wyart, P. Nassoy and M. H. Li, *Proc. Natl. Acad. Sci. U. S. A.*, 2009, **106**, 7294–7298.
46. H. J. Choi and C. D. Montemagno, *Nano Lett.*, 2005, **5**, 2538–2542.
47. D. M. Vriezema, P. M. L. Garcia, N. S. Oltra, N. S. Hatzakis, S. M. Kuiper, R. J. M. Nolte, A. E. Rowan and J. C. M. van Hest, *Angew. Chem. Int. Ed.*, 2007, **46**, 7378–7382.
48. N. Ben-Haim, P. Broz, S. Marsch, W. Meier and P. Hunziker, *Nano Lett.*, 2008, **8**, 1368–1373.
49. E. G. Bellomo, M. D. Wyrsta, L. Pakstis, D. J. Pochan and T. J. Deming, *Nat. Mater.*, 2004, **3**, 244–248.

50. E. P. Holowka, V. Z. Sun, D. T. Kamei and T. J. Deming, *Nat. Mater.*, 2007, **6**, 52–57.
51. C. Schatz, S. Louguet, J. F. Le Meins and S. Lecommandoux, *Angew. Chem. Int. Ed.*, 2009, **48**, 2572–2575.
52. J. Huang, C. Bonduelle, J. Thevenot, S. Lecommandoux and A. Heise, *J. Am. Chem. Soc.*, 2012, **134**, 119–122.
53. K. K. Upadhyay, J. F. Le Meins, A. Misra, P. Voisin, V. Bouchaud, E. Ibarboure, C. Schatz and S. Lecommandoux, *Biomacromolecules*, 2009, **10**, 2802–2808.
54. Z. Pang, L. Feng, R. Hua, J. Chen, H. Gao, S. Pan, X. Jiang and P. Zhang, *Mol. Pharmaceutics*, 2010, **7**, 1995–2005.
55. D. Demirgoz, T. O. Pangburn, K. P. Davis, S. Lee, F. S. Bates and E. Kokkoli, *Soft Matter*, 2009, **5**, 2011–2019.
56. T. O. Pangburn, F. S. Bates and E. Kokkoli, *Soft Matter*, 2012, **8**, 4449–4461.
57. J. S. Lee, T. Groothuis, C. Cusan, D. Mink and J. Feijen, *Biomaterials*, 2011, **32**, 9144–9153.
58. K. K. Upadhyay, A. N. Bhatt, A. K. Mishra, B. S. Dwarakanath, S. Jain, C. Schatz, J.-F. Le Meins, A. Farooque, G. Chandraiah, A. K. Jain, A. Misra and S. Lecommandoux, *Biomaterials*, 2010, **31**, 2882–2892.
59. K. K. Upadhyay, A. K. Mishra, K. Chuttani, A. Kaul, C. Schatz, J.-F. Le Meins, A. Misra and S. Lecommandoux, *Nanomedicine (Philadelphia, PA, U. S.)*, 2012, **8**, 71–80.

CHAPTER 7

Polymeric Micelle-Based Nanomedicine for siRNA Delivery

XI-QIU LIU, XIAN-ZHU YANG AND JUN WANG*

CAS Key Laboratory of Brain Function and Disease, and School of Life Sciences, University of Science and Technology of China, Hefei, Anhui 230027, P. R. China
*E-mail: jwang699@ustc.edu.cn

7.1 Introduction

The RNA interference (RNAi) machinery was first discovered in plants; this post-transcriptional gene-silencing mechanism was subsequently demonstrated by Fire and Mello in *Caenorhabditis elegans*.[1,2] Two approaches utilize RNAi to inhibit target genes: short hairpin RNA (shRNA) and small interfering RNA (siRNA).[3] Sequence-specific 21–23 base-pair double-stranded siRNA selectively degrades complementary messenger RNAs (mRNAs).[4,5] Inside the cell, siRNA is incorporated into the RNA-induced silencing complex (RISC), which separates the strands of the RNA duplex and discards the sense strand. The antisense RNA strand then guides the activated RISC to anneal and cleave the target mRNA.[6] An endonuclease, Argonaute 2, plays a key role in unwinding the duplex (sense and antisense siRNA strands) and degrading the target mRNA.[7] siRNA molecules can block specific expression of endogenous and heterologous genes in various mammalian cell lines.[5] In addition to biological research and drug development, siRNA has tremendous therapeutic

RSC Polymer Chemistry Series No. 3
Functional Polymers for Nanomedicine
Edited by Youqing Shen
© The Royal Society of Chemistry 2013
Published by the Royal Society of Chemistry, www.rsc.org

potential for the treatment of cancer and macular degeneration by inhibiting oncogene or angiogenic growth factor overexpression, respectively. The scientific community's commitment to RNAi technology was evidenced by the 2006 Nobel Prize in Medicine and by high-profile startup biotech companies as well as billion-dollar investments from established pharmaceutical companies.[8]

Only 10 years after the discovery of the RNAi mechanism, some siRNA candidate drugs have entered clinical trials.[9] Most siRNAs in clinical trials are administered by local delivery such as the intravitreal or intranasal routes.[10] Opko Health produced the first siRNA therapeutic (Bevasiranib) to reach a Phase III clinical trial for the treatment of wet neovascular age-related macular degeneration (AMD). Bevasiranib is a 21 nt-long siRNA with two deoxythymidine (dTs) on both 3′-ends and targets *VEGFA* mRNA.[11] Although the Phase I and II trials demonstrated the safety, tolerability, and efficacy of the siRNA, Opko terminated the Phase III study in March 2009, based on the recommendation of the Independent Data Monitoring Committee. The committee reasoned that the siRNA drug was unlikely to achieve its primary endpoint of reducing vision loss in the AMD patients.[12] Sirna Therapeutics developed the drug AGN-745 (formerly known as Sirna-027), a chemically modified siRNA targeting VEGF receptor 1. As the second RNAi drug to reach clinical trials, early data were promising, considering its efficacy and safety in a relevant subset of patients. However, the trial was terminated in May 2009, due to the lack of improvement in visual acuity.[12]

For systemic drug administration by intravenous injection, most clinically tested siRNA therapeutics require synthetic carriers for the payload. Tekmira has been advancing internal RNA interference (RNAi) product candidates using the stable nucleic acid lipid particle (SNALP) carrier. TKM-ApoB was developed for the treatment of hypercholesterolemia by a siRNA designed to silence ApoB. In preclinical models, TKM-ApoB was delivered with high efficiency into hepatocytes producing ApoB, where the siRNA acts to knock down the precursor mRNA coding for the ApoB protein. The decrease in circulating very low-density lipoproteins (VLDLs) and low-density lipoproteins (LDLs) resulted in significant reductions in LDLs and triglycerides. Tekmira initiated a Phase 1 human clinical trial for TKM-ApoB in July 2009 but terminated it in January 2010. Seventeen subjects received a single dose of TKM-ApoB at one of seven different dosing levels and six subjects received a placebo. Of the two subjects treated at the highest dose level, one subject experienced flu-like symptoms consistent with stimulation of the immune system caused by the ApoB siRNA payload (http://www.tekmirapharm.com/Programs/Products.asp). Calando Pharmaceuticals has developed a siRNA therapeutic (CALAA-01), which is a cyclodextrin-based polymeric nanoparticle containing a siRNA that targets the M2 subunit of ribonucleotide reductase (RRM2). The nanoparticle was decorated with the human transferrin (TF) protein and poly(ethylene glycol) (PEG) for stability.[13] Given the recent successful investigational new drug (IND) filing from the U.S.

Food and Drug Administration (FDA), CALAA-01 has entered a Phase I trial for safety study to treat solid tumor cancers; this study is currently recruiting participants.

In recent years, development of delivery vehicles capable of administering siRNA efficiently and safely both *in vitro* and *in vivo* has been strongly needed. An ideal carrier for systemic siRNA administration should have the following properties: (a) be nontoxic and non-immunogenic for systemic human administration; (b) condense siRNA efficiently; (c) maintain the integrity of its content before reaching the target site and avoid rapid elimination from the blood circulation; (d) reach diseased tissue, then specifically interact and become internalized by target cells; and (e) dissociate in intracellular compartments of the target cell to release the entrapped siRNA, making it accessible to mRNA.[14] Although viral vectors have high transfection efficiency in cancer cells or primary cells and sustained gene knockdown capacity,[15] safety concerns, including possible oncogenicity, inflammation, and immunogenicity, hamper their clinical application.[16,17] Hence, nanomedicne-based vectors have been widely investigated as potential candidates for effective siRNA delivery, such as polymeric micelles, quantum dots, liposomes, polymer–drug conjugates, dendrimers, biodegradable nanoparticles, inorganic nanoparticles, and other materials in the nanometer size range.[18] In this chapter, we focus on the barriers impeding siRNA delivery and its therapeutic effects, the development of different designs in polymeric micelles of siRNA, their successful gene-silencing evaluation, and advanced applications in the co-delivery of drugs and siRNA based on the micellar core–shell structure.

7.2 Barriers to the Efficacy of siRNA Therapeutics

siRNA is a type of anionic and hydrophilic double-stranded small RNA. The plasma membrane is a significant barrier to siRNA uptake. Despite their small size, the hydrophilicity and negative charge of siRNA molecules prevent them from readily crossing biological membranes (Figure 7.1). The intracellular trafficking of siRNAs delivered by different reagents begins in early endosomal vesicles. These early endosomes subsequently fuse with sorting endosomes, which in turn transfer their contents to late endosomes. ATP-mediated proton accumulation makes the endosomal compartments of cells significantly more acidic (pH 5.0–6.2) than the cytosol or intracellular space (pH $\approx$ 7.4).[19] The endosomal content is then relocated to the lysosomes, which are further acidified (pH $\approx$ 4.5) and contain various nucleases that promote the degradation of siRNAs.[20]

In contrast to the direct accessibility of localized targets, many tissues can only be reached through systemic administration in the bloodstream. siRNA formulations for systemic application face a series of hurdles *in vivo* before reaching the cytoplasm of the target cell.[21] Post-injection, the siRNA complex must navigate the circulatory system of the body while avoiding kidney filtration, uptake by phagocytes, aggregation with serum proteins, and

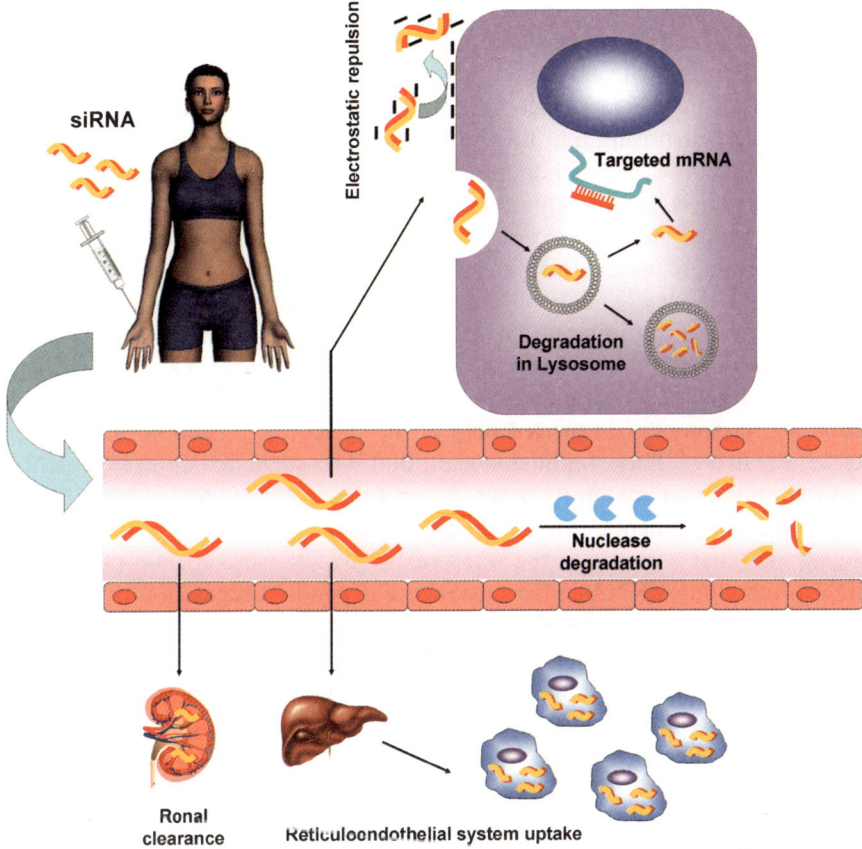

Figure 7.1 Barriers to siRNA uptake following systemic administration. siRNAs may be degraded in the blood, removed by renal excretion or by macrophages. siRNAs may not reach their target cells because of electrostatic repulsion. Following internalization, siRNAs may be prevented from reaching their intracellular targeted mRNA due to an inability to escape from the endosome–lysosome pathway.

enzymatic degradation by endogenous nucleases.[22] One of the first biological barriers encountered by administered siRNAs is presented by the nuclease activity in plasma and tissues. The major activity in plasma is a 3′-exonuclease; however, cleavage of internucleotide bonds can also take place. The reported half-life for unmodified siRNAs in serum ranges from several minutes to around an hour.[23–25] In addition, the kidney plays a key role in siRNA clearance and several studies in animals have reported that the biodistribution of siRNA shows the highest uptake in the kidney.[26–28] In addition to circulating nuclease degradation and renal clearance, a major barrier to *in vivo* delivery of siRNAs is uptake by the reticuloendothelial system (RES). The RES is composed of phagocytic cells, including circulating monocytes and tissue macrophages, whose physiological function is to clear foreign pathogens and to remove cellular debris and apoptotic cells.[29] Tissue macrophages are most abundant in the liver (where they are called Kupffer cells) and the spleen, tissues that also receive high blood flow and exhibit a fenestrated vasculature. Thus, it is not surprising that these organs accumulate high concentrations of siRNAs following systemic administration.[30] Four hours after hydrodynamic i.v. tail vein injection with fluorescently labeled siRNA, a strong fluorescent signal was noted in the liver, spleen, and bone marrow. Similarly, siRNA uptake after standard i.v. tail vein injection or i.p. injection was noted in the liver, spleen, kidney, and bone marrow at 4 h, but the overall signal was weaker.[31] Even after different chemical modifications, high concentrations of phosphodiester (PO), phosphorothioate (PS), 2'-OMe, and locked nucleic acid (LNA)-modified siRNA duplexes were found in large nucleated cells within the red bone marrow,[32–34] in the red pulp of the spleen,[32,33,35] and in the Kupffer cells of the liver.[36] These findings suggest that RES-mediated uptake of siRNA plays an important role in its tissue distribution.

siRNAs can generate off-target effects and can lead to unanticipated phenotypes that complicate the interpretation of the therapeutic benefits of siRNA, including siRNA-induced sequence-dependent regulation of unintended transcripts through partial sequence complementarity to their 3'-UTRs, as well as widespread effects on microRNA processing and function through saturation of the endogenous RNAi machinery by exogenous siRNAs.[37] Off-target effects are used in RNAi screens to identify novel components of signal transduction pathways in a variety of organisms. All siRNA hits, whatever their intended direct target, reduce the mRNA levels of two known upstream pathway components, the TGF-β receptors 1 and 2 (TGFBR1 and TGFBR2), *via* micro (mi)RNA-like off-target effects. The scale of these off-target effects is remarkable, with at least 1% of the sequences in the unbiased siRNA library having measurable off-target effects on one of these two genes.[38] Transfection of small RNAs can globally perturb gene regulation by endogenous miRNAs. Targets of endogenous miRNAs are expressed at significantly higher levels after specific siRNA transfection, consistent with impaired effectiveness of endogenous miRNA repression, which results in unexpected gene expression changes.[39]

Another challenge for siRNA therapy is immune stimulation, as introducing too much siRNA is known to result in nonspecific events owing to the activation of innate immune responses.[40] siRNAs can activate the cells of the immune system and induce the production of cytokines *in vitro* and *in vivo*.[41–43] Mammalian immune cells express a sub-family of pattern-recognition receptors called Toll-like receptors (TLRs) that recognize pathogen-associated molecular patterns, including unmethylated CpG DNA and viral dsRNA.[44] Several TLRs are involved in the recognition of siRNAs, including TLR3, TLR7, and TLR8.[45,46] TLR3 is the receptor for dsRNA and TLR3 overexpressed in cultured human embryonic kidney (HEK) 293 cells is capable of recognizing siRNAs.[47] siRNAs have been shown to activate TLR3 signaling in a sequence-independent manner.[45] TLR7 and TLR8 were initially shown to mediate the recognition of RNA viruses[48,49] and small synthetic antiviral compounds referred to as imidazoquinolines.[50] It has been shown that TLR7 is absolutely required for the induction of cytokines using the appropriate knockout mice in murine immune cells in response to siRNAs.[42,43,51] siRNAs can be recognized by human plasmacytoid dendritic cells (pDCs) through TLR7[42,51,52] and human monocytes likely *via* TLR8.[52,53] TLR7 and TLR8 mediate the recognition of siRNAs in a sequence-dependent manner, and RNA sequences including UG dinucleotides and the 5'-UGU-3' motif are preferentially recognized.[54,55] Thus, the sequence issue of siRNA-mediated immune stimulation still needs to be further investigated.

7.3 Polymeric Micelles for siRNA Delivery

To overcome the barriers to siRNA therapy, it is necessary to deliver siRNAs by suitable systemic carriers to treat most cancers and other diseases. With rapid advances in nanotechnology, many delivery systems have been developed, including polymer–drug conjugates, polymeric micelles, dendrimers, liposomes, and inorganic particles with a size range of 1–1000 nm.

Polymeric micelles are nanosized particles that are made up of polymer chains and are usually spontaneously formed by self-assembly in a liquid, generally as a result of hydrophobic or ion pair interactions between polymer segments. Micelles typically have a so-called core–shell structure. Polymeric micelles as drug delivery systems were introduced in the early 1990s by Kataoka's group by the development of doxorubicin-conjugated block copolymer micelles.[56] They have been used for drug or gene delivery for cancer therapy in preclinical and clinical studies.

Self-assembly of polymeric micelles starts as the concentration of the polymer reaches the critical micelle concentration (CMC). Below the CMC, the number of amphiphilic polymer molecules at the air/water interface increases with increasing concentration; at the CMC, both the bulk and the interface become saturated with monomers. Expulsion of ordered water molecules into the bulk aqueous phase brings about micelle association, which is entropically driven[57] as shown in Figure 7.2. The CMC of polymeric micelles can be as low as 10^{-6} M,[58] which increases the stability of the micellar structure with extreme dilutions after intravenous administration to patients.[59] Polymeric micelles possess other properties requested by ideal drug carriers, such as biodegradability, small particle size, high loading capacity, prolonged circulation, and accumulation in tumors.[60] The characteristic size in the range of several tens to hundreds of nanometers and the surrounding hydrophilic polymer layers endow them with the ability to avoid undesirable foreign-body recognition by the RES in biological entities and to achieve effective extravasation from the bloodstream into tumors through the enhanced permeation and retention (EPR) effect, resulting in high drug accumulation at the target site. On the other hand, inner cores play a pivotal role in the stability of micelles as well as

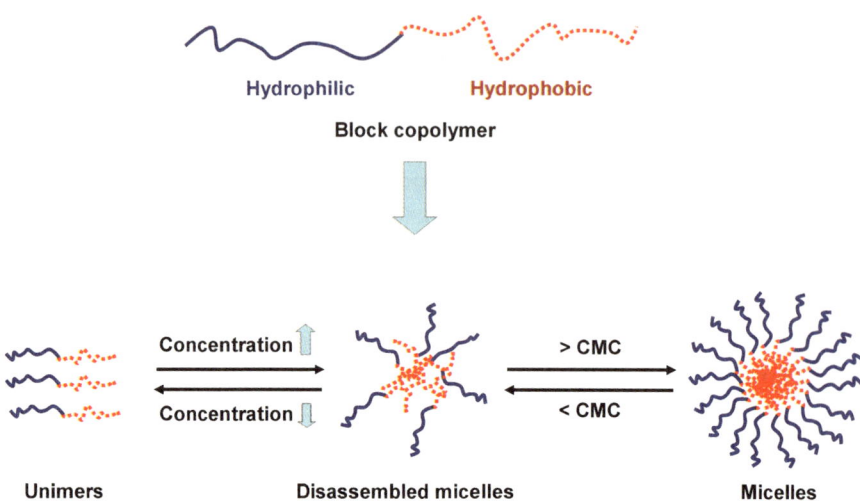

Figure 7.2   The formation procedure of polymeric micelles with increased concentrations of block copolymers.

the accommodation of cargo molecules.[61] Furthermore, various ligands have been attached to the hydrophilic shell for active targeting, including different sugars or peptides, transferrin, and folate. Polymeric micelles can also be engineered by means of the addition of pH- or redox-sensitive moieties according to the biological characteristics of the tumor environment.[59]

In recent years, significant progress has been made in the design of polymeric siRNA carriers. Polymeric micelles of siRNA can be classified as following: (a) polymeric micelles formed by the complexation of synthetic amphiphilic polymers containing polycation segments with siRNA, followed by micellization of polymer/siRNA complexes; (b) smart polymeric micelles having pH-responsive capacity or degradable linkages to facilitate siRNA release intracellularly. Incorporation of targeting ligands on the surface of carriers under each category has also been tried.

7.3.1 Polymeric Micelles Based on Amphiphilic Polymers for siRNA Delivery

Kakizawa *et al.* first developed engineered block copolymers to construct polyplex micelles useful for siRNA delivery. The micellization behavior of block copolymers was utilized to prepare organic–inorganic hybrid nanoparticles with a core–shell structure based on the self-assembly of poly(ethylene glycol)-*block*-poly(aspartic acid) block copolymers [PEG-P(Asp)] with calcium phosphate. The nanoparticles had diameters in the range of several hundreds of nanometers, depending on the PEG-PAA concentration, and revealed excellent colloidal stability due to the steric repulsion effect of the PEG layer surrounding the calcium phosphate core. The loading capacities for siRNA were fairly high, reaching almost 100% under optimal conditions. The calcium phosphate core dissociated in the intracellular environment with appreciably lowered calcium ion concentration compared to the exterior, allowing the release of the incorporated siRNA in a controlled manner. Appreciable silencing of GL3 luciferase gene expression (up to about 60%) was observed for the siRNA-incorporated nanoparticles prepared over the polymer concentration range from 420 to 700 μg mL^{-1}.[61]

Other different micelle assembling block copolymers were further designed for siRNA delivery. Compared with the water-soluble polycations, such micellar nanoparticles exhibited their own advantages. Instead of mixing water-soluble polycations and siRNA to form complexes, which would result in the formation of uncontrollable large particles, micellar nanoparticles with a positive charge allow nucleic acid loading after the formation of nanoparticles.[62] This is believed to be favorable for the construction of size-controllable and monodispersed nucleic acid loaded nanoparticles, which may display unique advantages for *in vivo* applications.[63] In addition, such a preparation method may be convenient for expansion to meet the large quantity requirements for therapeutic applications.[62] The common copolymers for siRNA delivery are those having polyester as core-forming blocks and PEG as the shell-forming blocks, with cationic moieties in the polyester or PEG blocks. The biocompatibility of both PEG and polyesters has been demonstrated. PEG has been extensively used for coating different pharmaceuticals to modify their pharmacokinetics, increase their safety, or lower their immunogenicity.[64,65] Polyesters, such as poly(ε-caprolactone), polylactide, and poly(lactide-*co*-glycolide), are proven biodegradable polymers and have a history of safe application in absorbable biomedical devices such as sutures.[66,67]

Xiong *et al.* developed a series of promising polymeric micelle carriers for siRNA delivery based on degradable poly(ethylene oxide)-*block*-poly(ε-caprolactone) (PEO-*b*-PCL) containing polycationic side chains on their polyester block, such as PEO-*b*-PCL with grafted spermine [PEO-*b*-P(CL-*g*-SP)], tetraethylenepentamine [PEO-*b*-P(CL-*g*-TP)], and *N*,*N*-dimethyldipropylenetriamine [PEO-*b*-P(CL-*g*-DP)]. Those polyamine-grafted PEO-*b*-PCL polymers were found to effectively bind siRNA, self-assemble into micelles, protect siRNA from degradation by nucleases, and be efficient in the delivery of MDR-1 siRNA to silence P-glycoprotein (P-gp) expression in the resistant human MDA435/LCC6 cancer cell line.[68] An advanced functional siRNA carrier was based on PEO-*b*-P(CL-*g*-SP) decorated with integrin $\alpha_v\beta_3$ ligand (RGD4C) and/or a cell-penetrating ligand (TAT) on the PEO shell, in order to overcome the steric effect of the hydrophilic PEO shell that impeded the attachment of the micellar carrier to cells (Figure 7.3). Increased cellular uptake and effective endosomal escape of siRNA were demonstrated in those delivered with the peptide-functionalized micelles, especially those with dual functionality (RGD/TAT micelles) compared to unmodified micelles in resistant MDA435/LCC6 cells. Transfection of MDR-1 siRNA formulated in peptide-modified micelles led to P-gp downregulation both at the mRNA

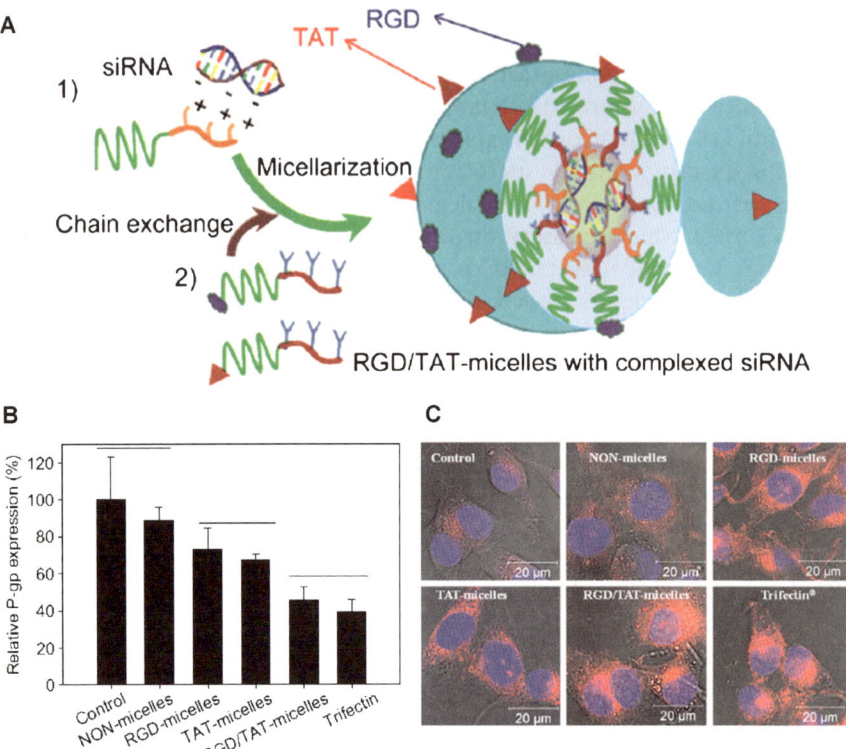

Figure 7.3   (A) Schematic illustration of RGD4C and/or TAT polymeric micellar siRNA complex formation. (B) P-gp expression by flow cytometry by comparing the P-gp related fluorescence intensity to untreated controls after 48 h of incubation. (C) Reversal of resistance to DOX in resistant MDA435/LCC6 cells after transfection with mdr1 siRNA formulations. After transfection, the cells were exposed to free DOX (5 mg mL^{-1}) and assessed for DOX cellular accumulation and distribution by fluorescence microscopy. (Adapted from Xiong and Uludag[69] with permission from Elsevier.)

and protein levels, leading to increased cellular accumulation of doxorubicin (DOX) in the cytoplasm and nucleus. RGD/TAT micellar siRNA complexes produced improved cellular uptake, P-gp silencing, DOX cellular accumulation, DOX nuclear localization, and DOX-induced cytotoxicity in MDA435/ LCC6 cells when compared to micelles decorated with the individual peptides. In the presence of DOX (5 μg mL^{-1}), the viability of cells with mdr1 siRNA transfection using various carriers were ranked as: control (no siRNA) (90%) > NON-micelles (53.4%) > RGD-micelles (29.5%) = TAT micelles (34.9%) > RGD/TAT micelles (18.9%) = Trifectin® (21.4%).[69]

   Sun *et al.* reported self-assembled micellar nanoparticles (MNPs) of monomethoxy poly(ethylene glycol)-*block*-poly(caprolactone)-*block* poly(2-aminoethyl ethylene phosphate) (PPEEA) (mPEG-*b*-PCL-*b*-PPEEA). The

MNPs took on a uniformly spherical morphology with a zeta potential of around 45 mV and were stabilized by hydrophobic–hydrophobic interactions in the PCL core, exhibiting a CMC at 2.7×10^{-3} mg mL^{-1}. Such MNPs allowed siRNA loading after nanoparticle formation without a change in uniformity. The siRNA loaded nanoparticles could be effectively internalized and subsequently released siRNA in cells, resulting in significant gene knockdown activity, which was demonstrated by delivering two siRNAs targeting green fluorescence protein (GFP) that effectively silenced GFP expression in 40–70% of GFP-expressing HEK293 cells.[62] Anticeramidase siRNA-loaded nanoparticles (micelleplex$_{siAC}$) showed significant gene knock-down activities toward the endogenous acid ceramidase (AC) gene *in vitro* and a significant inhibition of tumor growth in a BT474 xenograft murine model *via* tail vein injection (Figure 7.4). In addition to AC gene silencing, administration of micelleplex$_{siAC}$ in animals inhibited proliferation and elevated apoptosis in breast tumor cells. An evaluation of immunotoxicity indicated that this delivery system did not induce an immune response.[70]

Qi *et al.* synthesized siRNA carriers consisting of biodegradable cationic copolymers of monomethoxy poly(ethylene glycol)-*block*-poly(ε-caprolactone)-*block*-poly(L-lysine) (mPEG-*b*-PCL-*b*-PLL, abbreviated as "M") with a PCL block of different lengths. Compared to mPEG-*b*-PLL (abbreviated as "P"), M micellar nanoparticles with siRNA had a smaller size. The diameters of the P-complexes ranged from 130 nm to 300 nm, depending on the N/P

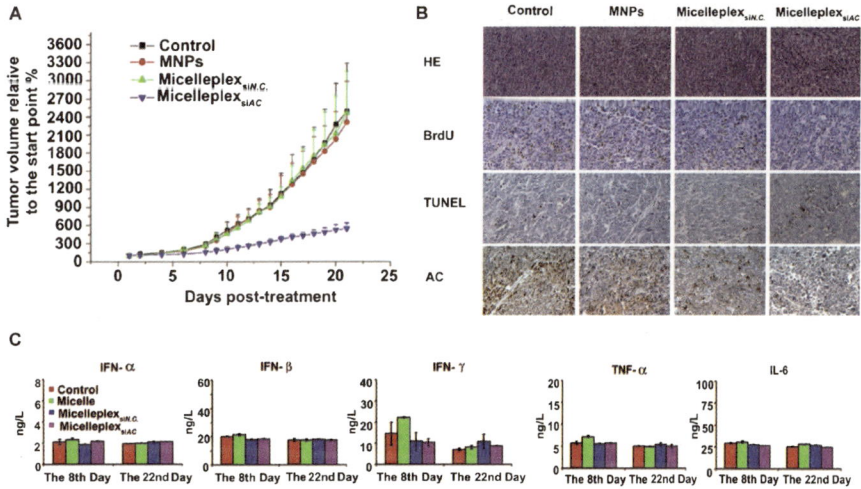

Figure 7.4   (A) Inhibition of BT474 xenograft tumor growth by micelleplex$_{siAC}$. (B) HE, TUNEL, BrdU, and anticeramidase (AC) analyses of tumor tissues after treatment for 22 days with various formulations. (C) Examination of mouse IFN-α, IFN-β, IFN-γ, tumor necrosis factor (TNF)-α, and interleukin (IL)-6 levels in the serum of BT474 xenograft-bearing mice after 8 or 22 days of intravenous injections of various formulations. (Adapted from Mao *et al.*[70] with permission from Elsevier.)

ratio. However, the diameters of the M1-complexes (48 $\pm$ 6 nm) and those of the M2-complexes (70 $\pm$ 10 nm) essentially did not change with N/P ratio. The higher particle density of the M-complexes led to easier internalization *via* endocytosis, higher siRNA delivery efficiency, and higher gene-silencing efficiency. During siRNA transfection experiments in Huh-7 Luc cells, the two M-complexes knocked the gene down by nearly 50% at both 48 h and 72 h when the N/P ratio was 60 or 75. This efficiency was comparable to that of Lipofectamine 2000 and much better than PEI-25 kDa. However, the P-complex exhibited little gene knockdown at any N/P ratio.[71]

Because polyelectrolyte complex nanoparticles suffer from poor stability under physiological conditions, an effective stabilization method is essential for the success of polycationic nanoparticle-mediated siRNA delivery. Nakanishi *et al.* used sodium triphosphate (TPP) as an ionic crosslinking agent to stabilize siRNA-containing nanoparticles by co-condensation. siRNA and TPP were co-encapsulated into a block copolymer, poly(ethylene glycol)-*b*-polyphosphoramidate (PEG-*b*-PPA), to form ternary nanoparticles. The PEG-*b*-PPA/siRNA/TPP ternary nanoparticles exhibited high uniformity with smaller size (80–100 nm) compared with PEG-*b*-PPA/siRNA nanoparticles and showed increased stability in physiological ionic strength and serum-containing medium, due to the stabilization effect from ionic crosslinks between the negatively charged TPP and cationic PPA segments. The transfection and gene-silencing efficiency of the TPP-crosslinked nanoparticles were markedly improved over PEG-*b*-PPA/siRNA complexes in serum-containing medium. No significant difference in cell viability was observed between the nanoparticles prepared with and without TPP co-condensation.[72]

High charge densities could cause the cytotoxicity of cationic polymers, peptides, and liposomes,[73,74] and low molecular weight carriers with fewer positive charges may be safer than high molecular weight ones with more positive charges. Thus, low molecular weight polycations have been developed for siRNA delivery. Ryu *et al.* evaluated amphiphilic peptides with arginine and valine residues as siRNA carriers. The peptides were composed of one to four arginine (R) blocks and six valine (V) blocks with a low charge density. In aqueous solution, the RV peptides formed micelles with hydrophobic cores composed of a valine block and a cationic surface composed of an arginine block. In their previous work, the RV peptides had much lower plasmid DNA delivery efficiency than PEI, suggesting that stable complexes could not form because the size of the plasmid DNA was much larger than that of the RV peptides.[75] However, the RV peptides have similar molecular weights to the siRNAs, and they may locally increase charge density in aqueous solution through the formation of micelles. The RV peptides formed more stable complexes with siRNA than they did with PEI25k; the results also showed that the R3V6 peptide was more efficient than the R1V6, R2V6, and R4V6 peptides in silencing reporter genes. R3V6 had the highest siRNA delivery efficiency at a 1:20 weight ratio and the siVEGF/R3V6 complex suppressed the VEGF expression by 35% compared with the untreated control. In addition, the suppression efficiency of the complex of the R3V6 peptide with siVEGF was similar to that of the siVEGF/PEI25k complex.[76]

Besides synthetic polymers, natural polymers are also utilized in siRNA therapy. Spermine is a tetraamine involved in cellular metabolism and present in all eukaryotic cells.[77] Hyaluronic acid (HA), a major constituent of the extracellular matrix, is a naturally occurring non-sulfated glycosaminoglycan (GAG) polysaccharide composed of N-acetyl-D-glucosamine and D-glucuronic acid,[78] and its receptors such as CD44 and RHAMM are abundantly presented in tumor cells.[79–81] Shen *et al.* developed a target-specific siRNA carrier using hydrophobized hyaluronic acid–spermine conjugates (HHSCs).[82] The polymers were able to effectively bind siRNA, self-assemble into micelles, protect siRNA from degradation by nucleases, and release complexed siRNA efficiently in the presence of low concentrations of polyanionic heparin. HHSCs demonstrated improved transfection efficiency in an HA receptor (CD44) overexpressing SGC-7901 cells compared to polyethylenimine with a high molecular weight of 25 000 Da (PEI 25k) and Lipofectamine due to HA receptor-mediated endocytosis of the complex.[83]

7.3.2 Smart Responsive Micelles for siRNA Delivery

Micelle-based siRNA vehicles can passively accumulate in solid tumor tissues owing to the EPR effect and thus improve siRNA therapeutic efficiency. However, the concentration of active siRNA within cancer cells is often insufficient due to the inefficient release of siRNA from the micelles into the cytoplasm, resulting in a requirement for higher doses. One promising approach to improving efficacy is to develop responsive carrier systems that can release loading agents triggered by intracellular stimuli, such as pH[84] or glutathione.[85,86] On reaching the targeted tumor, such carriers can be subsequently provoked by these stimuli to release siRNA after internalization, hence inducing aggressive activity within tumor cells and leading to maximal therapeutic efficacy with reduced side effects.

7.3.2.1 *pH-Responsive Polymeric Micelles*

Once siRNA is endocytosed, the predominant fate is enzymatic degradation in the lysosome and extracellular clearance.[87] Escape from the endosomal compartment is believed to be a major rate-limiting step for many delivery approaches.[88] In order to circumvent this fate, several pH-sensitive polymers have been employed to enhance endosomal escape, including block copolymers, conjugates, and dendrimers. The high efficiency of siRNA delivery by protonable cationic polymers is mainly attributed to their high buffering capacity, which has been hypothesized to mediate endosomal release by acting as "proton sponges",[88] as shown in Figure 7.5. Proton absorbance by buffering polymers prevents acidification of endosomal vesicles, thereby increasing the ATPase-mediated influx of protons and counterions. Increased

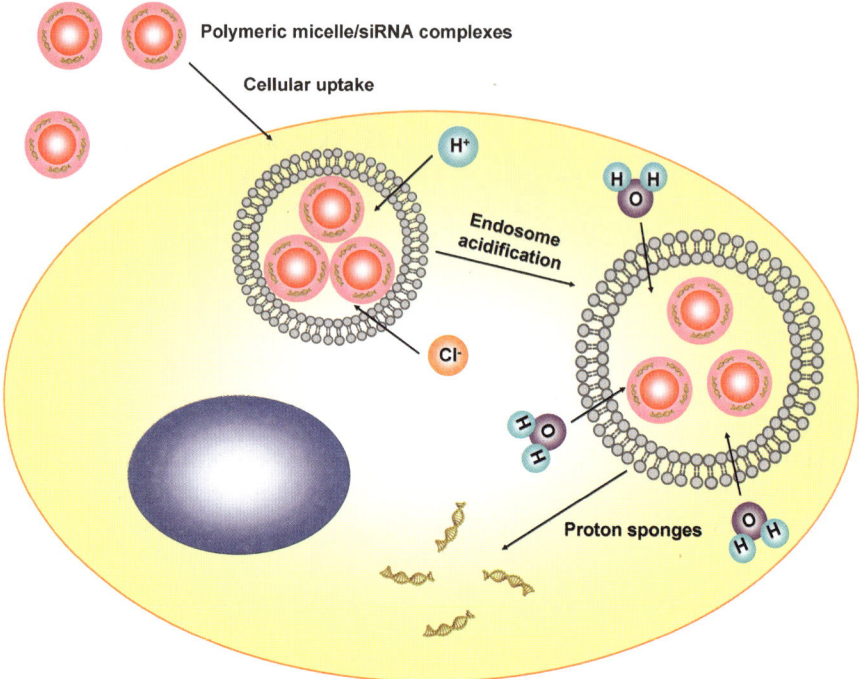

Figure 7.5   pH-sensitive polymers can enhance siRNA endosomal escape. The initial
step is endocytosis of the polymeric micelle/siRNA complexes, followed
by acidic endosome buffering, which leads to increased osmotic pressure
and finally to lysis and release of siRNA into the cytosol.

counterion concentrations inside the endosome lead to osmotic swelling,
endosomal membrane rupture, and the eventual leakage of nucleic acids into
the cytosol.[89]

Itaka *et al.* prepared PEG-polycation diblock copolymers possessing
diamine side chains with distinctive pK_a values for siRNA encapsulation into
polyplex micelles with high endosomal escape ability. PEG-poly{3-[(3-
aminopropyl)amino]propylaspartamide} (PEG-DPT; PEG, 12 000 g mol^{-1};
polymerization degree of the DPT segment, 68), carrying a diamine side chain
with a distinctive pK_a value, was newly synthesized by a side-chain aminolysis
reaction of PEG-poly(β-benzyl-L-aspartate) block copolymer (PEG-PBLA)
with dipropylenetriamine (DPT) (Figure 7.6). The unique feature of PEG-
DPT is the regulated location of primary and secondary amino groups in the
side chain. The former groups, with higher pK_a values, are settled at the distal
end of the side chain to participate in ion complex formation with phosphate
groups in the siRNA molecule, whereas the latter groups, with lower pK_a
values, are located closer to the polymer backbone, where they are expected to
have a substantial fraction of the unprotonated form even in the complex,
presumably due to the lower protonation power and the spatial restriction,

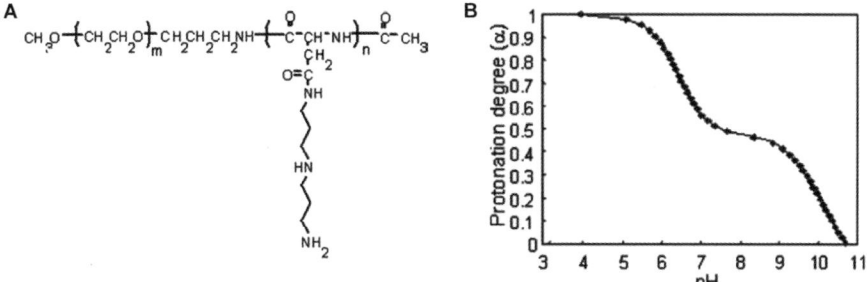

Figure 7.6   (A) Chemical structure of PEG-DPT. (B) Change in protonation degree
(α) with pH for Boc-Asp(DPT)-Pr. (Adapted from Itaka *et al.*[90] with
permission from the American Chemical Society.)

directing to the enhanced intracellular activity of siRNA through the buffering
capacity in the endosomal compartment. Notably, the gene knockdown
abilities of the siRNA/PEG-DPT complex were remarkable, especially at N/P
≥ 10: it showed more than an 80% knockdown, which exceeded commercially
available RNAiFect. Meanwhile, The siRNA/PEG-DPT complexes showed
comparable abilities of gene knockdown, even after co-incubation with serum
for 30 min, which indicated the excellent feasibility of the PEG-DPT/siRNA
complex, particularly under physiological conditions.[90]

In a separate study, the same group constructed pH-sensitive and targetable
polyion complex (PIC) micelles by electrostatic assembly of poly(L-lysine) and
lactosylated PEG-siRNA conjugates with acid-labile linkages (Lac-PEG-
siRNA), which exhibited significant gene silencing for firefly luciferase
expression in HuH-7 hepatoma cells. The PIC micelles achieved far more
effective RNAi activity than the Lac-PEG-siRNA conjugate alone, *viz.* the
50% inhibitory concentration (IC$_{50}$) was found to be 1.3 nM and 91.4 nM for
the PIC micelle and Lac-PEG-siRNA conjugate, respectively. There was
almost a 100-fold increase in RNAi activity with the PIC micelles. Several
important factors were likely to be synergistically involved in the pronounced
RNAi activity of the PIC micelles, such as improved stability against
enzymatic degradation, minimal interaction with serum proteins, enhancement
of cellular uptake through asialoglycoprotein (ASGP) receptor-mediated
endocytosis, and the effective transport of free siRNA from endosomes into
the cytoplasm due to the cleavage of the acid-labile linkage allowing the release
of hundreds of free PEG strands to increase the colloidal osmotic pressure.[91]

Convertine *et al.* synthesized a family of diblock copolymer siRNA carriers
using controlled reversible addition–fragmentation chain transfer (RAFT)
polymerization. The carriers were composed of a positively charged block of
dimethylaminoethyl methacrylate (DMAEMA) to mediate siRNA binding
and a second pH-responsive endosome-releasing block composed of
DMAEMA and propylacrylic acid (PAA) in roughly equimolar ratios along
with butyl methacrylate (BMA). However, the polymers could not form

micelles or any higher order structures at concentrations from 0.1 to 10 mg mL^{-1} at pH 7.4.[92] Then, a new generation of siRNA delivery polymers based on this design was developed which exhibited enhanced transfection efficiency and low cytotoxicity. This design incorporated a longer endosomolytic block with increased hydrophobic content to induce micelle formation. These polymers spontaneously formed spherical micelles in the size range of 40 nm with CMC values of approximately 2 μg mL^{-1}. siRNA binding to the cationic shell block did not perturb micelle stability or significantly increase particle size. The self-assembly of the diblock copolymers into particles was shown to provide a significant enhancement in mRNA knockdown at siRNA concentrations as low as 12.5 nM.[93]

Based on the same linear polymer, Palanca-Wessels *et al.* further developed a targeted delivery system comprising (i) an internalizing streptavidin-conjugated monoclonal antibody (mAb-SA) directed against CD22 and (ii) a biotinylated diblock copolymer, poly(DMAEMA)$_m$-*b*-(BMA$_x$-*co*-DMAEMA$_y$-*co*-PAA$_z$)$_n$. Treatment of lymphoma DoHH2 cells with CD22-targeted polymeric micelles exhibited enhanced siRNA uptake and gene knockdown, with a 70% reduction in gene expression at the 15 nmol L^{-1} siRNA concentration. Thus, this CD22-targeted polymer carrier might be useful for siRNA delivery to lymphoma cells.[94]

For sustained local delivery of siRNA, Nelson *et al.* provided a means to package and protect siRNA within the pH-responsive, endosomolytic micellar nanoparticles (si-NPs) formed by poly(DMAEMA)$_m$-*b*-(BMA$_x$-*co*-DMAEMA$_y$-*co*-PAA$_z$)$_n$ that can be incorporated into nontoxic, biodegradable, and injectable polyurethane (PU) tissue scaffolds. The si-NPs were homogeneously incorporated throughout the porous PUR scaffolds, and they were shown to be released *via* a diffusion-based mechanism for over three weeks. The PUR scaffold releasate collected *in vitro* in PBS at 37 °C for 1–4 days was able to achieve dose-dependent siRNA-mediated silencing with approximately 50% silencing of the model gene GAPDH achieved in NIH3T3 mouse fibroblasts.[95]

Elsabahy *et al.* developed pH-responsive polyion complex micelles (PICMs) consisting of a poly(amidoamine) (PAMAM) dendrimer–nucleic acid core and a detachable poly(ethylene glycol)-*block*-poly(propyl methacrylate-*co*-methacrylic acid) [PEG-*b*-P(PrMA-*co*-MAA)] shell. The micelles displayed a mean hydrodynamic diameter ranging from 50 to 70 nm, a narrow size distribution, and a nearly neutral surface charge. The resulting nanocomplexes could accommodate siRNA in their core. Anti-CD71 Fab' was conjugated at the extremity of the PEG segment *via* a disulfide linkage to promote entry of the micelles into cancer cells mediated by specific ligand–receptor recognition. Upon cellular uptake, the acidic pH in the endosomal compartment protonates the carboxylate groups of the MAA, thus causing the displacement of PEG-*b*-P(PrMA-*co*-MAA) from the PICM and the remaining unshielded PAMAM nucleic acid core can then promote endosomal escape (Figure 7.7). When these pH-responsive targeted PICMs were loaded with siRNA targeting the oncoprotein Bcl-2, they exhibited a greater transfection activity than non-targeted PICMs or commercial PAMAM dendrimers.[96] Because the coupling

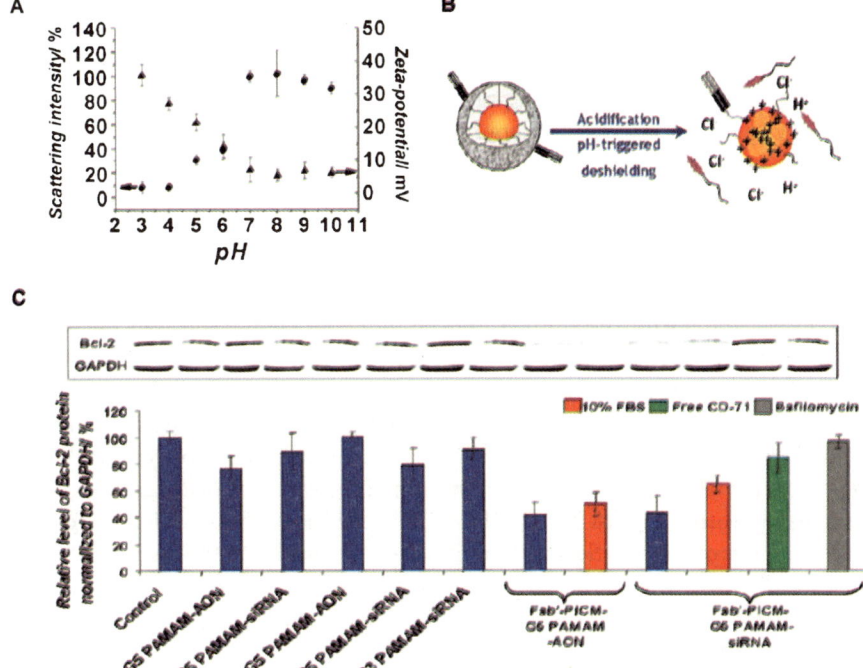

Figure 7.7 (A) Effect of pH on the scattering intensity (•) and zeta-potential (▲) of PEG$_{115}$-*b*-(PrMA$_{28}$-*co*-MAA$_{53}$)/AON/G5 PAMAM-based PICMs. (B) Proposed mechanism of shell dissociation. (C) Bcl-2 gene silencing in PC-3 cells transfected for 5 h with AON (400 nM) or siRNA (25 nM) complexed to G5 PAMAM or entrapped in plain or Fab′-PEG$_{115}$-*b*-P(PrMA$_{28}$-*co*-MAA$_{53}$)/G5 or G3 PAMAM PICMs at an N/(P+COO$^-$) ratio of 1.5 (AON: antisense oligonucleotides). (Adapted from Elsabahy *et al.*[96] with permission from Wiley.)

procedure (disulfide linkage) employed to attach the targeting ligand to PEG-*b*-P(PrMA-*co*-MAA) was relatively inefficient and potentially subject to cleavage in the blood, then anti-CD71 Fab′ was conjugated to a modified amino-PEG-*b*-P(PrMA-*co*-MAA) *via* a maleimide/activated ester bifunctional linker with a more stable sulfide bond.[97]

7.3.2.2 Redox-Responsive Polymeric Micelles

Chemical reactions involving the reductive degradation of disulfide bonds of polymers by intracellular glutathione (GSH) have been widely investigated for responsive drug and gene delivery.[98–100] GSH is a thiol-containing tripeptide and reduces disulfide bonds in the cytoplasm. It has been demonstrated that the intracellular concentration of GSH (*ca.* 10 mM) is significantly higher than the extracellular level (*ca.* 2 μM).[99] Therefore, different disulfide-linked redox delivery systems have been designed for intracellular siRNA release triggered by GSH in tumor cells, classified mainly under two categories of siRNA conjugates with disulfide linkages and micelles with disulfide crosslinked cores (Figure 7.8).

siRNA conjugates can improve the *in vivo* pharmacokinetics, increase the half-life, and increase the delivery efficiency of siRNA,[101–104] but these chemical modifications affect its activity and specificity. Therefore, "reversible" siRNAs have been developed by attaching siRNAs to polymers *via* disulfide linkages. Park's group successfully developed polyelectrolyte complex (PEC) micelle delivery systems for antisense oligonucleotides (ODN) based on the chemical conjugation of ODN to PEG,[105,106] and they extended the ODN PEC micelle strategy to a siRNA delivery system for effective silencing of the VEGF gene in human prostate carcinoma cells. VEGF siRNA was conjugated to PEG *via* a disulfide linkage in order to be cleaved in a reductive cytosolic environment. Branched polyethylenimine (PEI) was used as a cationic core condensing agent (Figure 7.9).[107] The PEC micelles self-assembled to form a core–shell structure *via* electrostatic interactions between the two oppositely

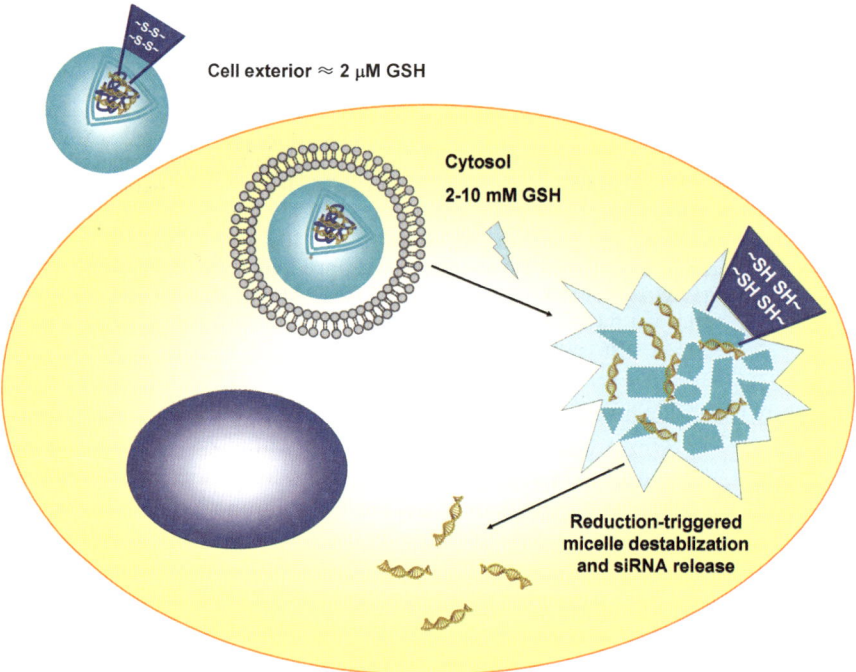

Figure 7.8   Intracellular siRNA release can be triggered by reductive cleavage of disulfide linkages in redox-sensitive polymeric micelles. GSH is a thiol-containing tripeptide and its intracellular concentration is 1000 times higher than the extracellular level.

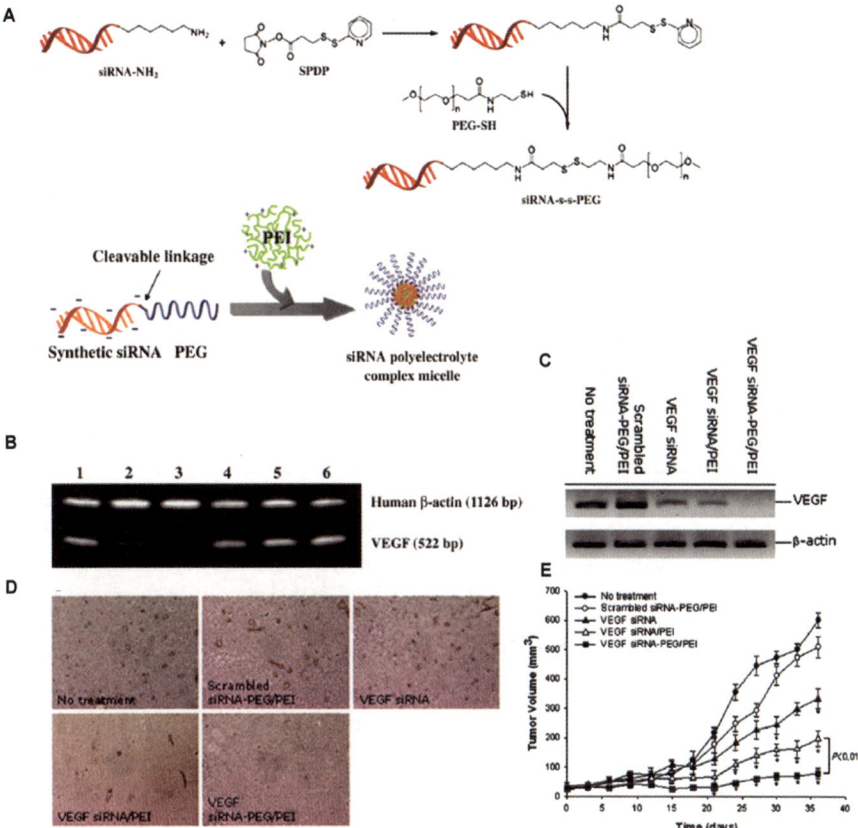

Figure 7.9 (A) Synthesis of the siRNA–PEG conjugate and preparation of the siRNA–PEG/PEI polyelectrolyte complex (PEC). (B) RT–PCR analysis showing downregulation of VEGF mRNA in PC-3 cells by VEGF siRNA–PEG/PEI PEC micelles: lane 1, no treatment; lane 2, VEGF siRNA/PEI complexes; lane 3, VEGF siRNA–PEG/PEI PEC micelles; lane 4, GFP siRNA/PEI complexes; lane 5, scrambled siRNA/PEI complexes; lane 6, scrambled siRNA–PEG/PEI PEC micelles. (C) Intratumoral VEGF mRNA transcript abundance level in PC-3 tumor xenografts at day 3 after the last treatment of intratumoral injection of VEGF siRNA–PEG/PEI PEC micelles. (D) Representative immuno-histochemical staining of factor VIII antibody in PC-3 tumor xenografts 36 days after the initial injection for microvessel density assay. (E) Tumor growth curves of PC-3 tumors; treatments were initiated when the average tumor volume reached around 50 mm³. (Panels A, C, D, and E adapted from Kim *et al.*[86] and panel B adapted from Kim *et al.*[107], both with permission from Elsevier.)

charged polyelectrolytes of siRNA and PEI, resulting in the formation of an insoluble core. At the same time, the siRNA/PEI inner core was stabilized by the PEG chains exposed on the surface.[108] The VEGF siRNA–PEG/PEI PEC micelles showed greater stability than naked VEGF siRNA against enzymatic degradation. Under reductive conditions similar to the cytosolic environment, an intact form of siRNA was released from the siRNA–PEG conjugate by cleavage of the disulfide linkage. The VEGF siRNA–PEG/PEI PEC micelles effectively silenced VEGF gene expression in prostate carcinoma cells (PC-3) by up to 96.5% under an optimized formulation condition.[107] Intravenous as well as intratumoral administration of the PEC micelles significantly inhibited VEGF expression at tumor tissues, leading to tumor growth suppression and an anti-angiogenesis effect in an animal tumor model without showing any detectable inflammatory responses in mice. Upon examination of the PEC micelle distribution and *in vivo* optical imaging following intravenous injection, enhanced accumulation of the PEC micelles was observed in the tumor region by the EPR effect.[86] The PEC micelle formulation also showed full penetration through multiple layers of cancer cells with a unique punctuated distribution pattern and significant suppression of mRNA and protein expression in a dose-dependent manner.[109]

Furthermore, the authors continued this work and generated targeted siRNA PEC micelles. A cell-specific targeting property was added to the PEC micellar delivery system by employing luteinizing hormone releasing hormone (LHRH) as a cancer targeting moiety. For the LHRH receptor overexpressing ovarian cancer cells (A2780), the PEC micelles with LHRH exhibited enhanced cellular uptake *via* receptor-mediated endocytosis, compared to those without LHRH. As a result, LHRH-modified PEC micelles containing 50 nM siRNA reduced VEGF expression by 63%, significantly better than the unmodified counterpart (siRNA-PEG/PEI) that reduced VEGF expression by 50%. In contrast, in SK-OV3 cells that are LHRH receptor negative, the same level of VEGF expression was observed after treatment with LHRH-modified and unmodified PEC micelles containing VEGF siRNA.[108]

In a separate study, PEC micelles were prepared by conjugating siRNA to PEG *via* a disulfide linkage and further condensing it with cationic fusogenic KALA peptide (WEAKLAKALAKALAKHLAKALAKALKACEA) instead of PEI. KALA is an amphipathic peptide that undergoes a conformational change from pH 5.0 to 7.5, resulting in effective disruption of the endosomal membrane.[110–112] siRNA-PEG/KALA PEC micelles inhibited gene expression as effectively as siRNA-PEG/PEI PEC micelles at a much reduced N/P ratio. As PEI's cytotoxicity might limit clinical applications, KALA could be utilized as a non-cytotoxic core condensing agent for siRNA-PEG conjugates to form PEC micelles.[113] Multifunctional PEC micelles based on this delivery system are continually being developed. siRNA was conjugated to a six-arm PEG derivative and the conjugate was also functionalized with Hph1, a cell penetrating peptide, to enhance its cellular uptake property (6PEG-siRNA-Hph1). The 6PEG-siRNA-Hph1 conjugate was electrostatically complexed with cationic self-crosslinked fusogenic KALA peptide (cl-KALA). The 6PEG-siRNA-Hph1/cl-KALA complexes showed superior physical stability, resistance to enzymatic degradation and enhanced gene-silencing efficiency in MDA-MB-435 cells under serum-containing conditions without eliciting cytotoxicity.[114]

Another type of reversible siRNA conjugate was designed by Musacchio *et al.* by attaching siRNA to phospholipids (PE) *via* a disulfide linkage. When inside the cell, the S–S linkage in such a conjugate would be reduced by intracellular glutathione (GSH) and liberate the native siRNA into the cytoplasm. To protect siRNA from nucleolytic degradation *en route* to the target, the siRNA-S–S-PE conjugate was incorporated *via* its hydrophobic PE moiety into stable PEG-PE-based micelles to form mixed polymeric micellar nanoparticles. In the obtained mixed siRNA-S-S-PE/PEG-PE micelles, siRNA was well-protected against degradation by nucleases for at least 24 h, and was released easily from these nanoparticles in free form in the presence of glutathione (GSH) at a concentration mimicking intracellular levels. In GFP-C166 endothelial cells, mixed GFP-siRNA-S–S-PE/PEG-PE micelles downregulated GFP production 50-fold more effectively than free siRNA.[115]

Micelles with disulfide crosslinked cores can maintain a stable micellar structure at physiological ionic strength but are disrupted under reductive conditions because of the cleavage of disulfide crosslinks, which is desirable for siRNA release in the reducing intracellular environment. A core–shell-type polyion complex (PIC) micelle with a disulfide crosslinked core was prepared by Matsumoto *et al.* through the assembly of iminothiolane-modified poly(ethylene glycol)-*block*-poly(L-lysine) [PEG-*b*-(PLL-IM)] and siRNA at a characteristic optimum mixing ratio. The PIC micelles showed a spherical shape of ∼60 nm in diameter with a narrow distribution, and achieved 100-fold higher siRNA transfection efficacy compared with non-crosslinked PICs prepared from PEG-*b*-poly(L-lysine) which were not stable at physiological ionic strength. However, PICs formed with PEG-*b*-(PLL-IM) at non-optimal ratios did not assemble into micellar structures and did not achieve gene silencing following siRNA transfection.[116] One possible explanation for this observed complexation behavior between PEG-*b*-PLL(IM) and siRNA could be related to the instability of amidines formed with 2-iminothiolane (2-IT). Rearrangement of 2-IT modified amines is known to occur following reactions with amino acids involving an intramolecular reaction between sulfur and the amidine carbon, with subsequent release of ammonia and formation of an N-substituted 2-iminothiolane ring.[117] This five-membered ring structure contains an imine bond ($pK_a \approx 6.7$), not an amidine bond ($pK_a \approx 12$), thus reducing the positive charge of the block copolymer at pH 7.4.[118,119] To further investigate the properties of micellar siRNA delivery vehicles prepared with PEG-*b*-PLL comprising lysine amines modified to contain amidine and thiol functionality, lysine modification was achieved using 2-IT [yielding PEG-*b*-PLL(N2IM-IM)] or dimethyl 3,3′-dithiobispropionimidate (DTBP) [yielding PEG-*b*-PLL(MPA)] (Figure 7.10). Amidines formed with 2-IT were unstable and rearranged into an uncharged ring structure lacking free thiol functionality, whereas amidines generated with DTBP were stable. Micelles formed with siRNA and PEG-*b*-PLL(N2IM-IM) at higher molar ratios of polymer/siRNA, while PEG-*b*-PLL(MPA) produced micelles only near stoichiometric molar ratios. *In vitro* gene silencing was highest for PEG-*b*-PLL(MPA)/siRNA micelles, which were also more sensitive to

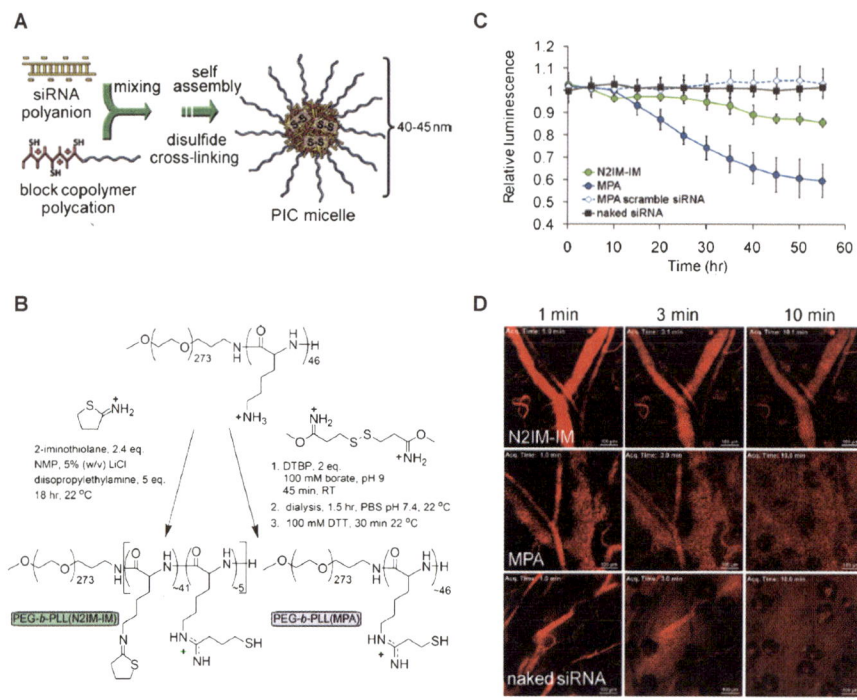

Figure 7.10   (A) Schematic illustration of the preparation of disulfide-crosslinked PIC micelles containing siRNA. (B) Schematic illustration of the synthesis of PEG-*b*-PLL derivatives. (C) *In vitro* gene knockdown of luciferase in B16F10-luc cells. (D) Behavior of siRNA incorporating micelles prepared with PEG-*b*-PLL(X) (X = N2IM-IM or MPA) following intravenous injection. Snapshots of the ear lobe dermis at 1, 3, and 10 min with Cy5 fluorescence shown as red. (Adapted from Christie *et al.*[117] with permission from the American Chemical Society.)

disruption under disulfide-reducing conditions. Blood circulation was most improved for PEG-*b*-PLL(N2IM-IM)/siRNA micelles, with a circulation half-life three-fold longer than naked siRNA.[117]

7.4   Co-delivery of siRNA and Drugs Based on Polymeric Micelles

In the past decade, siRNA therapy has achieved *in vitro* reduction in target gene expression and promising activity against a wide variety of tumors. However, because of the multigenic alterations of tumors, the use of siRNAs as single agents does not seem to be effective in the treatment of malignancies. Thus, siRNA therapy that interferes with signaling pathways in cell proliferation and apoptosis are particularly promising in combination with conventional anticancer treatment. Spänkuch *et al.* identified that treatment of breast cancer cells with siRNAs or antisense oligonucleotides targeting polo-like kinase 1 (Plk1) improved the sensitivity of cancer cells to paclitaxel.[120,121] Macdiarmid *et al.* reported that sequential administration of targeted minicells containing specific siRNA or a cytotoxic drug showed more significant therapeutic efficiency in the treatment of drug-resistant tumors than single administration of minicells loaded with siRNA or chemotherapeutics.[122] Nevertheless, to exert their maximal effect *in vivo*, it is expected the chemotherapeutic drug and the siRNA should be simultaneously delivered to the same tumor cell after systemic administration and, ideally, be distributed in the cells at an optimized ratio for maximal intracellular cooperation.[123] Polymeric micelles are self-assembled nanoparticles composed of amphiphilic materials with a structure that includes a hydrophobic core and a hydrophilic corona. The amphiphilic structure of micelles can provide advantages for multifunctional tasks, such that the hydrophilic shell modified with cationic charges can electrostatically interact with siRNA or DNA, and the hydrophobic core can serve as a payload for hydrophobic drugs.[124]

Zhu *et al.* have reported cationic micelles based on well-defined poly(-dimethylaminoethyl methacrylate)-poly(ε-caprolactone)-poly(dimethylami-noethyl methacrylate) (PDMAEMA-PCL-PDMAEMA) triblock copolymers for the combinatorial delivery of siRNA and paclitaxel. The molecular weights of PDMAEMA blocks varied at 2700, 4800, and 9100 (denoted as polymer 1, 2, and 3, respectively). A gel retardation assay showed that micelle 1 could effectively complex with siRNA at and above N/P ratios of 4:1 and achieved over 70% GFP gene-silencing efficiency at an N/P ratio of 36:1, more efficiently than 25 kDa bPEI and the 20 kDa PDMAEMA homopolymer due to its improved siRNA condensation and endosomal escaping ability. Paclitaxel-loaded micelle 1 with a drug loading content of 6.8 wt% demonstrated clearly higher drug efficacy compared to free paclitaxel. The combinatorial delivery of VEGF siRNA and paclitaxel using PDMAEMA-PCL-PDMAEMA micelles resulted in significantly lower VEGF expression compared to delivery of VEGF siRNA alone, reaching a high silencing efficiency of *ca.* 85%.[125]

Polyethylenimine (PEI) can be modified by grafting stearic acid (SA) and further formulated into polymeric micelles (PEI-SA) with a positive surface charge. Huang's work showed that PEI-SA micelles provided high siRNA cellular uptake efficiency and improved post-transcriptional gene-silencing capacity. The transfection results demonstrated that 65% and 25% reduction of VEGF expression were observed for siVEGF delivered with 10k PEI-SA and 1.8k PEI-SA, respectively, compared with cells which received no treatment. In contrast, siVEGF delivered by complexing with 10k and 1.8k PEI only resulted in 20% and 10% inhibition of VEGF expression, respectively. Loading of doxorubicin to 10k PEI-SA could achieve a high encapsulation efficiency (EE) of 91.9% with a drug loading (DL) percentage of 4.3% without a significant change in particle size and zeta potential. In the animal Huh-7 cancer cell intratumoral model study, at day 30 after injection the relative tumor volumes for groups injected with the PEI-SA/DOX/siVEGF and PEI-SA/siVEGF complexes and PEI-SA/DOX nanoparticles were reduced to 13.0%, 33.7%, and 56.7%, respectively, compared with the untreated group. This result suggests a significant combined effect of co-delivery of DOX and VEGF siRNA by PEI-SA nanoparticles.[124]

Cao *et al.* developed a multifunctional targeting co-delivery system based on PEG, PCL, and linear PEI. Diblock copolymers of PEI-PCL were synthesized and assembled into biodegradable nanocarriers for the combined delivery of BCL-2 siRNA and DOX. The self-assembled micelles could completely inhibit siRNA migration at an N/P ratio around 5, with the highest loading content of DOX around 14.6%. Folic acid as a tumor-targeting ligand was conjugated to a polyanion, poly(ethylene glycol)-*block*-poly(glutamic acid) (FA-PEG-PGA). Driven by electrostatic interactions, FA-PEG-PGA was coated onto the surface of the cationic PEI-PCL nanoparticles pre-loaded with siRNA and DOX, potentiating ligand-directed delivery to human hepatic cancer Bel-7402 cells. The folate-targeted delivery of BCL-2 siRNA resulted in more significant gene suppression, inducing cancer cell apoptosis and improved the therapeutic efficacy of the co-administered DOX. Targeted DOX-loaded micelles complexed with BCL-2 siRNA incubation in the highest level of apoptosis (84.2 $\pm$ 5.4%), which was significantly higher than either targeted blank micelles complexed with BCL-2 siRNA (23.4 $\pm$ 2.5%) or targeted DOX-loaded micelles complexed with scrambled siRNA (42.8 $\pm$ 4.2%). Simultaneously delivering siRNA and the drug into the same cancer cells could yield a synergistic effect of RNA interference and chemotherapy in cancer.[126]

The co-delivery polymeric micellar system was also utilized to overcome multidrug resistance. Xiong *et al.* modified PEO-*b*-PCL block copolymers with functional groups on both blocks. The functional group on the PCL block was used to incorporate short polyamines for complexation with siRNA or to chemically conjugate DOX *via* a pH-sensitive hydrazone linkage. The PEO shell was conferred by attaching two ligands, *i.e.* the integrin $\alpha_v\beta_3$-specific ligand (RGD4C) for active cancer targeting and the cell-penetrating peptide TAT for membrane activity. This system was used to improve the efficacy of DOX in multidrug-resistant MDA-MB-435 human tumor models that overexpress P-glycoprotein (P-gp), by simultaneous intracellular delivery of DOX and siRNA against P-gp expression. RGD/TAT-micelles with DOX and MDR-1 siRNA demonstrated a maximum of $\sim 70\%$ of cell growth inhibition. Dy677-labeled siRNA was also used to assess the *in vivo* stability of the siRNA carrier. Significant Dy677-siRNA fluorescence was observed in tumors 24 h after injection, suggesting that these micelles could stably complex and protect siRNA in the micelle core and deliver the siRNA into tumor tissue, while NON-micelles/Dy677-siRNA did not show obvious fluorescence accumulation in the tumor, which indicated this multifunctional polymeric micellar system could target $\alpha_v\beta_3$-positive tumors *in vivo*.[127]

Although the combination of chemotherapy and siRNA-based therapy for cancer treatment has received increasing attention, the ideal delivery system that can maximize the efficiency of the active agents is a great challenge. It has been proposed that such delivery vehicles should: (i) be able to encapsulate different payloads with tunable doses; (ii) endow similar pharmacokinetics of the payloads and be able to systemically deliver them into the same tumor cells; (iii) synergistically inhibit tumor growth following systemic injection; and (iv) be safe for *in vivo* applications.[123] Sun *et al.* further used the micellar nanoparticle of a biodegradable mPEG-*b*-PCL-*b*-PPEEA triblock copolymer to systemically deliver siRNA and a chemotherapeutic drug (Figure 7.11). Efficient siRNA binding occurred at a molar ratio of nitrogen in the N/P ratio of 5:1 and paclitaxel could be entrapped in the hydrophobic PCL core with high encapsulation efficiency (>90%) *via* a hydrophobic–hydrophobic interaction. The micelleplex showed compact and spherical morphology with a mean diameter of 50 nm. The micelleplex was capable of delivering siRNA and paclitaxel simultaneously to the same tumor cells both *in vitro* and *in vivo*. In the MDA-MB-435s xenograft murine model, simultaneous delivery of the same doses of paclitaxel and si*Plk1* by ^{paclitaxel}micelleplex$_{\text{si}Plk1}$ (0.667 µg kg^{-1} paclitaxel and 0.223 mg kg^{-1} si*Plk1* per injection) exhibited particularly significant inhibition of tumor growth compared with PBS treatment ($p <$ 0.0001). A synergistic inhibitory effect of the two therapeutic agents on tumor growth was demonstrated (*c.i.* < 1), requiring 1000-fold less paclitaxel than needed for paclitaxel monotherapy delivered by the micelleplex. In contrast, combinatorial delivery of separate si*Plk1* and paclitaxel by micelleplex$_{\text{si}Plk1}$ and ^{paclitaxel}micelleplex$_{\text{si}Nonsense}$ only showed moderate inhibition of tumor growth and no synergistic effect was observed. Delivery of si*Plk1* by the micelleplex did not cause an elevation in human or mouse IFN-α, IFN-β, IFN-γ, tumor necrosis factor (TNF)-α, or interleukin (IL)-6 levels. Additionally, simultaneous delivery of si*Plk1* and paclitaxel with the micelleplex did not result in any sign of toxicity to the mice, as shown by normal heart, liver, and kidney functions following treatment. The results demonstrate that systemic administration of a micelleplex carrying *Plk1*-specific siRNA and paclitaxel could induce a synergistic tumor suppression effect without activation of the innate immune response or generation of carrier-associated toxicity.[123]

7.5 Future Perspectives

RNAi is a post-transcriptional gene-silencing mechanism that can be induced by 21–23 nt siRNA to degrade a target mRNA in a highly sequence-dependent manner. Since the discovery of RNAi, many oncogenic targets, including proliferation, anti-apoptosis, angiogenesis, and drug resistance, have been studied for siRNA-mediated cancer therapy. Compared to traditional chemical drug discovery, the identification of highly selective and inhibitory sequences of siRNA is much faster. Meanwhile, it is relatively easy to synthesize and manufacture siRNA on a large scale.[128] Although a number of reports have

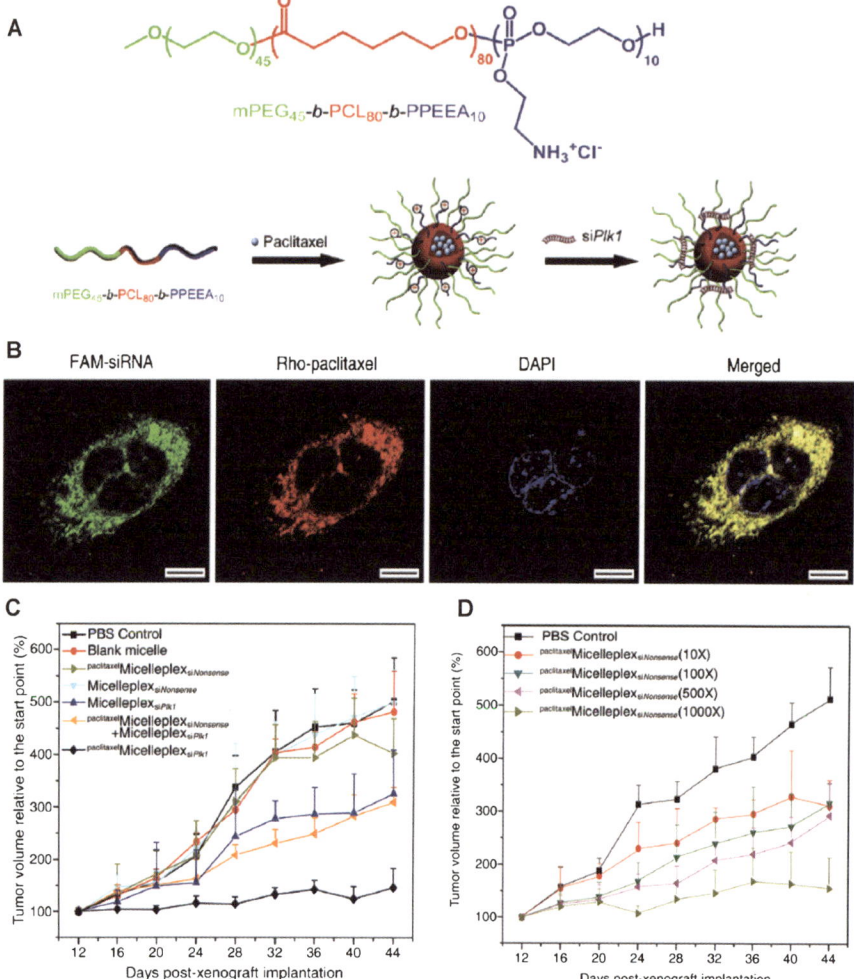

Figure 7.11 (A) The chemical structure of mPEG$_{45}$-*b*-PCL$_{80}$-*b*-PPEEA$_{10}$ and schematic illustration of micellar nanoparticle formation and the loading of paclitaxel and siRNA. (B) Confocal laser scanning microscope (CLSM) image of the intracellular distribution of $^{\text{Rho-paclitaxel}}$micelleplex$_{\text{FAM-siRNA}}$ in MDA-MB-435s cells after incubation for 2 h (630 ×). The scale bar is 10 μm. Paclitaxel and siRNA were labeled with rhodamine (*red*) and fluorescein (*green*), respectively. Cell nuclei were stained with 4′,6-diamidino-2-phenylindole (DAPI; *blue*). (C) Inhibition of MDA-MB-435s xenograft tumor growth by ^{paclitaxel}micelleplex$_{\text{siPlk1}}$ in comparison with various formulations (*n* = 6). (D) Dose–response study of paclitaxel delivered by ^{paclitaxel}micelleplex$_{\text{siNonsense}}$ on the inhibition of MDA-MB-435s xenograft tumor growth (*n* = 6). Paclitaxel doses were 10- to 1000-fold higher (10 × to 1000 ×) compared to those used in the ^{paclitaxel}micelleplex$_{\text{siPlk1}}$. (Adapted from Sun *et al.*[123] with permission from the American Chemical Society.)

demonstrated the great potential of siRNA in cancer treatment, safe and efficient delivery of siRNA into target cells or organs still remains a considerable challenge for therapeutic applications.

Nanoparticles are rapidly emerging as siRNA delivery systems both *in vitro* and *in vivo*. A successful siRNA delivery carrier should have several characteristics, including effective siRNA protection, high transfection efficacy, improved immune evasion, few off-target effects, and the ability to bypass intracellular and extracellular barriers. Polymeric micelles in the range of several tens to hundreds of nanometers can serve as siRNA delivery systems due to their structure, which can be easily changed and modified. The outer hydrophilic segments can evade RES recognition to achieve effective extravasation from the bloodstream to tumors through the EPR effect, resulting in high siRNA accumulation, and can also be modified with polycations to interact electrostatically with siRNA and different moieties to target specific cells. The inner hydrophobic cores improve the stability of particles and allow loading of hydrophobic drugs. Smart polymeric micelles triggered by intracellular stimuli can be further developed to improve siRNA efficacy. Several types of polymeric micelles have already exhibited great therapeutic potential in xenograft cancer models, and simultaneous delivery of siRNA and drugs into the same cancer cells by polymeric micellar delivery systems have demonstrated the synergistic effects of RNA interference and chemotherapy in cancer.

In the future, a significant amount of work will have to be performed to turn polymeric micelles/siRNA complexes into clinically acceptable therapeutic drugs. (1) There is still a need for better polymers for micellar formulations and more highly effective methods of siRNA loading and protection during *in vivo* delivery. (2) Methods for preparing polymeric micelles/siRNA complexes on a larger scale for ongoing production should be developed, and attention must also be paid to developing more stable polymeric micelles/siRNA complexes for long-term storage because siRNA is easily degraded under physiological conditions. (3) siRNA pharmacokinetic and pharmacodynamic parameters should be clearly defined after systemic administration, which is very important to provide necessary information for the optimization of an siRNA dosing regimen. Convenient analysis methods to detect siRNA in biological samples should also be developed. (4) *In vivo* safety needs to be further studied to eliminate unexpected immune stimulation and toxicity of both polymeric micelle materials and their possible degradation products. (5) Although enhanced efficacy and specificity of cancer treatment has been found using a combined approach with siRNA and traditional chemotherapy, the optimized ratio of the two distinct drugs needs to be further investigated to obtain the maximum synergistic effect.

7.6 Conclusion

Polymeric micelles as siRNA delivery systems will remain an active area in future research because the specific core–shell structure provides great potential for loading siRNAs and protecting them to overcome many *in vitro* and *in vivo* hurdles to achieve targeted delivery to tissues/cells. Meanwhile, advanced applications for the simultaneous delivery of both siRNA and hydrophobic drugs can be developed using polymeric micellar delivery systems to enhance therapeutic efficacy. The coming years represent a critical time in the field as the concept requires further optimization to become a suitable clinical delivery system for cancer therapy.

References

1. M. K. Montgomery, S. Q. Xu and A. Fire, *Proc. Natl. Acad. Sci. U. S. A.*, 1998, **95**, 15502.
2. A. Fire, S. Xu, M. K. Montgomery, S. A. Kostas, S. E. Driver and C. C. Mello, *Nature*, 1998, **391**, 806.
3. G. Liu, F. Wong-Staal and Q. X. Li, *Histol. Histopathol.*, 2007, **22**, 211.
4. C. S. Gondi and J. S. Rao, *J. Cell Physiol.*, 2009, **220**, 285.
5. S. M. Elbashir, J. Harborth, W. Lendeckel, A. Yalcin, K. Weber and T. Tuschl, *Nature*, 2001, **411**, 494.
6. S. M. Hammond, E. Bernstein, D. Beach and G. J. Hannon, *Nature*, 2000, **404**, 293.
7. S. M. Hammond, S. Boettcher, A. A. Caudy, R. Kobayashi and G. J. Hannon, *Science*, 2001, **293**, 1146.
8. Q. Leng, M. C. Woodle, P. Y. Lu and A. J. Mixson, *Drugs Future*, 2009, **34**, 721.
9. I. Melnikova, *Nat. Rev. Drug Discovery*, 2007, **6**, 863.
10. Y. K. Oh and T. G. Park, *Adv. Drug Delivery Rev.*, 2009, **61**, 850.
11. N. S. Dejneka, S. H. Wan, O. S. Bond, D. J. Kornbrust and S. J. Reich, *Mol. Vision*, 2008, **14**, , 997.
12. J. C. Burnett, J. J. Rossi and K. Tiemann, *Biotechnol. J.*, 2011, **6**, 1130.
13. M. E. Davis, J. E. Zuckerman, C. H. J. Choi, D. Seligson, A. Tolcher, C. A. Alabi, Y. Yen, J. D. Heidel and A. Ribas, *Nature*, 2010, **464**, 1067.
14. A. Falamarzian, X. B. Xiong, H. Uludag and A. Lavasanifar, *J. Drug Delivery Sci. Technol.*, 2012, **22**, 43.
15. Y. Dorsett and T. Tuschl, *Nat. Rev. Drug Discovery*, 2004, **3**, 318.
16. A. K. Zaiss and D. A. Muruve, *Curr. Gene Ther.*, 2005, **5**, 323.
17. D. Xu, D. McCarty, A. Fernandes, M. Fisher, R. J. Samulski and R. L. Juliano, *Mol. Ther.*, 2005, **11**, 523.
18. S. Nimesh, N. Gupta and R. Chandra, *Nanomedicine (London, U. K.)*, 2011, **6**, 729.
19. S. Ohkuma and B. Poole, *Proc. Natl. Acad. Sci. U. S. A.*, 1978, **75**, 3327.
20. M. Dominska and D. M Dykxhoorn, *J. Cell Sci.*, 2010, **123**, 1183.
21. K. A. Whitehead, R. Langer and D. G. Anderson, *Nat. Rev. Drug Discovery*, 2009, **8**, 129.
22. F. Alexis, E. Pridgen, L. K. Molnar and O. C. Farokhzad, *Mol. Pharmaceutics*, 2008, **5**, 505.

23. D. M. Dykxhoorn and J. Lieberman, *Annu. Rev. Biomed. Eng.*, 2006, **8**, 377.

24. J. M. Layzer, A. P. McCaffrey, A. K. Tanner, Z. Huang, M. A. Kay and B. A. Sullenger, *RNA*, 2004, **10**, 766.

25. D. V. Morrissey, K. Blanchard, L. Shaw, K. Jensen, J. A. Lockridge, B. Dickinson, J. A. McSwiggen, C. Vargeese, K. Bowman, C. S. Shaffer, B. A. Polisky and S. Zinnen, *Hepatology*, 2005, **41**, 1349.

26. D. A. Braasch, Z. Paroo, A. Constantinescu, G. Ren, O. K. Oz, R. P. Mason and D. R. Corey, *Bioorg. Med. Chem. Lett.*, 2004, **14**, 1139.

27. F. M. van de Water, O. C. Boerman, A. C. Wouterse, J. G. Peters, F. G. Russel and R. Masereeuw, *Drug Metab. Dispos.*, 2006, **34**, 1393.

28. O. M. Merkel, D. Librizzi, A. Pfestroff, T. Schurrat, M. Behe and T. Kissel, *Bioconjugate Chem.*, 2009, **20**, 174.

29. D. M. Mosser and J. P. Edwards, *Nat. Rev. Immunol.*, 2008, **8**, 958.

30. R. Juliano, J. Bauman, H. Kang and X Ming, *Mol. Pharmaceutics*, 2009, **6**, 686.

31. S. D. Larson, L. N. Jackson, L. A. Chen, P. G. Rychahou and B. M. Evers, *Surgery*, 2007, **142**, 262.

32. M. Butler, K. Stecker and C. F. Bennett, *Lab. Invest.*, 1997, **77**, 379.

33. G. Lendvai, I. Velikyan, M. Bergstrom, S. Estrada, D. Laryea, M. Valila, S. Salomaki, B. Langstrom and A. Roivainen, *Eur. J. Pharm. Sci.*, 2005, **26**, 26.

34. G. Lendvai, I. Velikyan, S. Estrada, B. Eriksson, B. Langstrom and M. Bergstrom, *Oligonucleotides*, 2008, **18**, 33.

35. R. S. Geary, R. Z. Yu, T. Watanabe, S. P. Henry, G. E. Hardee, A. Chappell, J. Matson, H. Sasmor, L. Cummins and A. A. Levin, *Drug Metab. Dispos.*, 2003, **31**, 1419.

36. M. J. Graham, S. T. Crooke, D. K. Monteith, S. R. Cooper, K. M. Lemonidis, K. K. Stecker, M. J. Martin and R. M. Crooke, *J. Pharmacol. Exp. Ther.*, 1998, **286**, 447.

37. A. L. Jackson and P. S. Linsley, *Nat. Rev. Drug Discovery*, 2010, **9**, 57.

38. N. Schultz, D. R. Marenstein, D. A. De Angelis, W. Q. Wang, S. Nelander, A. Jacobsen, D. S. Marks, J. Massague and C. Sander, *Silence*, 2011, **2**, 3.

39. A. A. Khan, D. Betel, M. L. Miller, C. Sander, C. S. Leslie and D. S Marks, *Nat. Biotechnol.*, 2009, **27**, 549.

40. Y. K. Oh and T. G. Park, *Adv. Drug Delivery Rev.*, 2009, **61**, 850.

41. V. Hornung, M. Guenthner-Biller, C. Bourquin, A. Ablasser, M. Schlee, S. Uematsu, A. Noronha, M. Manoharan, S. Akira, A. de Fougerolles, S. Endres and G. Hartmann, *Nat. Med.*, 2005, **11**, 263.

42. A. D. Judge, V. Sood, J. R. Shaw, D. Fang, K. McClintock and I. MacLachlan, *Nat. Biotechnol.*, 2005, **23**, 457.

43. M. Sioud, *J. Mol. Biol.*, 2005, **348**, 1079.

44. A. L. Jackson and P. S. Linsley, *Nat. Rev. Drug Discovery*, 2010, **9**, 57.

45. K. Kariko, P. Bhuyan, J. Capodici and D. Weissman, *J. Immunol.*, 2004, **172**, 6545.
46. J. T. Marques and B. R. Williams, *Nat. Biotechnol.*, 2005, **23**, 1399.
47. K. Kariko, P. Bhuyan, J. Capodici and D. Weissman, *J. Immunol.*, 2004, **172**, 6545.
48. S. S. Diebold, T. Kaisho, H. Hemmi, S. Akira and C. R. E. Sousa, *Science*, 2004, **303**, 1529.
49. F. Heil, H. Hemmi, H. Hochrein, F. Ampenberger, C. Kirschning, S. Akira, G. Lipford, H. Wagner and S. Bauer, *Science*, 2004, **303**, 1526.
50. H. Hemmi, T. Kaisho, O. Takeuchi, S. Sato, H. Sanjo, K. Hoshino, T. Horiuchi, H. Tomizawa, K. Takeda and S. Akira, *Nat. Immunol.*, 2002, **3**, 196.
51. V. Hornung, M. Guenthner-Biller, C. Bourquin, A. Ablasser, M. Schlee, S. Uematsu, A. Noronha, M. Manoharan, S. Akira, A. de Fougerolles, S. Endres and G. Hartmann, *Nat. Med.*, 2005, **11**, 263.
52. A. Iwasaki and R. Medzhitov, *Nat. Immunol*, 2004, **5**, 987.
53. Y. J. Liu, *Annu. Rev. Immunol.*, 2005, **23**, 275.
54. S. S. Diebold, T. Kaisho, H. Hemmi, S. Akira and C. Reis e Sousa, *Science*, 2004, **303**, 1529.
55. F. Heil, H. Hemmi, H. Hochrein, F. Ampenberger, C. Kirschning, S. Akira, G. Lipford, H. Wagner and S. Bauer, *Science*, 2004, **303**, 1526.
56. M. Yokoyama, G. S. Kwon, T. Okano, Y. Sakurai, T. Seto and K. Kataoka, *Bioconjugate Chem.*, 1992, **3**, 295.
57. M. L. Adams, A. Lavasanifar and G. S. Kwon, *J. Pharm. Sci.*, 2003, **92**, 1343.
58. V. P. Torchilin, *J. Controlled Release*, 2001, **73**, 137.
59. U. Kedar, P. Phutane, S. Shidhaye and V. Kadam, *Nanomedicine (Philadelphia, U. S.)*, 2010, **6**, 714.
60. R. Gref, Y. Minamitake, M. T. Peracchia, V. Trubetskoy, V. Torchilin and R. Langer, *Science*, 1994, **263**, 1600.
61. Y. Kakizawa, S. Furukawa and K. Kataoka, *J. Controlled Release*, 2004, **97**, 345.
62. T. M. Sun, J. Z. Du, L. F. Yan, H. Q. Mao and J. Wang, *Biomaterials*, 2008, **29**, 4348.
63. J. M. Zhu, A. G. Tang, L. P. Law, M. Feng, K. M. Ho, D. K. L. Lee, F. W. Harris and P. Li, *Bioconjugate Chem.*, 2005, **16**, 139.
64. M. Chinol, P. Casalini, M. Maggiolo, S. Canevari, E. S. Omodeo, P. Caliceti, F. M. Veronese, M. Cremonesi, F. Chiolerio, E. Nardone, A. G. Siccardi and G. Paganelli, *Br. J. Cancer*, 1998, **78**, 189.
65. J. M. Harris and R. B. Chess, *Nat. Rev. Drug Discovery*, 2003, **2**, 214.
66. K. A. Athanasiou, G. G. Niederauer and C. M. Agrawal, *Biomaterials*, 1996, **17**, 93.
67. B. G. Amsden, M. Y. Tse, N. D. Turner, D. K. Knight and S. C. Pang, *Biomacromolecules*, 2006, **7**, 365.
68. X. B. Xiong, H. Uludag and A. Lavasanifar, *Biomaterials*, 2009, **30**, 242.

69. X. B. Xiong, H. Uludag and A. Lavasanifar, *Biomaterials*, 2010, **31**, 5886.
70. C. Q. Mao, J. Z. Du, T. M. Sun, Y. D. Yao, P. Z. Zhang, E. W. Song and J. Wang, *Biomaterials*, 2011, **32**, 3124.
71. R. Qi, S. Liu, J. Chen, H. Xiao, L. Yan, Y. Huang and X. Jing, *J. Controlled Release*, 2012, **159**, 251.
72. M. Nakanishi, R. Patil, Y. Ren, R. Shyam, P. Wong and H. Q. Mao, *Pharm. Res.*, 2011, **28**, 1723.
73. D. Fischer, T. Bieber, Y. X. Li, H. P. Elsasser and T. Kissel, *Pharm. Res.*, 1999, **16**, 1273.
74. K. Kunath, A. von Harpe, D. Fischer, H. Peterson, U. Bickel, K. Voigt and T. Kissel, *J. Controlled Release*, 2003, **89**, 113.
75. D. W. Ryu, H. A. Kim, H. Song, S. Kim and M. Lee, *J. Cell. Biochem.*, 2011, **112**, 1458.
76. D. W. Ryu, H. A. Kim, J. H. Ryu, D. Y. Lee and M. Lee, *J. Cell. Biochem.*, 2012, **113**, 619.
77. T. Azzam, H. Eliyahu, A. Makovitzki, M. Linial and A. J. Domb, *J. Controlled Release*, 2004, **96**, 309.
78. T. Laurent, *Ciba Found. Symp.*, 1989, **143**, 1.
79. N. Yerushalmi, A. Arad and R. Margalit, *Arch. Biochem. Biophys.*, 1994, **313**, 267.
80. Y. Luo, M. R. Ziebell and G. D. Prestwich, *Biomacromolecules*, 2000, **1**, 208.
81. D. Coradini, C. Pellizzaro, G. Miglierini, M. G. Daidone and A. Perbellini, *Int. J. Cancer*, 1999, **81**, 411.
82. Y. Shen, Q. Li, J. S. Tu and J. B. Zhu, *Carbohydr. Polym.*, 2009, **77**, 95.
83. Y. Shen, B. H. Wang, Y. Lu, A. Ouahab, Q. Li and J. S. Tu, *Int. J. Pharm.*, 2011, **414**, 233.
84. E. S. Lee, Z. G. Gao and Y. H. Bae, *J. Controlled Release*, 2008, **132**, 164.
85. S. Takae, K. Miyata, M. Oba, T. Ishii, N. Nishiyama, K. Itaka, Y. Yamasaki, H. Koyama and K. Kataoka, *J. Am. Chem. Soc.*, 2008, **130**, 6001.
86. S. H. Kim, J. H. Jeong, S. H. Lee, S. W. Kim and T. G. Park, *J. Controlled Release*, 2008, **129**, 107.
87. L. K. Medina-Kauwe, J. Xie and S. Hamm-Alvarez, *Gene Ther.*, 2005, **12**, 1734.
88. M. Dominska and D. M. Dykxhoorn, *J. Cell Sci.*, 2010, **123**, 1183.
89. N. D. Sonawane, F. C. Szoka and A. S. Verkman, *J. Biol. Chem.*, 2003, **278**, 44826.
90. K. Itaka, N. Kanayama, N. Nishiyama, W. D. Jang, Y. Yamasaki, K. Nakamura, H. Kawaguchi and K. Kataoka, *J. Am. Chem. Soc.*, 2004, **126**, 13612.
91. M. Oishi, Y. Nagasaki, K. Itaka, N. Nishiyama and K. Kataoka, *J. Am. Chem. Soc.*, 2005, **127**, 1624.
92. A. J. Convertine, D. S. W. Benoit, C. L. Duvall, A. S. Hoffman and P. S. Stayton, *J. Controlled Release*, 2009, **133**, 221.

93. A. J. Convertine, C. Diab, M. Prieve, A. Paschal, A. S. Hoffman, P. H. Johnson and P. S. Stayton, *Biomacromolecules*, 2010, **11**, 2904.

94. M. C. Palanca-Wessels, A. J. Convertine, R. Cutler-Strom, G. C. Booth, F. Lee, G. Y. Berguig, P. S. Stayton and O. W. Press, *Mol. Ther.*, 2011, **19**, 1529.

95. C. E. Nelson, M. K. Gupta, E. J. Adolph, J. M. Shannon, S. A. Guelcher and C. L. Duvall, *Biomaterials*, 2012, **33**, 1154.

96. M. Elsabahy, N. Wazen, N. Bayo-Puxan, G. Deleavey, M. Servant, M. J. Damha and J. C. Leroux, *Adv. Funct. Mater.*, 2009, **19**, 3862.

97. A. E. Felber, B. Castagner, M. Elsabahy, G. F. Deleavey, M. J. Damha and J. C. Leroux, *J. Controlled Release*, 2011, **152**, 159.

98. L. Y. Tang, Y. C. Wang, Y. Li, J. Z. Du and J. Wang, *Bioconjugate Chem.*, 2009, **20**, 1095.

99. G. Saito, J. A. Swanson and K. D. Lee, *Adv. Drug Delivery Rev.*, 2003, **55**, 199.

100. C. Lin, C. J. Blaauboer, M. M. Timoneda, M. C. Lok, M. van Steenbergen, W. E. Hennink, Z. Y. Zhong, J. Feijen and J. F. J. Engbersen, *J. Controlled Release*, 2008, **126**, 166.

101. J. Soutschek, A. Akinc, B. Bramlage, K. Charisse, R. Constien, M. Donoghue, S. Elbashir, A. Geick, P. Hadwiger, J. Harborth, M. John, V. Kesavan, G. Lavine, R. K. Pandey, T. Racie, K. G. Rajeev, I. Rohl, I. Toudjarska, G. Wang, S. Wuschko, D. Bumcrot, V. Koteliansky, S. Limmer, M. Manoharan and H. P. Vornlocher, *Nature*, 2004, **432**, 173.

102. S. A. Moschos, S. W. Jones, M. M. Perry, A. E. Williams, J. S. Erjefalt, J. J. Turner, P. J. Barnes, B. S. Sproat, M. J. Gait and M. A. Lindsay, *Bioconjugate Chem.*, 2007, **18**, 1450.

103. T. C. Chu, K. Y. Twu, A. D. Ellington and M. Levy, *Nucleic Acids Res.*, 2006, **34**, 73.

104. K. Nishina, T. Unno, Y. Uno, T. Kubodera, T. Kanouchi, H. Mizusawa and T. Yokota, *Mol. Ther.*, 2008, **16**, 734.

105. J. H. Jeong, S. H. Kim, S. W. Kim and T. G. Park, *Bioconjugate Chem.*, 2005, **16**, 1034.

106. J. H. Jeong, S. W. Kim and T. G. Park, *J. Controlled Release*, 2003, **93**, 183.

107. S. H. Kim, J. H. Jeong, S. H. Lee, S. W. Kim and T. G. Park, *J. Controlled Release*, 2006, **116**, 123.

108. S. H. Kim, J. H. Jeong, S. H. Lee, S. W. Kim and T. G. Park, *Bioconjugate Chem.*, 2008, **19**, 2156.

109. A. M. Al-Abd, S. H. Lee, S. H. Kim, J. H. Cha, T. G. Park, S. J. Lee and II. J. Kuh, *J. Controlled Release*, 2009, **137**, 130.

110. E. Wagner, *Adv. Drug Delivery Rev.*, 1999, **38**, 279.

111. C. Plank, B. Oberhauser, K. Mechtler, C. Koch and E. Wagner, *J. Biol. Chem.*, 1994, **269**, 12918.

112. T. Niidome, N. Ohmori, A. Ichinose, A. Wada, H. Mihara, T. Hirayama and H. Aoyagi, *J. Biol. Chem.*, 1997, **272**, 15307.
113. S. H. Lee, S. H. Kim and T. G. Park, *Biochem. Biophys. Res. Commun.*, 2007, **357**, 511.
114. S. W. Choi, S. H. Lee, H. Mok and T. G. Park, *Biotechnol. Prog.*, 2010, **26**, 57.
115. T. Musacchio, O. Vaze, G. D'Souza and V. P. Torchilin, *Bioconjugate Chem.*, 2010, **21**, 1530.
116. S. Matsumoto, R. J. Christie, N. Nishiyama, K. Miyata, A. Ishii, M. Oba, H. Koyama, Y. Yamasaki and K. Kataoka, *Biomacromolecules*, 2009, **10**, 119.
117. R. J. Christie, K. Miyata, Y. Matsumoto, T. Nomoto, D. Menasco, T. C. Lai, M. Pennisi, K. Osada, S. Fukushima, N. Nishiyama, Y. Yamasaki and K. Kataoka, *Biomacromolecules*, 2011, **12**, 3174.
118. M. Mokotoff, Y. M. Mocarski, B. L. Gentsch, M. R. Miller, J. H. Zhou, J. Chen and E. D. Ball, *J. Pept. Res.*, 2001, **57**, 383.
119. R. Singh, L. Kats, W. A. Blattler and J. M. Lambert, *Anal. Biochem.*, 1996, **236**, 114.
120. B. Spankuch, S. Heim, E. Kurunci-Csacsko, C. Lindenau, J. P. Yuan, M. Kaufmann and K. Strebhardt, *Cancer Res.*, 2006, **66**, 5836.
121. B. Spankuch, E. Kurunci-Csacsko, M. Kaufmann and K. Strebhardt, *Oncogene*, 2007, **26**, 5793.
122. J. A. MacDiarmid, N. B. Amaro-Mugridge, J. Madrid-Weiss, I. Sedliarou, S. Wetzel, K. Kochar, V. N. Brahmbhatt, L. Phillips, S. T. Pattison, C. Petti, B. Stillman, R. M. Graham and H. Brahmbhatt, *Nat. Biotechnol.*, 2009, **27**, 643.
123. T. M. Sun, J. Z. Du, Y. D. Yao, C. Q. Mao, S. Dou, S. Y. Huang, P. Z. Zhang, K. W. Leong, E. W. Song and J. Wang, *ACS Nano*, 2011, **5**, 1483.
124. H. Y. Huang, W. T. Kuo, M. J. Chou and Y. Y. Huang, *J. Biomed. Mater. Res., A*, 2011, **97**, 330.
125. C. H. Zhu, S. Jung, S. B. Luo, F. H. Meng, X. L. Zhu, T. G. Park and Z. Y. Zhong, *Biomaterials*, 2010, **31**, 2408.
126. N. Cao, D. Cheng, S. Y. Zou, H. Ai, J. M. Gao and X. T. Shuai, *Biomaterials*, 2011, **32**, 2222.
127. X. B. Xiong and A. Lavasanifar, *ACS Nano*, 2011, **5**, 5202.
128. D. Bumcrot, M. Manoharan, V. Koteliansky and D. W. Y. Sah, *Nat. Chem. Biol.*, 2006, **2**, 711.

CHAPTER 8

Polysaccharide/Polynucleotide Complexes for Cell-Specific DNA Delivery

SHINICHI MOCHIZUKI[a] AND KAZUO SAKURAI*[a,b]

[a] Department of Chemistry and Biochemistry, The University of Kitakyushu, 1-1, Hibikino, Wakamatsu-ku, Kitakyushu, Fukuoka, 808-0135, Japan; [b] CREST, Japan Science and Technology Agency, 4-1-8, Honcho, Kawaguchi-shi, Saitama, 332-0012, Japan
*E-mail: sakurai@kitakyu-u.ac.jp

8.1 Introduction

Recent studies have shown that synthetic oligonucleotides, including antisense, CpG DNAs, and siRNA, are useful in the treatment of various incurable diseases. The first generation of antisense oligonucleotides has reached clinical testing[1–4] in various sites, while many groups propose the second generation with higher binding affinities, greater stability, and lower toxicity as clinical candidates.[5–7] However, there are a number of problems to overcome for *in vivo* use, such as rapid excretion *via* the kidneys, degradation in serum, uptake by phagocytes of the reticuloendothelial system, and inefficient endocytosis by target cells. In order to put these therapeutic oligonucleotides to practical use, the development of an efficient drug currier is deemed inevitable.[8–10] A variety of supramolecular nanocarriers, including liposomes,[11] cationic polymer complexes,[12,13] and various polymeric nanoparticles,[14] have been used to deliver antisense and siRNA oligonucleotides.[3,15–20] Complexation of oligonucleotides

RSC Polymer Chemistry Series No. 3
Functional Polymers for Nanomedicine
Edited by Youqing Shen
© The Royal Society of Chemistry 2013
Published by the Royal Society of Chemistry, www.rsc.org

with polycations is a common approach for intracellular delivery; this includes PEGylated polycations,[21] polyethylenimine (PEI) complexes,[22,23] cationic block copolymers,[24] and dendrimers.[25–27] However, the large size and/or considerable toxicity[28,29] of cationic lipid particles and cationic polymers may cause problems for in vivo utilization.

Some members of a family of natural polysaccharides, called β-1,3-glucans, are approved for clinical use in Japan and have been used for the treatment of uterine cancer.[30] It has been demonstrated that β-1,3-glucans can form novel complexes with homo-oligodeoxynucleotides (homo-ODNs) *via* a combination of hydrogen bonding and hydrophobic interactions (Figure 8.1).[31–33] The complexation is accomplished in a rather peculiar manner. Thus the β-1,3-glucans adopt a triple helix in neutral water and the helix dissociates into single chains in dimethyl sulfoxide or alkaline solution (>0.25 N $NaOH_{aq}$).[34] When the solvent of the single chains is substituted with neutral water, the original triple helix is regenerated.[35] When a particular polynucleotide such as poly(C) or poly(dA) is present in this process, two main-chain glucoses of β-1,3-glucans and one ODN base form a stoichiometric complex instead of retrieving the original triple helix. The complexation not only provides scientifically interesting issues but also a new tool to deliver biologically functional polynucleotides to specific targets. Among the β-1,3-glucans, we have focused on schizophyllan (SPG; Figure 8.1), because SPG in a medically approved grade is available in Japan and its molecular characteristics are well understood.[30]

The fungi containing β-1,3-glucans have been used in Chinese herbal medicine for a long time[36] and, among others, SPG and lentinan[37] have been commercially distributed as medicines for various cancers and health foods for

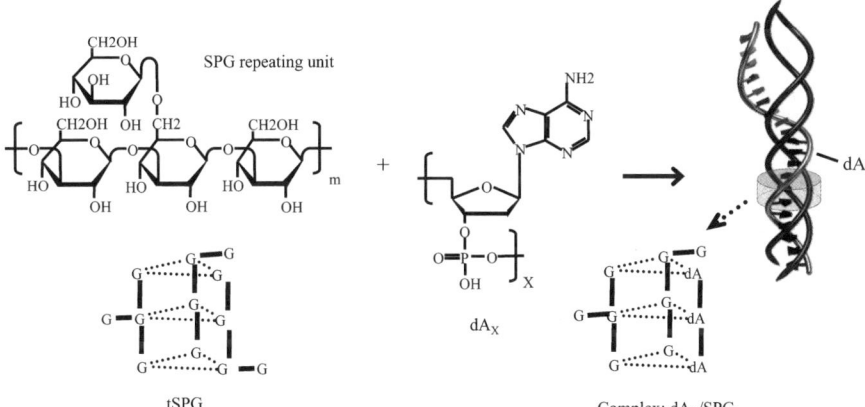

Figure 8.1 Repeating unit of schizophyllan (SPG) and poly(deoxyadenylic acid) (dA_x) and an illustration of their complexation and stoichiometric structures of the triple helix of SPG and the complex, denoted tSPG and dA_x/SPG.

activating gut immunity in Japan. Recent work has revealed that the immunostimulatory nature of β-glucans is related to the effector functions of leukocytes as well as the inflammatory processes. Recently, dectin-1 was identified as a major receptor involved in the recognition of β-glucans.[38,39] Dectin-1 was originally found as a dendritic cell (DC)-specific receptor and thus it was named as "dendritic-cell-associated C-type lectin-1". However, it was revealed that dectin-1 is expressed by many other antigen presenting cells (APCs), including macrophages, monocytes, neutrophils, and a subset of T cells.[40] The APC binding ability of β-1,3-glucans implies that the glucans can specifically deliver the bound oligonucleotides to APCs.

We are investigating the fundamental properties of this novel polysaccharide/polynucleotide complex as well as developing a new method for DNA delivery using this complex. The present chapter reviews the characterization of the complex, and its application to delivery of CpG and antisense oligonucleotides to the APCs *in vitro* and *in vivo*.

8.2   Characterization of the SPG/DNA Complex

8.2.1   Preparation of the SPG/DNA Complex

SPG ($M_w = 1.5 \times 10^5$ as the single chain, determined using gel-permeation chromatography coupled to multi-angle light scattering analysis) was dissolved in 0.25 N $NaOH_{aq}$ for more than 2 days to completely dissociate the triple helix to single chains. All DNA samples were purified with high-performance liquid chromatography. In this work, we used not only a phosphodiester but also a phosphorothioate backbone. Among various nucleotide analogs, phosphorothioate, in which one oxygen atom of the phosphodiester moiety is replaced by sulfur, have improved nuclease resistance and cellular uptake.[6,41] The SPG solution, ODN (dA_{60}) in water, and phosphate buffer solution (330 mM NaH_2PO_4, pH = 4.7) were mixed. After mixing, the pH was controlled around 7 and the mixture was stored at 4 °C overnight.

8.2.2   Solution Properties and Characterization

Figure 8.2A shows size exclusion chromatography (SEC) chromatograms for triple-helix SPG (tSPG), dA_{60}/SPG, and dA_{60} detected with a light scattering (LS) photodiode at $2\theta = 90°$ (upper left), the refractive index (RI) refractometer (middle left), and the UV spectrometer at $\lambda = 260$ nm (bottom right).[42] The weight-averaged molar mass was determined for each fraction and plotted against the elution time (upper right). Since tSPG has no absorption at 260 nm, whereas dA has, UV chromatograms were only observed for dA_{60}/SPG and dA_{60}. For the main peaks of dA_{60}/SPG in the RI and UV chromatograms, the peak-top positions and their shapes were very similar but not identical: the RI peak was eluted more forward (*i.e.*, higher molecular weight) than the UV peak and showed tailing to the low molar-mass

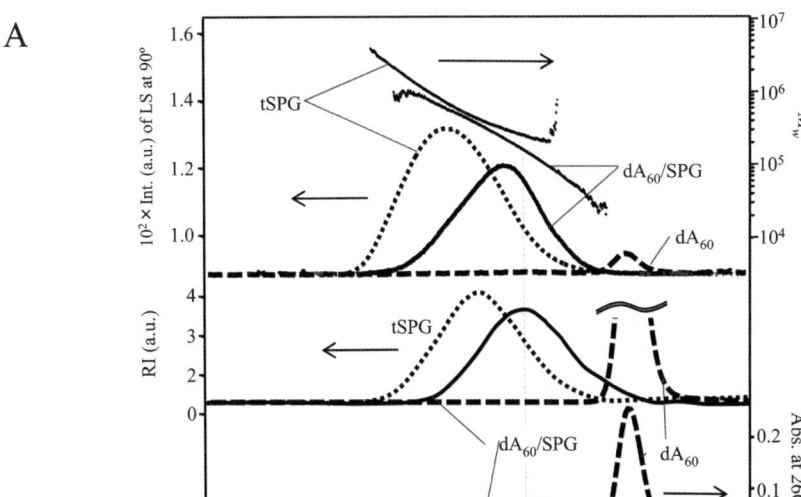

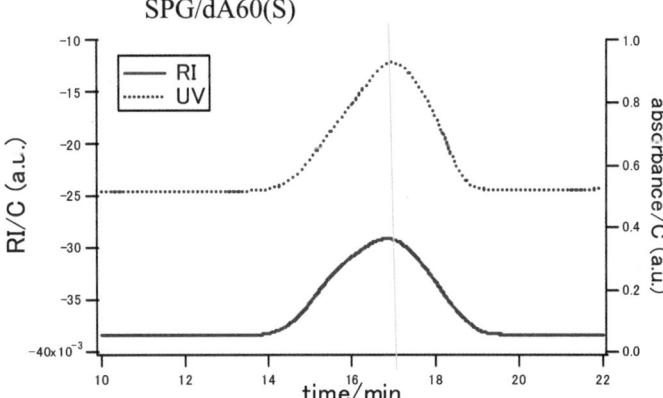

Figure 8.2 (A) SEC chromatograms for tSPG ($M_w = 4.5 \times 10^5$), dA_{60}, and dA_{60}/SPG prepared at the stoichiometric composition. From the top, the weight-averaged molecular weight, the scattering intensity at 90°, the refractive index signal, and the UV absorbance at 260 nm are shown. The UV chromatogram for dA_{60}/SPG was deconvoluted into two Gaussians to represent the complex and free DNA, and each deconvoluted line is shown as a *dotted line* (indicated by *arrow*). (B) Comparison of the RI and UV chromatograms for S-dA_{60}/SPG. The stoichiometric sample had no free S-dA_{60} and the complexation yield was almost 100%. The peak-top appeared at 17 min, becoming more forward than dA_{60}/SPG (18 min) and close to tSPG (16.2 min).

side. This fact indicates that uncomplexed SPG was mixed with the dA_{60}/SPG complex in the main peak. The UV showed a small secondary peak at 20 min with the same position as dA60 alone, indicating the presence of uncomplexed dA_{60}. When this bimodal peak was deconvoluted with two Gaussians representing dA_{60}/SPG and dA_{60} (dotted lines in the UV chart of Figure 8.2A), the complexation yield was estimated to be 96%. When the sample was prepared at a dA_{60} excess composition of 3:1, the UV and RI became less different, while the amount of uncomplexed dA was increased along with the composition (data not shown). This result can be interpreted as the main peak contained less free SPG and the composition was almost uniform in the main peak. Therefore, we used the dA excess composition for further analysis for the all phosphodiester complexes. When we carried out the same experiment for S-dA_{60}/SPG, we observed that the stoichiometric sample had no free S-dA_{60} and the complexation yield was almost 100% (Figure 8.2B). The peak top appeared at 17 min, becoming more forward than dA_{60}/SPG (18 min) and close to tSPG (16.2 min), indicating that S-dA_{60}/SPG shows a higher complexation yield than dA_{60}/SPG.

8.2.3 Thermal Stability of the Complexes

Since the above result can be interpreted by the S-dA_{60} having a higher affinity than dA_{60}, we examined the composition dependence of the phosphorothioated backbone. We prepared six dA_{40} samples with different phosphorothioate (PS) contents: 0/39 PS, 8/39 PS, 13/39 PS, 20/39 PS, 26/39 PS, and 39/39 PS (see Table 8.1 for the substitution sites), and made the complexes with SPG.[43] Figure 8.3A shows the CD spectra in the range 240–320 nm for all PS and phosphodiester (PO) dA_{40} samples at 20 °C. The overall spectral shapes were almost same for all samples, although with increasing the PS content the negative peak at 248 nm slightly increased in intensity. Since this wavelength range is related to the induced CD between the base moieties, the similarity in CD indicates that all dA_{40} samples take the same conformation. Upon complexation, the CD spectra are drastically changed, as presented in Figure 8.3B, where for convenience the naked and complexed PO-dA_{40} are compared. This means that dA adopts a new conformation to form the SPG

Table 8.1 Sample codes and the sequences.

Sample code	DNA sequence[a]
0/39	A_{40}
8/39	$(AsAAAA)_8$
13/39	$(AsAA)_{13}A$
20/39	$(AsA)_{20}$
26/39	$(AsAsA)_{13}A$
39/39	$(As)_{39}A$

[a]The lower-case letter "s" indicates the phosphorothioate (PS) linkage.

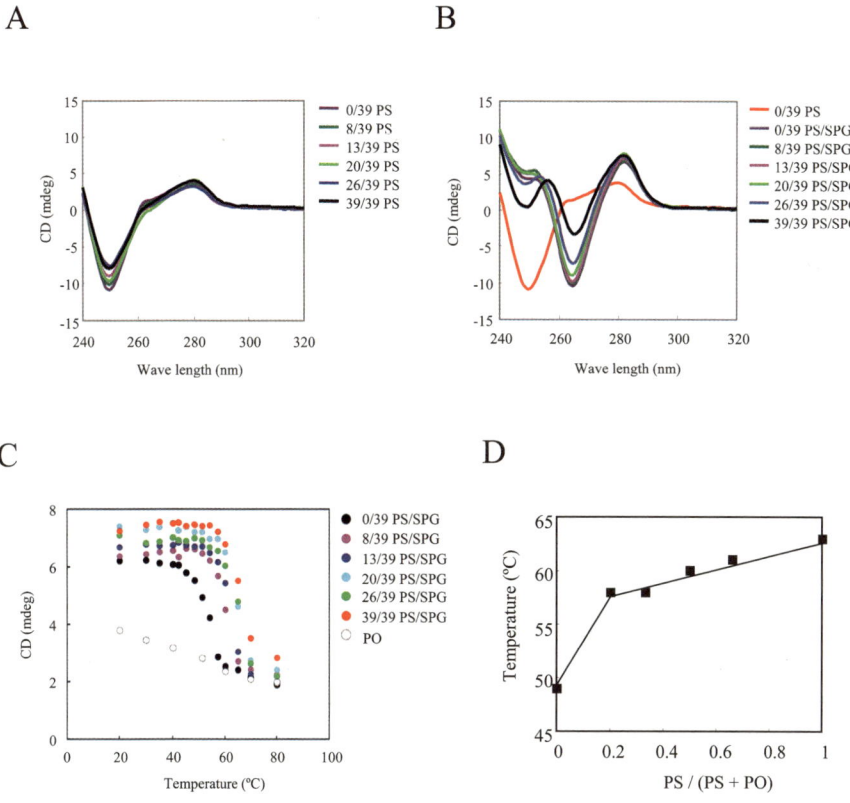

Figure 8.3 PS content dependence of the CD spectrum of the naked (A) and complexed (B) dA$_{40}$ measured at 20 °C. (C) Temperature dependence of the CD intensity at 282 nm for the complex. (D) T_m for dA$_{40}$ with different PS contents.

complex, as reported previously.[31,44] When compared among the complexed spectra, the four complexes from 0/39 PS to 20/39 PS were almost identical. With increasing the S content to 26/39 and 39/39, they showed different patterns from the others. For 39/39 PS, the CD value at 248 nm decreased and that of 264 nm increased. These results agree qualitatively with the above-mentioned speculation that the PS can provide a more preferable conformation for the complexation than PO. However, quantitatively, there is a discrepancy in the CD and PAGE results, where 33% is enough to reach 100% yield while more than 66% is necessary to change the CD.

Figure 8.3C plots the CD intensity at 282 nm against the temperature. As mentioned above, at low temperatures the CD intensities for the complexes are larger than the naked ones. Within a certain temperature range, the intensities for the complexes decreased and merged into those for the corresponding dA$_{40}$ samples. From this transition the dissociation (or melting) temperature, T_m, is plotted in Figure 8.3D. With increasing the PS content from 0/39 to 8/39 PS,

T_m increased from 49 to 59 °C. Further increase of the PS content did not increase T_m. These results confirm that the PS backbone modification induces a more stable complex than the PO backbone, while only 20 mol% modification should be enough to give the same increment in T_m as the 100% one. Our results demonstrate that the complex with the phosphorothioate dA_{40} was thermally more stable than phosphodiester dA_{40}, and the stability increased to the same level as the fully phosphorothioated one by only replacing 20–30 mol% of the oxygen of the phosphodiester moiety.

8.3   Application of the Complex to ODN Delivery

8.3.1   Uptake of the Complex by Macrophages

We used RT-PCR to analyze the expression of dectin-1 mRNA in thioglycolate-elicited macrophages (thio-Mφ; Figure 8.4A).[45] The expression doubled after 24 h and stayed at this level to 48 h (data not shown), which is consistent with the drastic increase at the protein level for the 24 h pre-cultured cells measured with a flow cytometer (Figure 8.4B). For Figure 8.4A, the expression at 0 h is due to elicitation by thioglycolate *in vivo*, and the further increment can be ascribed to additional stimulation generated by cellular adhesion to the culture plate.[46] The expression of CD11b, a macrophage-specific antigen, was unaffected by 24 h culture (data not shown), indicating that the macrophage's essential nature was unaltered during the pre-culturing. We examined the uptake of ODN/SPG by 3-, 24-, and 48-h thio-Mφ (Figure 8.4C). Even some naked ODN was taken up, regardless of the dectin-1 expression level. This result can be interpreted in the knowledge that phosphorothioate ODN can enter cells *via* an unidentified pathway.[47] The uptake of ODN/SPG increased from 3-h to 24-h thio-Mφ. This increment can be ascribed to dectin-1-mediated uptake, because dectin-1 expression at 24 and 48 h was more enhanced than at 3 h, as mentioned above. After 24 h culture we treated thio-Mφs with TAMRA-labeled ODN or its complex with FITC-labeled SPG, and examined the distributions of these two markers by fluorescence microscopy (Figure 8.4D). The naked ODN gave no fluorescence, while the complex gave strong fluorescence with both FITC and TAMRA, appearing as small dots scattered and sprinkled over the cells. This distribution suggests that they are localized inside vesicles, most likely endocytosis vesicles. Superimposition of these two images showed that ODN and SPG were co-localized in the cells. These results demonstrate that SPG can act a pilot molecule to transport the bound ODN to APCs by means of dectin-1.

8.3.2   IL-12 Secretion Due to Administration of CpG-ODN/SPG Complexes

Peritoneal macrophages express TLR9 to induce CpG DNA-mediated IL-12.[49] Accordingly, the complexes with various compositions were applied to the

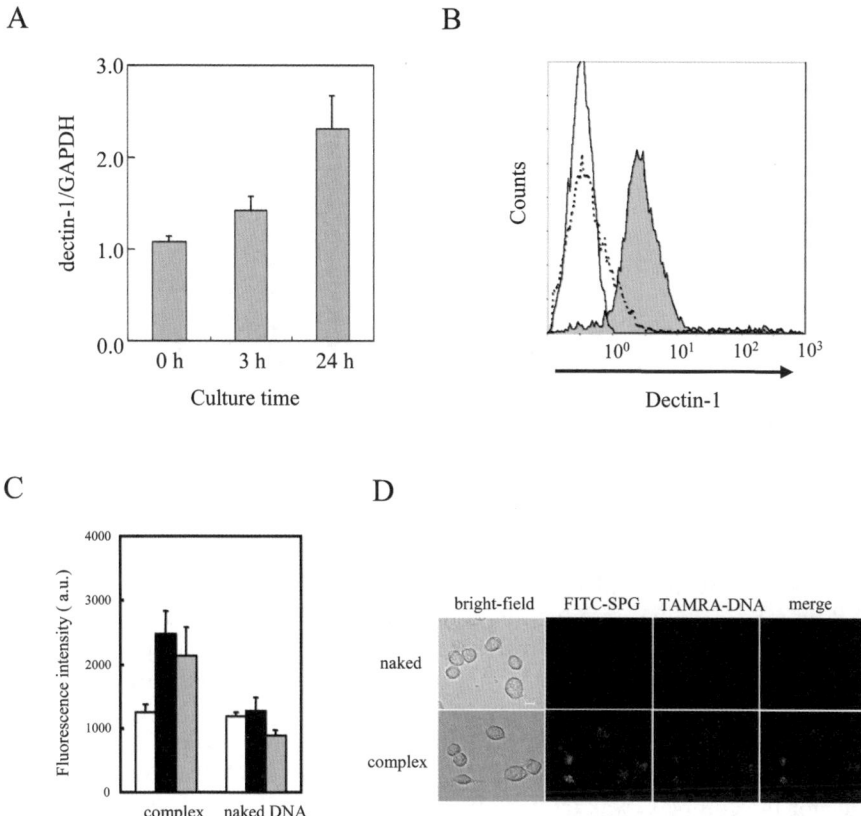

Figure 8.4 Dectin-1 expression on thio-Mɸs and cellular uptake of DNA/SPG complexes due to dectin-1. Pre-culturing time dependence of the amount of mRNA expression expressed in thio-Mɸs. After seeding of 1.0×10^6 cells/well in six-well plates, the levels of dectin-1 and GAPDH were assayed by RT-PCR. Data represent the mean $\pm$ SD of three experiments. (B) Flow cytometry analysis profiles for 0 h (*dashed line*) and 24 h pre-cultured thio-Mɸs (*shaded in gray*) after anti-dectin-1 antibody staining. The *solid line* histogram shows unstained control. (C) Comparison of the FITC fluorescence intensity from the cell lysates, after FITC-ODN was applied to the different pre-culturing cells: 3 h (*white*), 24 h (*black*), or 48 h (*gray*) with or without complexation. Data represent the mean $\pm$ SD of triplicate wells. (D) Fluorescence images for FITC and TAMRA after TAMRA-ODN was applied with or without complexation with FITC-SPG to 48 h pre-cultured thio-Mɸs. The bar represents 10 µm.

thio-Mɸs, and the amount of secreted IL-12 was measured with ELISA (Figure 8.5).[48] The naked CpG-dA$_{60}$ produced low levels of IL-12 (0.8 ng mL^{-1}) in macrophages, and this value was consistent with other results.[49] In the range of [mG]/[dA] = 0−1.0, the production slightly increased with an increase of the amount of SPG, and dramatically enhanced at [mG]/

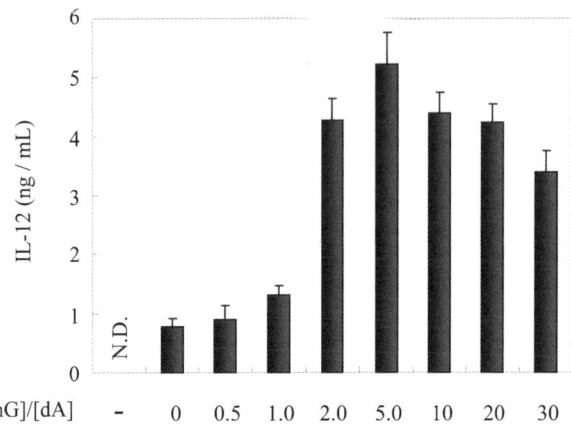

Figure 8.5 The complex composition dependence (CpG; 0.5 μM) of IL-12 production from the thio-Mφs after incubation for 24 h ($n = 3$); N.D. = not detectable.

[dA] = 2. At [mG]/[dA] = 5, the complex induced about seven-fold (5.2 ng mL^{-1}) higher production than that of naked CpG-dA60, and the production decreased with further increase of the amount of SPG. Interestingly, between [mG]/[dA] = 1 and 5 there was a significant difference in the IL-12 production from cells, while the amount of CpG-dA60 taken up by the cells differed only slightly among them (data not shown). In addition, this drastic increase did not occur when the mixture of CpG-dA60 and SPG, but not the complex, was treated. When multiple CpG DNAs are conjugated on one substrate and this substrate is taken up by macrophages, the cytokine production is dramatically (or allosterically) induced.[50–53] This is explained by the observation that the multiple DNAs increase local concentrations of CpG sequences in endolysosomes and, subsequently, TLR9 recognition of CpG DNA and further signals are enhanced. The allostericity is probably related to the fact that dimerization of TLR9 is necessary for its activation.[54] In addition, our recent study has shown that the flexibility of the polymer containing CpG DNA is a key to induce the large amount of IL-12.[55] Once one CpG branch binds to one TLR9 dimer, the polymer is locked adjacently to the vesicle surface where other TLR9 dimers are present. When the structure is flexible enough to turn, the second binding can easily occur, and the adjacent second binding seems to be critical for the allosteric effect. At [mG]/[dA] < 2 in Figure 8.5, the complex is considered to have the maximum number of CpG-dA$_{60}$ due to the DNA-rich composition. On the other hand, at [mG]/[dA] > 2, most of the complexes have fewer CpG-dA$_{60}$ than at [mG]/[dA] < 2 due to the SPG-rich composition. With increasing [mG]/[dA], the flexibility of the complex is considered to increase. The complex with a rigid structure could only access the adjacent TLR9 dimers with difficulty, while the complex with a flexible structure could access them easily. Therefore, the cytokine secretion drastically increased at [mG]/[dA] > 2 because of the easy accessibility of the complex to TLR9 dimers. These results suggested the existence of a particular higher-order structure to activate TLR9 more efficiently. The present work shows a novel strategy to deliver CpG DNA to immune cells by the use of β-1,3-glucan/DNA complexes.

8.3.3 LPS-Induced TNF-α Suppression by the AS-ODN/SPG Complex *in vitro* and *in vivo*

Lipopolysaccharide (LPS) stimulation leads thio-Mφ to secrete a large amount of TNF-α.[56] We evaluated how complexed AS-ODN can silence TNF-α secretion (Figure 8.6A–C).[45] The complex suppressed TNF-α secretion in a dose-dependent manner, while the naked dose did not cause any suppression (Figure 8.6A). The suppression was appreciable even at 10 nM, lower than the dose usually reported for antisense assays *in vitro*.[57] As the dA length increased from 30 to 60, the complex suppressed TNF-α more, but the naked AS-ODN did not (Figure 8.6B). This relation can be related to the fact that the complexation yield increased with the tail length (data not shown). Only the antisense sequence reduced the secretion (Figure 8.6C), showing that the sequence specificity was maintained after the complexation. The dectin-1-mediated host defense response to a fungus such as *C. albicans* or *S. cerevisiae* is reported to induce TNF-α production.[58] Since SPG is also produced by the fungus *Schizophyllum commune*, we suspected a similar effect. However, the dose we used did not induce any such activity due to SPG itself (Figure 8.6C), suggesting that SPG itself is inert.

We examined whether AS-ODN/SPG could cure LPS/D-GalN-induced hepatitis. In mice, 600 mg kg^{-1} D-GalN + 10 μg kg^{-1} LPS gave 80% mortality within 12 h (Figure 8.7A). Treatment with 5 mg kg^{-1} of the complex gave 80% survival, but the naked dose gave no improvement. Even 0.5 mg kg^{-1} of the complex gave 60% survival. This dose level is much lower (1/20 to 1/100) than reported values.[59,60] The complex significantly reduced the serum levels of TNF-α at 1 h compared with that treated with PBS (Figure 8.7B). Histological analysis of the livers from mice treated with LPS/D-GalN showed serious damage, death of a large amount of cells, and severe hemorrhage (Figure 8.7C). The naked dose did not give any improvement. In contrast, the livers of mice treated with the complex did not show any of these abnormalities. These histological results are consistent with the survival test and the reduction of the serum TNF-α.

8.3.4 A New Therapy for Inflammatory Bowel Disease Using Antisense Macrophage-Migration Inhibitory Factor

The pathogenesis role of macrophage-migration inhibitory factor (MIF) has been shown in inflammatory bowel disease (IBD), including animal models.[62] Indeed, MIF, which was mainly produced from CD11b$^+$ macrophages, significantly increased in dextran sodium sulfate (DSS)-treated mice. MIF is

A

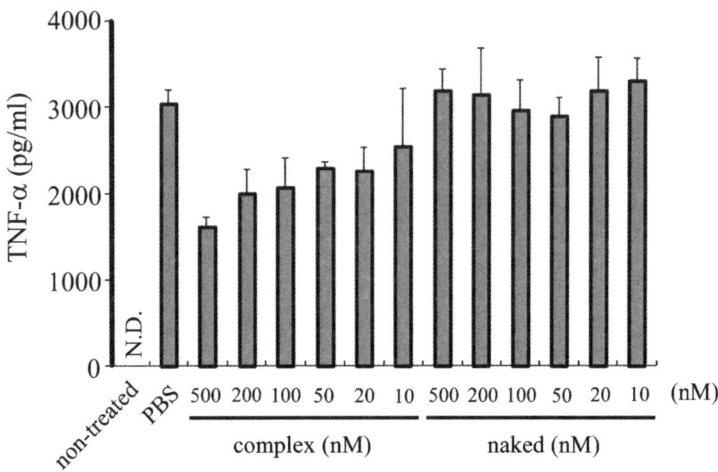

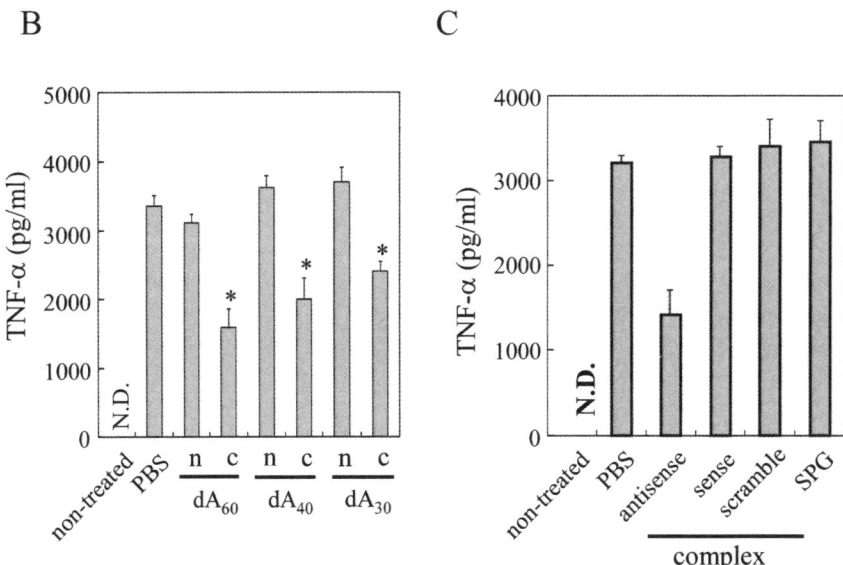

Figure 8.6   Silencing TNF-α secretion determined with ELISA due to AS-ODN administration to LPS-stimulated 48 h pre-cultured thio-Mφs for different dose conditions. (A) AS-ODN dose dependence of the silencing, after incubation for 30 min at 37 °C. (B) dA tail length dependence of the silencing for the complexed (c) and naked (n) AS-ODN (ASODN; 0.5 μM); *$P < 0.01$ *versus* the cells in PBS. (C) Comparison of AS-ODN/ SPG, sense-ODN/SPG, and scrambled-ODN/SPG (AS-ODN, 0.5 μM); N.D. = not detected. Data represent the mean ± SD of triplicate wells.

A

B

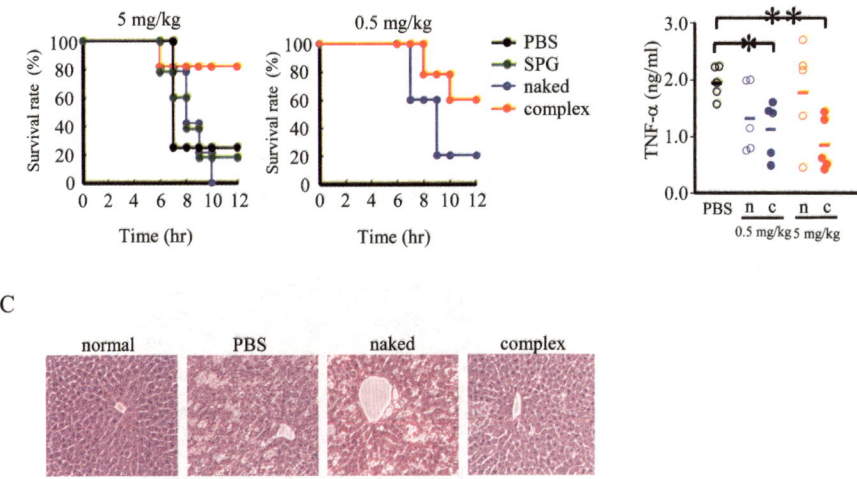

C

Figure 8.7 AS-ODN/SPG-mediated TNF-α inhibition on the progression of LPS/D-GalN-induced fulminant hepatitis. (A) Survival rate of the mice treated i.p. with PBS, SPG, AS-ODN, or AS-ODN/SPG before hepatitis induction. The animals were protected with the preceding treatment; at 24 h later they were i.p. co-injected with LPS/D-GalN and SPG, AS-ODN, or AS-ODN/SPG, and the survival rate to 12 h was obtained ($n =$ 5). (B) TNF-α level in the serum was measured at 1 h after LPS/D-GalN administration. The blood was obtained by retro-orbital bleeding ($n = 5$). n and c represent naked AS-ODN and complexed AS-ODN/SPG administration, respectively; $*P < 0.05$, $**P < 0.01$. (C) A representative liver histology of LPS/D-GalN-induced hepatitis in mice injected with PBS, AS-ODN, or AS-ODN/SPG (5 mg kg^{-1}). At 6 h after LPS/D-GalN injection the livers were stained with hematoxylin and eosin (H&E). Original magnification $\times 200$.

now known to be more abundantly expressed by macrophages themselves, although it was first reported as a T lymphokine acting on macrophages. MIF is constitutively expressed in macrophages, with MIF protein preformed in cytoplasmic stores.[63] The experimental approach to antagonize MIF actions was usage of neutralizing anti-MIF antibodies. This has proven therapeutically effective in several models of autoimmune disease, such as IBD,[64] experimental autoimmune encephalomyelitis (EAE),[65] and arthritis.[66] Antibody-based anti-MIF drugs have significant associated risks and limitations, including their potential immunogenicity, their short half-life *in vivo*, possible side effects, and the high cost of their application. Therefore, if there is a nucleotide base therapy such as antisense DNA and siRNA available, it would be a great advantage.

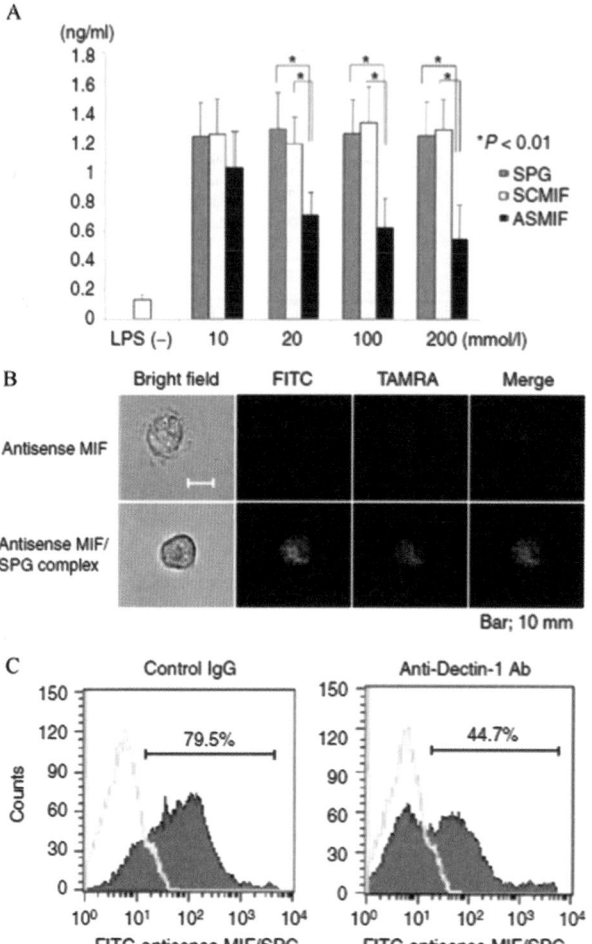

Figure 8.8   The antisense MIF/SPG complex inhibited MIF production induced by LPS *in vitro*. (A) CD11b⁺ cells from lamina propria (LP) and mesenteric lymph node (MLN) were cultured with several concentrations of antisense MIF/SPG complex (ASMIF), scramble control DNA/SPG complex (SCMIF), and SPG as a control. After 10 h, 1 μg mL⁻¹ of LPS was added under each condition and the cells were cultured for 24 h. MIF expressions were measured using an enzyme-linked immunosorbent assay (ELISA) ($n = 5$ per group). The data presented are the means of the cytokine concentration $\pm$ SD. (B) Immunofluorescence in CD11b⁺ cells was performed by labeling antisense MIF with TAMRA and the SPG with FITC. CD11b⁺ cells took up the antisense MIF/SPG complex effectively compared with the antisense MIF alone. (C) Neutralizing anti-dectin-1 antibodies inhibited the uptake of FITC-labeled antisense MIF/SPG complex into CD11b⁺ cells compared with rat IgG2b. The data are presented as the mean of three independent experiments.

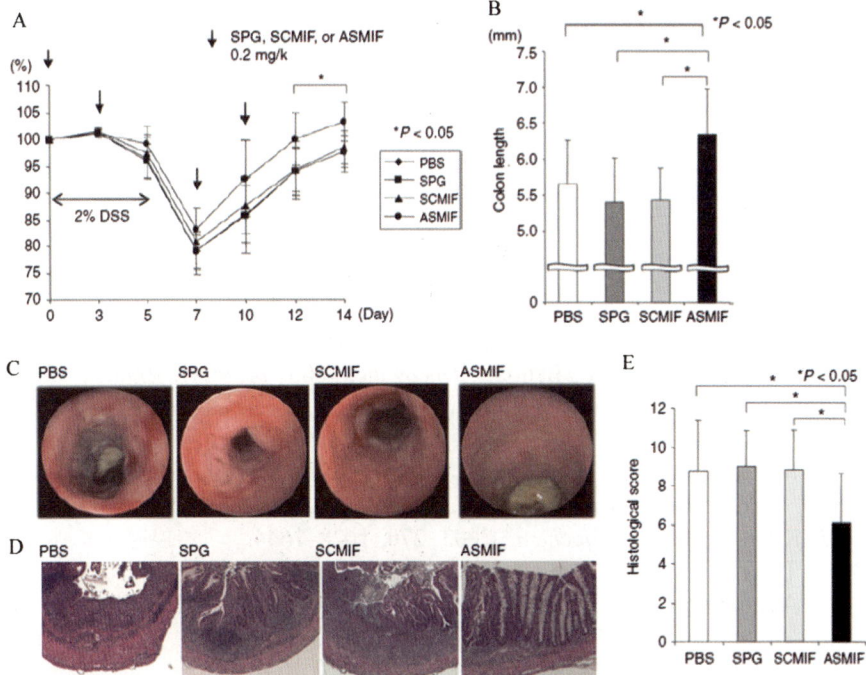

Figure 8.9 Attenuation of DSS-induced colitis by the administration of the antisense MIF/SPG complex. A total of 0.2 mg kg^{-1} of the antisense MIF/SPG complex (ASMIF), scramble control DNA/SPG complex (SCMIF), SPG, or PBS as a control were injected intraperitoneally (i.p.) twice weekly into mice receiving DSS ($n = 8$ per group). (A) Body weights as a percentage of the initial weight on day 0 are shown. (B) Colon length from the terminal ileum to the rectum. (C) Endoscopic findings for the colon. (D) H&E staining of the colon (original magnification ×100). (E) The histological scores were evaluated.

We examined the biological function and the therapeutic effect of SPG as a new delivery system formed with antisense MIF against colitis. MIF production *in vitro* was significantly suppressed by administering the antisense MIF/SPG complex, which was taken into the CD11b^{+} macrophage more effectively than antisense MIF alone (Figure 8.8). Administration of the antisense MIF/SPG complex safely and effectively ameliorated the inflammation of the colon in the DSS-induced colitis model (Figure 8.9). Our results demonstrated the possibility of a new therapeutic approach, namely administration of the antisense MIF/SPG complex, for inflammatory bowel disease.[61]

8.4 Conclusion

In this review, we have demonstrated that β-1,3-glucans have a novel feature to complex with polynucleotides, distinguishing them from other polysaccharides.

The resultant complexes have well-defined helical superstructures composed of two polysaccharide strands and one polynucleotide strand. This unique property of β-1,3-glucans has made it possible to utilize these polysaccharides as oligo-DNA carriers. Especially, it was possible to deliver the polynucleotides into the APCs selectively without using modified SPG because of the presence of a β-1,3-receptor, dectin-1, on these cells. These findings indicate that β-1,3-glucans are very attractive and useful materials in biotechnology.

References

1. R. Nahta and F. J. Esteva, *Semin. Oncol.*, 2003, **30**, 143–149.
2. N. M. Dean and C. F. Bennett, *Oncogene*, 2003, **22**, 9087–9096.
3. J. H. Chan, S. Lim and W. S. Wong, *Clin. Exp. Pharmacol. Physiol.*, 2006, **33**, 533–540.
4. F. M. Coppelli and J. R. Grandis, *Curr. Pharm. Des.*, 2005, **11**, 2825–2840.
5. M. Manoharan, *Antisense Nucleic Acid Drug Dev.*, 2002, **12**, 103–128.
6. J. Kurreck, *Eur. J. Biochem.*, 2003, **270**, 1628–1644.
7. S. T. Crooke, *Annu. Rev. Med.*, 2004, **55**, 61–95.
8. S. Akhtar and I. F. Benter, *J. Clin. Invest.*, 2007, **117**, 3623–3632.
9. A. Mescalchin, A. Detzer, M. Wecke, M. Overhoff, W. Wunsche and G. Sczakiel, *Expert Opin. Biol. Ther.*, 2007, **7**, 1531–1538.
10. V. Russ and E. Wagner, *Pharm. Res.*, 2007, **24**, 1047–1057.
11. D. C. Drummond, O. Meyer, K. Hong, D. B. Kirpotin and D. Papahadjopoulos, *Pharmacol. Rev.*, 1999, **51**, 691–743.
12. T. Suma, K. Miyata, T. Ishii, S. Uchida, H. Uchida, K. Itaka, N. Nishiyama and K. Kataoka, *Biomaterials*, 2012, **33**, 2770–2779.
13. H. J. Kim, A. Ishii, K. Miyata, Y. Lee, S. Wu, M. Oba, N. Nishiyama and K. Kataoka, *J. Controlled Release*, 2010, **145**, 141–148.
14. E. M. Pridgen, R. Langer and O. C. Farokhzad, *Nanomedicine (London, U. K.)*, 2007, **2**, 669–680.
15. A. Inoue, S. Y. Sawata and K. Taira, *J. Drug Targeting*, 2006, **14**, 448–455.
16. D. De Paula, M. V. Bentley and R. I. Mahato, *RNA*, 2007, **13**, 431–456.
17. I. R. Gilmore, S. P. Fox, A. J. Hollins and S. Akhtar, *Curr. Drug Delivery*, 2006, **3**, 147–145.
18. W. Li and F. C. Szoka, Jr., *Pharm. Res.*, 2007, **24**, 438–449.
19. E. Fattal, P. Couvreur and C. Dubernet, *Adv. Drug Delivery Rev.*, 2004, **56**, 931–946.
20. S. V. Vinogradov, E. V. Batrakova and A. V. Kabanov, *Bioconjugate Chem.*, 2004, **15**, 50–60.
21. M. Meyer, A. Philipp, R. Oskuee, C. Schmidt and E. Wagner, *J. Am. Chem. Soc.*, 2008, **130**, 3272–3273.
22. M. Grzelinski, B. Urban-Klein, T. Martens, K. Lamszus, U. Bakowsky, S. Hobel, F. Czubayko and A. Aigner, *Hum. Gene Ther.*, 2006, **17**, 751–766.

23. R. M. Schiffelers, A. Ansari, J. Xu, Q. Zhou, Q. Tang, G. Storm, G. Molema, P. Y. Lu, P. V. Scaria and M. C. Woodle, *Nucleic Acids Res.*, 2004, **32**, e149.
24. J. DeRouchey, C. Schmidt, G. F. Walker, C. Koch, C. Plank, E. Wagner and J. O. Radler, *Biomacromolecules*, 2008, **9**, 724–732.
25. H. Kang, R. DeLong, M. H. Fisher and R. L. Juliano, *Pharm. Res.*, 2005, **22**, 2099–2106.
26. T. Tsutsumi, F. Hirayama, K. Uekama and H. Arima, *J. Controlled Release*, 2007, **119**, 349–359.
27. H. Yoo and R. L. Juliano, *Nucleic Acids Res.*, 2000, **28**, 4225–4231.
28. H. Lv, S. Zhang, B. Wang, S. Cui and J. Yan, *J. Controlled Release*, 2006, **114**, 100–109.
29. S. Akhtar and I. Benter, *Adv. Drug Delivery Rev.*, 2007, **59**, 164–182.
30. K. Noda, S. Takeuchi, A. Yajima, K. Akiya, T. Kasamatsu, Y. Tomoda, M. Ozawa, K. Sekiba, H. Sugimori, S. Hashimoto, *et al.*, *Jpn. J. Clin. Oncol.*, 1992, **22**, 17–25.
31. K. Sakurai and S. Shinkai, *J. Am. Chem. Soc.*, 2000, **122**, 4520–4521.
32. K. Miyoshi, K. Uezu, K. Sakurai and S. Shinkai, *Chem. Biodivers.*, 2004, **1**, 916–924.
33. K. Sakurai, M. Mizu and S. Shinkai, *Biomacromolecules*, 2001, **2**, 641–650.
34. T. Norisuye, T. Yanaki and H. Fujita, *J. Polym. Sci., Polym. Phys. Ed.*, 1980, **18**, 547–558.
35. T. Sato, T. Norisuye and H. Fujita, *Macromolecules*, 1983, **16**, 185–189.
36. C. Hobbs, Medical Mushrooms, Botanical Press, Santa Cruz, CA, 1991, pp 158–159.
37. G. Chihara, Y. Maeda, J. Hamuro, T. Sasaki and F. Fukuoka, *Nature*, 1969, **222**, 687–688.
38. G. D. Brown and S. Gordon, *Nature*, 2001, **413**, 36–37.
39. G. D. Brown, P. R. Taylor, D. M. Reid, J. A. Willment, D. L. Williams, L. Martinez-Pomares, S. Y. Wong and S. Gordon, *J. Exp. Med.*, 2002, **196**, 407–412.
40. P. R. Taylor, G. D. Brown, D. M. Reid, J. A. Willment, L. Martinez-Pomares, S. Gordon and S. Y. Wong, *J. Immunol.*, 2002, **169**, 3876–3882.
41. A. A. Levin, *Biochim. Biophys. Acta*, 1999, **1489**, 69–84.
42. Y. Sanada, T. Matsuzaki, S. Mochizuki, T. Okobira, K. Uezu and K. Sakurai, *J. Phys. Chem. B*, 2012, **116**, 87–94.
43. S. Mochizuki and K. Sakurai, *Bioorg. Chem.*, 2010, **38**, 260–264.
44. M. Mizu, T. Kimura, K. Koumoto, K. Sakurai and S. Shinkai, *Chem. Commun.*, 2001, 429–430.
45. S. Mochizuki and K. Sakurai, *J. Controlled Release*, 2011, **151**, 155–161.
46. A. Friedman and D. I. Beller, *Immunology*, 1987, **61**, 469–474.
47. A. R. Thierry and A. Dritschilo, *Nucleic Acids Res.*, 1992, **20**, 5691–5698.
48. J. Minari, S. Mochizuki, T. Matsuzaki, Y. Adachi, N. Ohno and K. Sakurai, *Bioconjugate Chem.*, 2011, **22**, 9–15.

49. K. Suzuki, T. Suda, T. Naito, K. Ide, K. Chida and H. Nakamura, *Am. J. Respir. Crit. Care Med.*, 2005, **171**, 707–713.

50. J. D. Marshall, E. M. Hessel, J. Gregorio, C. Abbate, P. Yee, M. Chu, G. Van Nest, R. L. Coffman and K. L. Fearon, *Nucleic Acids Res.*, 2003, **31**, 5122–5133.

51. C. C. Wu, J. Lee, E. Raz, M. Corr and D. A. Carson, *J. Biol. Chem.*, 2004, **279**, 33071–33078.

52. M. Kerkmann, L. T. Costa, C. Richter, S. Rothenfusser, J. Battiany, V. Hornung, J. Johnson, S. Englert, T. Ketterer, W. Heckl, S. Thalhammer, S. Endres and G. Hartmann, *J. Biol. Chem.*, 2005, **280**, 8086–8093.

53. S. Rattanakiat, M. Nishikawa, H. Funabashi, D. Luo and Y. Takakura, *Biomaterials*, 2009, **30**, 5701–5706.

54. E. Latz, A. Verma, A. Visintin, M. Gong, C. M. Sirois, D. C. Klein, B. G. Monks, C. J. McKnight, M. S. Lamphier, W. P. Duprex, T. Espevik and D. T. Golenbock, *Nat. Immunol.*, 2007, **8**, 772–779.

55. J. Minari, S. Mochizuki and K. Sakurai, *Oligonucleotides*, 2008, **18**, 337–344.

56. B. Beutler, N. Krochin, I. W. Milsark, C. Luedke and A. Cerami, *Science*, 1986, **232**, 977–980.

57. E. C. LaCasse, G. G. Cherton-Horvat, K. E. Hewitt, L. J. Jerome, S. J. Morris, E. R. Kandimalla, D. Yu, H. Wang, W. Wang, R. Zhang, S. Agrawal, J. W. Gillard and J. P. Durkin, *Clin. Cancer Res.*, 2006, **12**, 5231–5241.

58. G. D. Brown, J. Herre, D. L. Williams, J. A. Willment, A. S. Marshall and S. Gordon, *J. Exp. Med.*, 2003, **197**, 1119–1124.

59. H. Zhang, J. Cook, J. Nickel, R. Yu, K. Stecker, K. Myers and N. M. Dean, *Nat. Biotechnol.*, 2000, **18**, 862–867.

60. L. Dong, L. Zuo, S. Xia, S. Gao, C. Zhang, J. Chen and J. Zhang, *J. Gene Med.*, 2009, **11**, 229–239.

61. H. Takedatsu, K. Mitsuyama, S. Mochizuki, T. Kobayashi, K. Sakurai, H. Takeda, Y. Fujiyama, Y. Koyama, J. Nishihira and M. Sata, *Mol. Ther.*, 2012, **20**, 1234–1241.

62. Y. P. de Jong, A. C. Abadia-Molina, A. R. Satoskar, K. Clarke, S. T. Rietdijk, W. A. Faubion, E. Mizoguchi, C. N. Metz, M. Alsahli, T. ten Hove, A. C. Keates, J. B. Lubetsky, R. J. Farrell, P. Michetti, S. J. van Deventer, E. Lolis, J. R. David, A. K. Bhan and C. Terhorst, *Nat. Immunol.*, 2001, **2**, 1061–1066.

63. T. Calandra, J. Bernhagen, R. A. Mitchell and R. Bucala, *J. Exp. Med.*, 1994, **179**, 1895–1902.

64. D. de Jong, C. J. Mulder and A. A. van Sorge, *Gut*, 2001, **49**, 874.

65. C. M. Denkinger, M. Denkinger, J. J. Kort, C. Metz and T. G. Forsthuber, *J. Immunol.*, 2003, **170**, 1274–1282.

66. L. Santos, P. Hall, C. Metz, R. Bucala and E. F. Morand, *Clin. Exp. Immunol.*, 2001, **123**, 309–314.

Design of Complex Micelles for Drug Delivery

RUJIANG MA AND LINQI SHI*

Key Laboratory of Functional Polymer Materials, Ministry of Education, and Institute of Polymer Chemistry, Nankai University, Tianjin 300071, P. R. China
*E-mail: shilinqi@nankai.edu.cn

9.1 Introduction

Polymeric micelles have been extensively studied as controlled drug delivery vehicles based on their inherent properties[1] such as tunable size in the nano range, stability against agglomeration, and long circulation time due to the protection by hydrophilic shells of poly(ethylene glycol) (PEG), enhancement of solubility of insoluble drugs, and decrease of toxicity of drugs. More importantly, polymeric micelles can be passively targeted to tumor issue by the enhanced permeability and retention (EPR) effect or actively targeted to cancer cells by conjugating with many ligands such as antibody fragments, epidermal growth factors, folate, *etc.*

Typical polymeric micelles for drug delivery are generally composed of a hydrophobically associated biodegradable core and a soluble PEG shell for biocompatibility. Water-insoluble drugs such as doxorubicin (DOX) and paclitaxel (PTX) can be loaded into the micelle core by hydrophobic interaction. The hydrophilic PEG shell protects the drug-loaded micelle not only from aggregation but also from nonspecific adsorption of proteins in the blood circulation. However, a simple core–shell structure is vulnerable under

RSC Polymer Chemistry Series No. 3
Functional Polymers for Nanomedicine
Edited by Youqing Shen
© The Royal Society of Chemistry 2013
Published by the Royal Society of Chemistry, www.rsc.org

complicated physiological conditions because of premature disintegration of the micelle structure due to enzymatic degradation and the burst release of drug absorbed on the surface of the micelle core. It is desirable to construct polymeric micelle-based drug carriers with superior structures that can enhance their ability to control the drug release profiles. Two kinds of complex micelles, either with a core–shell–corona (CSC) structure or with surface channels, were developed recently for the above purpose. Though a hydrophobic drug can be easily encapsulated in the micelle core during micellization, it is difficult to encapsulate hydrophilic or ionic drugs in hydrophobically associated micelles due to the lack of strong interaction. Polyion complex (PIC) micelles were developed for the incorporation of hydrophilic or ionic drugs and a series of studies on drug delivery by PIC micelles has been reported. This chapter will discuss recent advances in drug delivery by complex micelles.

9.2   Core–Shell–Corona Micelles for Drug Delivery

Core–shell–corona micelles have an onion structure and can be obtained by self-assembly of a triblock copolymer or two diblock copolymers. Gohy *et al.*[2] reported the formation of three-layer micelles by the triblock copolymer polystyrene-*block*-poly(2-vinylpyridine)-*block*-poly(ethylene oxide) (PS-*b*-P2VP-*b*-PEO) in aqueous solution. These micelles are referred to as core–shell–corona (CSC) micelles with PS as the core, P2VP as the shell, and PEO as the corona. Zhang *et al.*[3] also reported CSC micelles with a thermo-responsive shell self-assembled by the triblock copolymer poly(ethylene glycol)-*b*-poly(*N*-isopropylacrylamide)-*b*-polystyrene (PEG-*b*-PNIPAM-*b*-PS) in aqueous solution. In addition to self-assembly of triblock copolymers, CSC micelles can also be easily prepared by adsorption of block copolymer chains onto the core–shell micelles through hydrogen bonding or electrostatic interaction (Figure 9.1).[4–6]

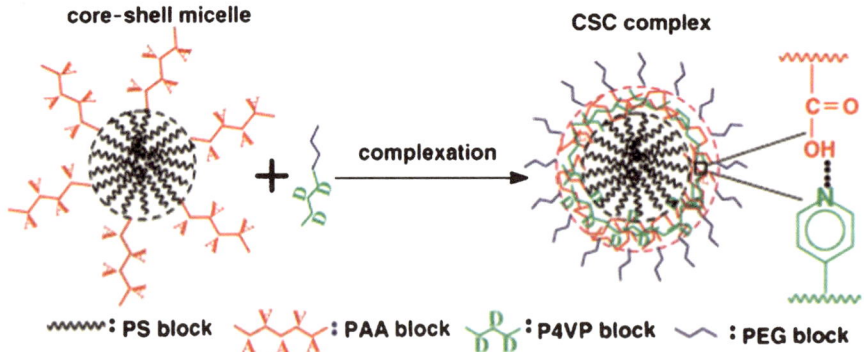

Figure 9.1   Schematic formation of CSC complex micelles formed by adsorption of a block copolymer onto core–shell micelles *via* hydrogen bonding.[4]

When used as drug delivery vehicles, different parts of the CSC micelles play important roles themselves. The micelle core is used as a reservoir for the loaded drug. The shell layer is usually endowed with a responsiveness, which is used to control and/or environmentally trigger drug release. The hydrophilic corona of PEG either stabilizes the micelles from aggregation or prevents them from premature clearance in blood circulation.

Shi *et al.*[7] reported CSC complex micelles as contractive "nanopumps" for thermo-sensitive release of drugs, as shown in Figure 9.2. The complex micelles were obtained from PS-*b*-PNIPAM-*b*-PAA [PAA = poly(acrylic acid)] micelles and PEG-*b*-P4VP block copolymers *via* the strong electrostatic interaction and hydrogen bonding between PAA and P4VP blocks in water. The PS block formed the core and the PAA/P4VP complex shell functioned as a semi-permeable membrane which could control the permeation of small molecules. Between the core and the shell, the large fluid-filled space that was formed with the thermo-responsive PNIPAM gel could retain the loaded drug for a long period of time. With increasing temperature, the shrinkage of the PNIPAM coils pumped the drug out of the complex micelles.

Other CSC complex micelles with surface channels were also reported by Shi *et al.*[8] for controlled drug release. An O/W emulsion with an oil phase of dichloroethane and a PAA surface was prepared and followed by deposition of PNIPAM-*b*-P4VP and PEG-*b*-P4VP through the interaction between PAA and P4VP. PEG channels were formed upon collapse of PNIPAM above its lower critical solution temperature (LCST). Ibuprofen was encapsulated by using a dichloroethane solution of the drug. The PAA/P4VP complex layer was swollen and permeable for small molecules, while the PEG channels played a key role in controlling drug release. As indicated in Figure 9.3, for the complex

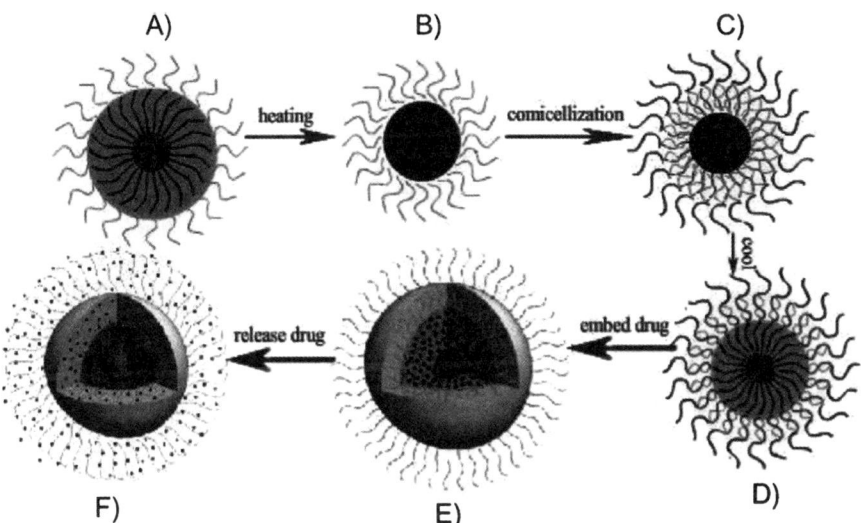

Figure 9.2 CSC micelles as contractive "nanopumps" for controlled drug release.[7]

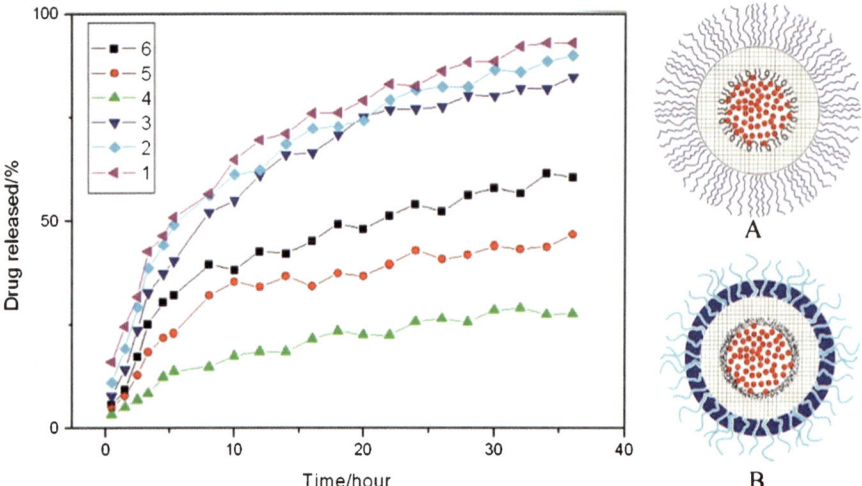

Figure 9.3   Ibuprofen release profiles of different micelles at 38 °C. Curves 1, 2, and 3 correspond to micelles with only PEG in the corona but different ratios of 4VP to AA units as 0.7, 1.0, and 1.4, respectively. Curves 4, 5, and 6 are the release profiles of complex micelles with a mixed corona; the weight content of PEG-*b*-P4VP in curves 4, 5, and 6 are 30, 50, and 70%, respectively.[8]

micelles without a collapsed PNIPAM shell the variation of 4VP/AA ratios did not have a marked influence on the release profiles, which demonstrates a good and similar permeability of the P4VP/PAA complex layer with different compositions. However, a distinct decrease of drug release rate was observed in the presence of collapsed PNIPAM, which covered the P4VP/PAA surface and blocked the diffusion of drugs. Only the PEG channels acted as a passageway for drug release. An increase in PNIPAM content resulted in a decrease of drug release rate due to increased coverage of the P4VP/PAA surface and a decreased amount of PEG channels.

Shell crosslinking of the drug-loaded micelles is frequently used to enhance their stability during blood circulation, prevent premature drug release, endow them with programmed disassembly of the nanocarriers, and also allow drug release at target sites. Wooley *et al.*[9] synthesized a series of shell-crosslinked knedel-like (SCK) nanoparticles for use as multifunctional imaging and therapeutic agents. The SCKs were constructed by a procedure that involved the supramolecular assembly of amphiphilic block copolymers into micelles, followed by covalent crosslinking throughout the shell layer. SCKs have been investigated as candidates for targeted biomedical applications due to their variable compositions, tunable sizes, and exceptional stability to avoid spontaneous disassembly at concentrations below the critical micelle concentration (CMC) or under challenging conditions as occur *in vivo*.

Wooley *et al.*[10] also reported cationic shell-crosslinked knedel-like (cSCK) nanoparticles (Figure 9.4) for highly efficient gene and oligonucleotide

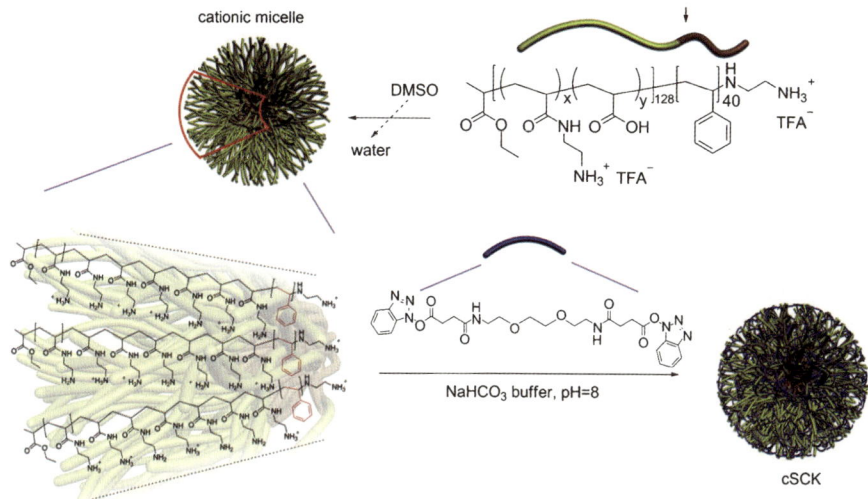

cationic micelle

DMSO

water

NH$_3^+$ TFA$^-$

NaHCO$_3$ buffer, pH=8

cSCK

Figure 9.4 Preparation of cationic shell-crosslinked knedel-like nanoparticles.[10]

transfection of mammalian cells. In this work, a robust synthetic nanostructure was designed for the effective packaging of DNA and it was shown to be an efficient agent for cell transfection. An amphiphilic block copolymer, poly(acrylamidoethylamine)-*b*-polystyrene (PAEA$_{128}$-*b*-PS$_{40}$), was synthesized, micellized in water, and shell-crosslinked using a diacid-derivatized crosslinker. A series of discrete complexes of the cSCKs with plasmid DNA (pDNA) was able to be formed over a broad range of polymer amine:pDNA phosphate ratios (N/P ratios), 2:1–20:1. Intracellular delivery of a splice-correcting phosphorothioate and genetic material, respectively, by the cSCKs led to much higher transfection efficiency than the performance of poly (amidoamine) (PAMAM) dendrimers.

Dual peptide nucleic acid- and peptide-functionalized shell-crosslinked nanoparticles were also designed by Wooley *et al.*[11] to target mRNA toward the diagnosis and treatment of acute lung injury. A well-designed amphiphilic block graft copolymer (Figure 9.5) was synthesized and used as the building blocks for the formation of well-defined SCKs, which were prepared by a series of robust, efficient, and versatile synthetic steps as indicated in Figure 9.6. The multifunctional biosynthetic hybrid nanostructures had potential utility in the recognition and inhibition of mRNA sequences for inducible nitric oxide synthase (iNOS), which are overexpressed at sites of inflammation, such as in cases of acute lung injury.

Lee *et al.*[12] reported a robust CSC micelle bearing redox-responsive shell-specific crosslinks as a carrier of docetaxel (DTX) for cancer therapy, as shown in Figure 9.7. The polymer micelles of poly(ethylene glycol)-*b*-poly(L-lysine)-*b*-poly(L-phenylalanine) (PEG-PLys-PPhe) in the aqueous phase provided the three distinct functional domains: the PEG outer corona for prolonged

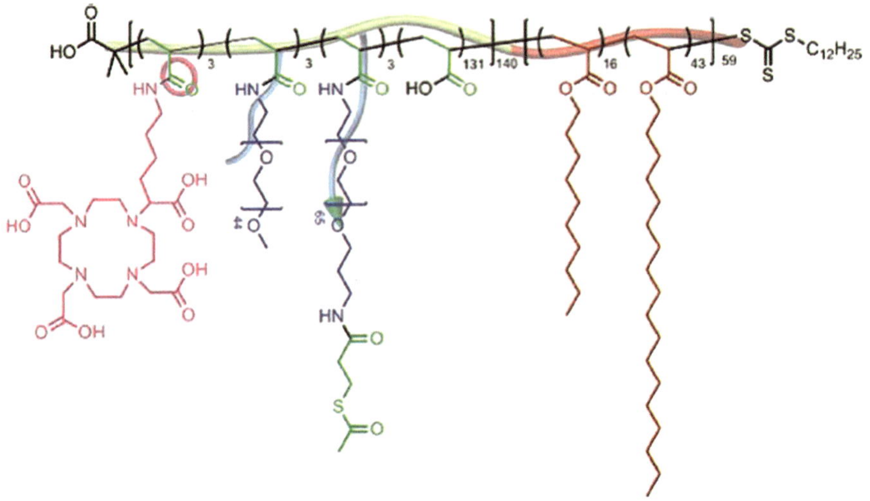

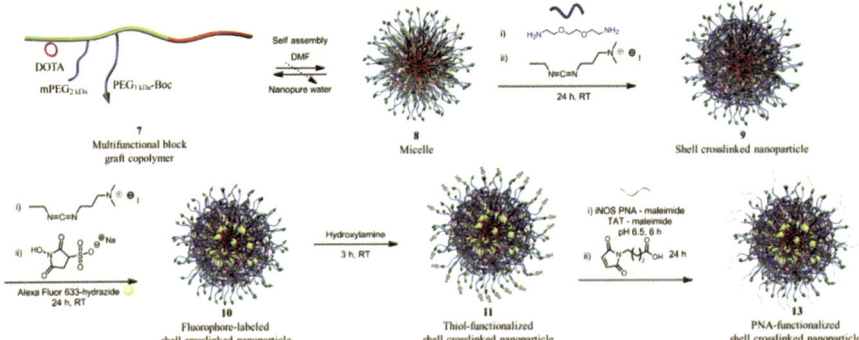

Figure 9.5   Amphiphilic functional block graft copolymer used for the preparation of SCK.[11]

circulation, the PLys middle shell for disulfide crosslinking, and the PPhe inner core for DTX loading. The shell crosslinking was performed by the reaction of disulfide-containing crosslinkers with Lys moieties in the middle shells. The DTX release from the DTX-loaded disulfide crosslinked micelles (DTX-SSCLM) was facilitated by increasing the concentration of glutathione (GSH). At an intracellular GSH level, DTX release was facilitated due to the reductive cleavage of the disulfide crosslinks in the shell domains. The DTX-SSCLM exhibited enhanced tumor specificity due to the prolonged stable circulation in blood and the EPR effect compared with the DTX-loaded non-crosslinked micelles (DTX-NCLM).

Figure 9.6   Self-assembly of the multifunctional block graft copolymer 7 into micelle 8, crosslinking into SCK 9, conjugation of alexa fluor 633 hydrazide 10, and functionalization with PNAs and TAT to give the final biosynthetic hybrid nanostructure 13.[11]

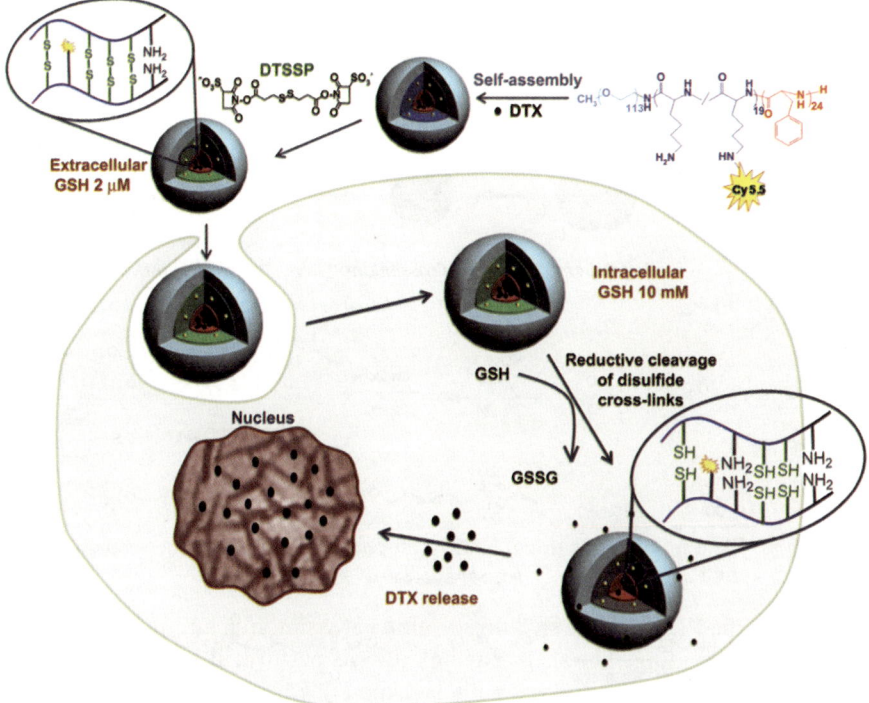

Figure 9.7   Illustration of shell crosslinking of DTX-loaded polymer micelles with redox-labile disulfide crosslinks and triggered release of DTX by intracellular GSH.[12]

Wang *et al.*[13] synthesized an amphiphilic triblock copolymer of poly(ε-caprolactone)-*block*-poly(mercaptoethyl ethylene phosphate)-*block*-poly(ethylene glycol) (PCL-*b*-PPE$_{SH}$-*b*-PEG), which could form CSC micelles with free thiols in the shell as indicated in Figure 9.8. Crosslinking of the micelles within the shell reduced their CMC and enhanced their stability against severe conditions. The redox-sensitive crosslinkage allowed the facilitated release of entrapped anticancer drugs in the cytoplasm in response to the intracellular reductive environment. With enhanced stability during circulation after administration, and accelerated intracellular drug release at the target site, the biocompatible and biodegradable shell-crosslinked polymeric micelle was proposed as a promising drug vehicle for cancer chemotherapy.

Polymeric micelle-based nanocarriers always encounter premature drug leakage in sample storage or during blood circulation and poor accumulation of drug-loaded micelles in tumor cells. Shuai *et al.*[14] described a novel multifunctional CSC micelle for tumor-targeted intracellular drug release and fluorescent imaging (Figure 9.9) to solve these problems. A triblock copolymer of poly(ethylene glycol)-poly[*N*-(*N'*,*N'*-diisopropylaminoethyl)aspartamide]-cholic acid [PEG-PAsp(DIP)-CA] was used to self-assemble into CSC micelles

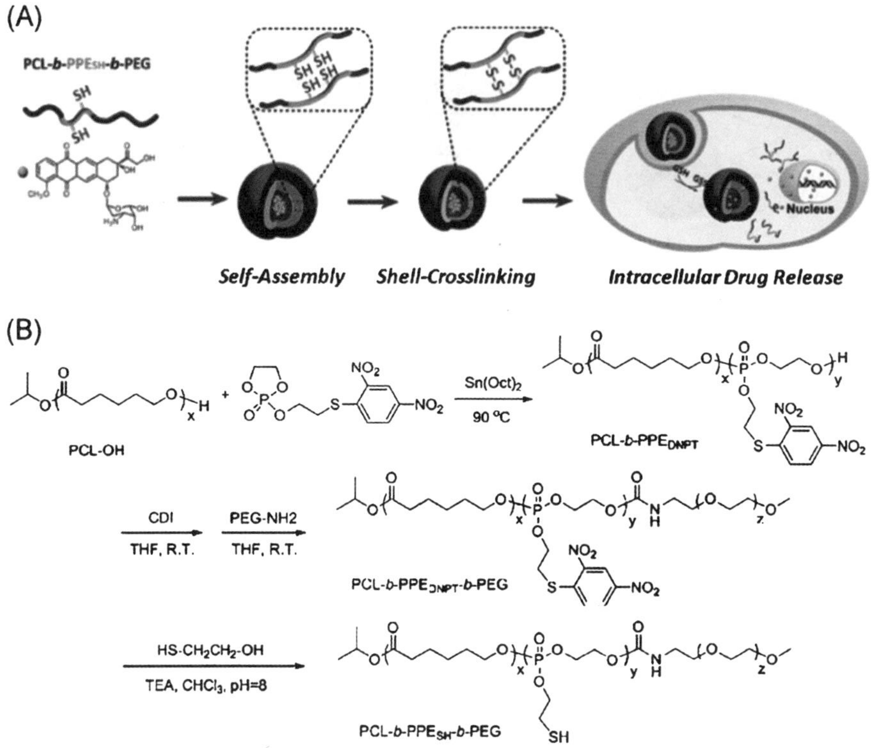

(A)

Self-Assembly   Shell-Crosslinking   Intracellular Drug Release

(B)

Figure 9.8   Schematic illustration of the formation of crosslinked micelles and intracellular drug release and synthetic pathway of the triblock copolymer PCL-*b*-PPE$_{SH}$-*b*-PEG.[13]

with CA as the core, the interlayer of PAsp(DIP) as the pH-sensitive shell, and PEG as the corona. The hydrophobic drug PTX was encapsulated into the micelle core while negatively charged quantum dots (QDs) were embedded in the shell layer as a fluorescent imaging probe for identifying tumor location. The pH-sensitive interlayer acted as an "on–off" module to turn the drug release "off" at neutral pH (*e.g.* in blood circulation) but "on" in the acidic lysosomal compartments (pH $\approx$ 5.0) of cells. Folic acid was used to functionalize the micelle for active tumor targeting, based on the fact that the folate receptor (FR) is overexpressed in many cancer cells yet rarely found in normal cells. Cell and animal studies revealed great potential for this unique micelle design in drug delivery applications.

A recent advance in nanomedicine-mediated cancer therapy is the development of multifunctional carriers for jointly delivering a chemotherapeutic drug and genetic material such as pDNA or siRNA.[15] Shuai *et al.*[16] developed CSC micelles as biodegradable nano-carriers for co-delivery of BCL-2 siRNA and DOX, as shown in Figure 9.10. An amphiphilic diblock copolymer of poly(ε-caprolactone)-*block*-poly(ethylenimine) (PCL-*b*-PEI) was

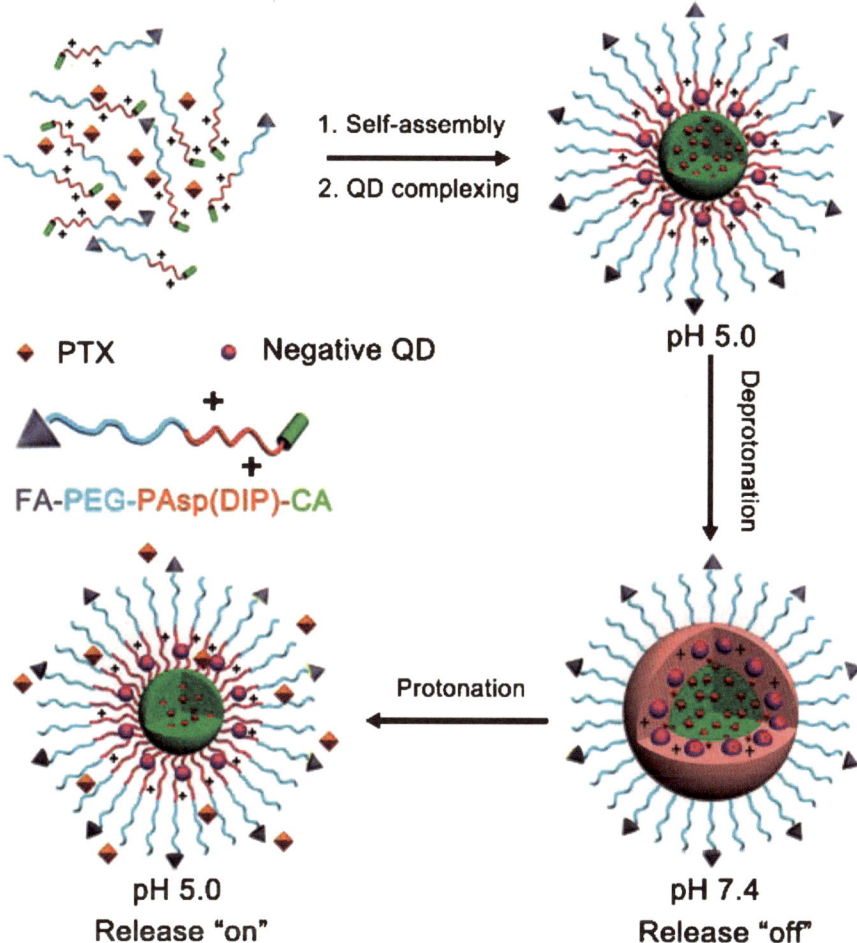

Figure 9.9 Illustrative preparation of PTX and QD-loaded micelle and pH-tunable drug release.[14]

first self-assembled into core–shell micelles with DOX loaded in the core. Then BCL-2 siRNA was incorporated into the micelles by complexation with the cationic shell. Finally, driven by the electrostatic interaction, folic acid-functionalized poly(ethylene glycol)-*block*-poly(glutamic acid) (FA-PEG-*b*-PGA) was coated onto the surface of the cationic PEI-PCL nanoparticles pre-loaded with siRNA and DOX, potentiating a ligand-directed delivery to the human hepatic cancer cells Bel-7402. The multifunctional hierarchical nano-assembly enabled simultaneously delivering siRNA and drug into the same cancer cells, yielding a synergistic effect of RNA interference and chemother-apy in cancer. At optimized composition, the nanoparticles exhibited not only high transfection efficiency but also ideally controlled release of drug. Compared to nonspecific delivery, the folate-targeted delivery of BCL-2

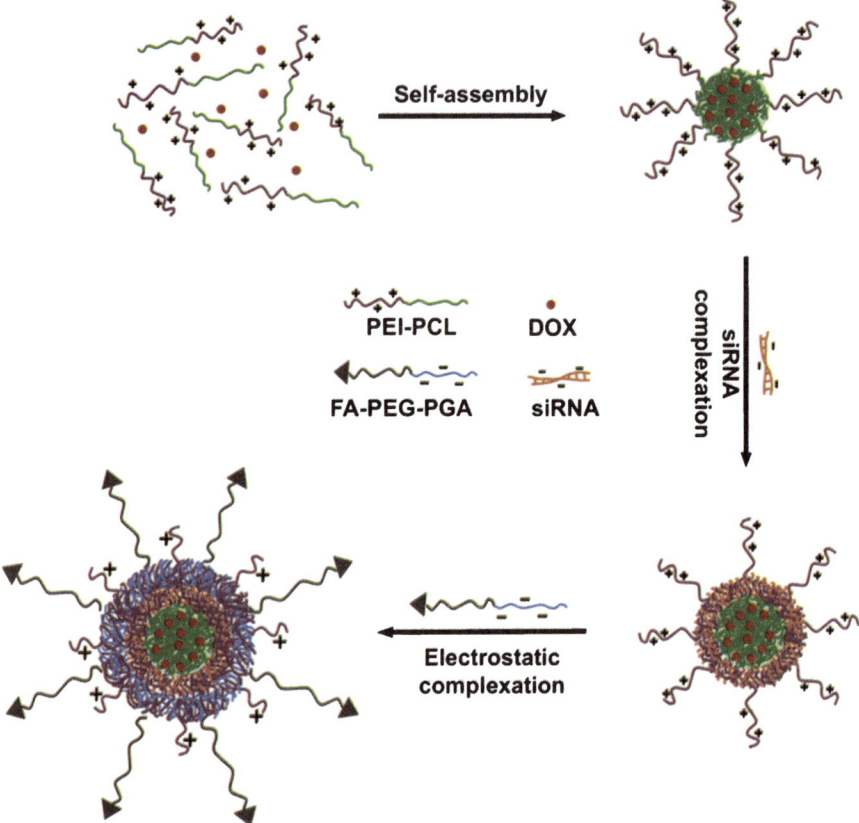

Figure 9.10   Formation of hierarchical nano-assemblies for combinatorial delivery of siRNA and anticancer drugs.[16]

siRNA resulted in more significant gene suppression at both the BCL-2 mRNA and protein expression levels, inducing cancer cell apoptosis and improving the therapeutic efficacy of the co-administered DOX.

9.3   Complex Micelles with Surface Channels for Drug Delivery

This kind of complex micelle refers to those that have a hydrophobically associated or electrostatically crosslinked core surrounded by two kinds of soluble polymer chains. Phase separation in the mixed shell can result in channels on the micelles' surface, which can be applied in controlled drug release. This section will cover the recent development of complex micelles with surface channels (CMSCs) in drug delivery.

Complex micelles with mixed shells (CMMSs) can be prepared by simultaneous self-assembly of two block copolymers either *via* hydrophobic or electrostatic interaction in aqueous solution. Using a mixed shell generally composed of neutral hydrophilic chains, *e.g.* PEG, and environment-sensitive chains such as PNIPAM, P4VP, *etc.*, a series of CMMSs has been developed by Shi *et al.* for the delivery of drugs under different conditions. Generally, the polymeric micelles have a common core surrounded by two different kinds of polymer chains as a mixed shell. External stimuli-induced collapse of one kind of polymer chain results in phase separation in the micelle shell. The collapsed polymer chains form hydrophobic domains around the micelle core and block mass exchange between the inner core and the outer milieu to some extent. Another kind of shell-forming polymer chain, which is soluble and connects the inner core and outer milieu, can act as channels for small molecules to pass through. As an example (Figure 9.11), a novel kind of complex micelle with a common P*t*BA core and a mixed and double-responsive P4VP/PNIPAM shell is prepared by co-micellization of two diblock copolymers, poly(*t*-butyl acrylate)-*b*-poly(*N*-isopropylacrylamide) (P*t*BA-*b*-PNIPAM) and poly(*t*-butyl acrylate)-*b*-poly(4-vinylpyridine) (P*t*BA-*b*-P4VP), in aqueous solution.[17] With an increase in temperature or pH value, the core–shell micelles can be converted into a CSC structure due to the collapse of PNIPAM or P4VP, while the soluble chains stretch out from the core through the collapsed shell and stabilize the micelles, resulting in the formation of P4VP or PNIPAM channels, respectively, as indicated in Figure 9.11. The channels, which connect the micelle core and the outer milieu, provide an effective way for mass exchange between the inner and outer micelles. Different channels may endow complex micelles with different functions in controlling mass exchange. The size and permeability of the channels can be regulated by manipulating the composition of the diblock copolymers or by changing the environmental conditions. By using similar methods, complex micelles with PEG channels are

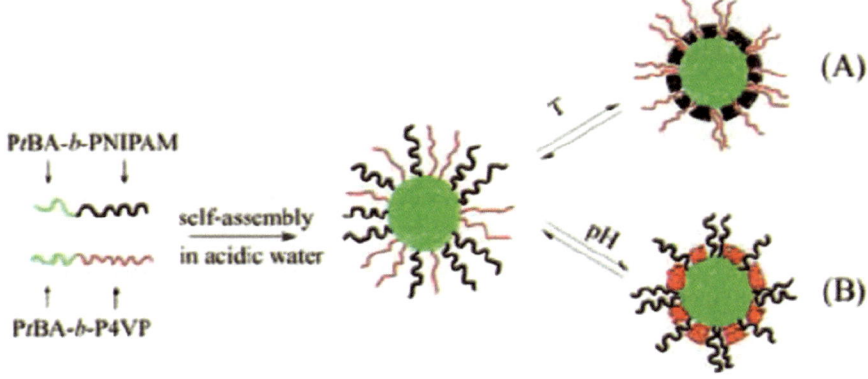

Figure 9.11 Formation of complex micelles with double responsive channels from a mixture of P*t*BA-*b*-PNIPAM and P*t*BA-*b*-P4VP.[17]

also constructed with different polymer compositions.[18] These new complex micelles with controllable channels may be promising candidates for use in controlled uptake/release processes.

In comparison with typical core–shell micelles as drug carriers, besides a well-defined core–shell structure, a nanoscale size (10–100 nm), the ability to solubilize water-insoluble drugs, and prolonged blood circulation times,[19] CMSCs also have other advantages, including facile manipulation of the drug release rate, efficient prevention of burst drug release, and degradation of the micelle core, due to their unique structures. CMSCs are promising nanocarriers for drug delivery due to their unique structure of an associated core as a drug reservoir, a collapsed inner shell as a barrier for burst drug release and enzyme invasion, and a soluble corona both as a stabilizer for the nanoparticles and channels for mass exchange between the core and the outer milieu.[20] As indicated in Figure 9.12, drug release from simple core–shell micelles (Figure 9.12a) may occur in all directions and burst release is often observed. For CMSCs (Figure 9.12b), drug release is expected mainly through the channels due to the protection of the shell layer of the PNIPAM, thus effectively reducing the risk of burst release. At the same time, the PNIPAM shell can also protect the micelle core and drug from invasion by enzymes, which enables the CMSCs to be used as nanocarriers for oral drug delivery. Several efforts have been made on the application of CMSCs in controlled drug release.

Shi *et al.*[21] have prepared complex micelles with a common PLA core and a mixed PEG/PNIPAM shell by simultaneous micellization of the diblock copolymers poly(L-lactide)-*b*-poly(*N*-isopropylacrylamide) (PLA-*b*-PNIPAM) and poly(ethylene glycol)-*b*-poly(L-lactide) (PEG-*b*-PLA) in aqueous solution at room temperature, as illustrated in Figure 9.13. Upon increasing the temperature above the LCST of PNIPAM, these complex micelles could be converted into a CSC structure composed of a PLA core, a collapsed PNIPAM shell, and a soluble PEG corona. The PEG chains stretched from the PLA core and penetrated through the PNIPAM shell to the outer milieu, leading to the formation of PEG channels. The PNIPAM block could collapse on the PLA core surface and the density of the PEG channels decreased on increasing the

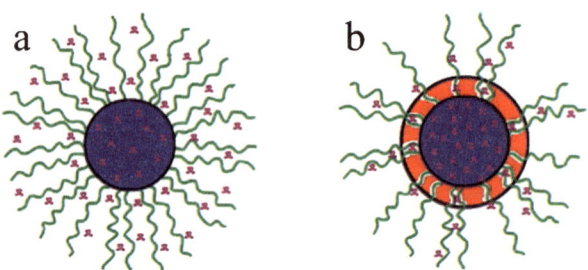

Figure 9.12   Schematic illustration of drug release from (a) core–shell micelles and (b) CMSC micelles with surface channels.[20]

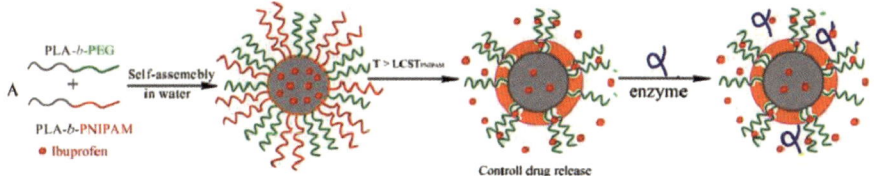

Figure 9.13   Schematic illustration of complex micelles with thermally induced PEG channels for controlled drug release and protecting PLA against enzymatic degradation.[21]

PNIPAM content. A decreased ibuprofen release rate with increasing PNIPAM content in the micelle shell was observed, which was ascribed to an enhanced retardation effect on drug diffusion from the micelle core by the collapsed PNIPAM and decreased amounts of PEG channels for drug release.

Lin *et al.*[18] have reported a new type of complex micelle with a poly(propylene oxide) (PPO) core and a mixed shell consisting of poly(L-glutamic acid) (PLGA) and PEG for drug delivery applications, as illustrated in Figure 9.14. The micelles were prepared through a co-micellization of PLGA-*b*-PPO-*b*-PLGA and PEG-*b*-PPO in water. In acidic conditions, PLGA undergoes a transformation from water-soluble random coils to a water-insoluble α-helix, leading to microphase separation in the mixed micelle shell. The PLGA chains collapse on the surface of the PPO core, forming patchy or continuous domains that are impermeable for drugs. The PEG chains with higher solubility connect the micelle core and the outer milieu, serving as channels for a rapid diffuse of loaded drugs. The drug release profile can be manipulated by tuning the composition of the mixed shell. An accelerated drug release rate is observed at the initial stage of increasing the PLGA in the mixed shell, which is suggested to be due to an increased stress of collapsed PLGA chains on the micelle core, which give rise to a core distortion inducing a leakage of loaded drugs. A higher content of PLGA chains can generate a higher stress force, but results in a smaller domain of PEG channels. Above

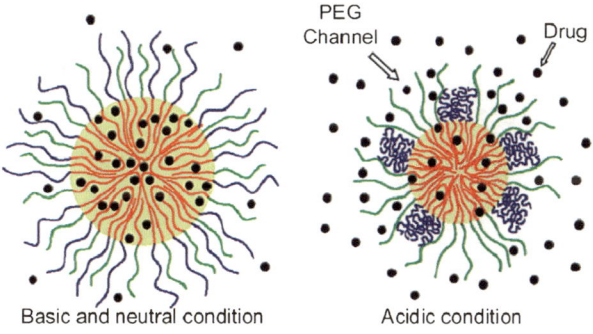

Figure 9.14   Schematic representation of the release mechanism proposed for the PLGA-*b*-PPO-*b*-PLGA/PEG-*b*-PPO drug-loaded complex micelles.[18]

50 wt% of PLGA in the micelle shell, a retardation of drug release is obvious because of increased coverage of the PPO core by PLGA domains and decreased PEG channels. The observed channel effect induced by the phase separation of micellar coronas is interesting, because the channels can be constructed and destructed within a narrow pH range without losing colloidal stability.

Oral administration of ionic drugs generally encounters significant fluctuation in plasma concentration due to the large variation of pH value in the gastrointestinal tract and the pH-dependent solubility of ionic drugs. Polymeric complex micelles with charged channels on the surface provided an effective way to reduce the difference in the drug release rate upon change in pH value. Shi *et al.*[22] reported complex micelles with charged channels for controlled release of ionic drugs (Figure 9.15). The complex micelles were prepared by self-assembly of PCL-*b*-PAsp and PCL-*b*-PNIPAM in water at room temperature, with PCL as the core and PAsp/PNIPAM as the mixed

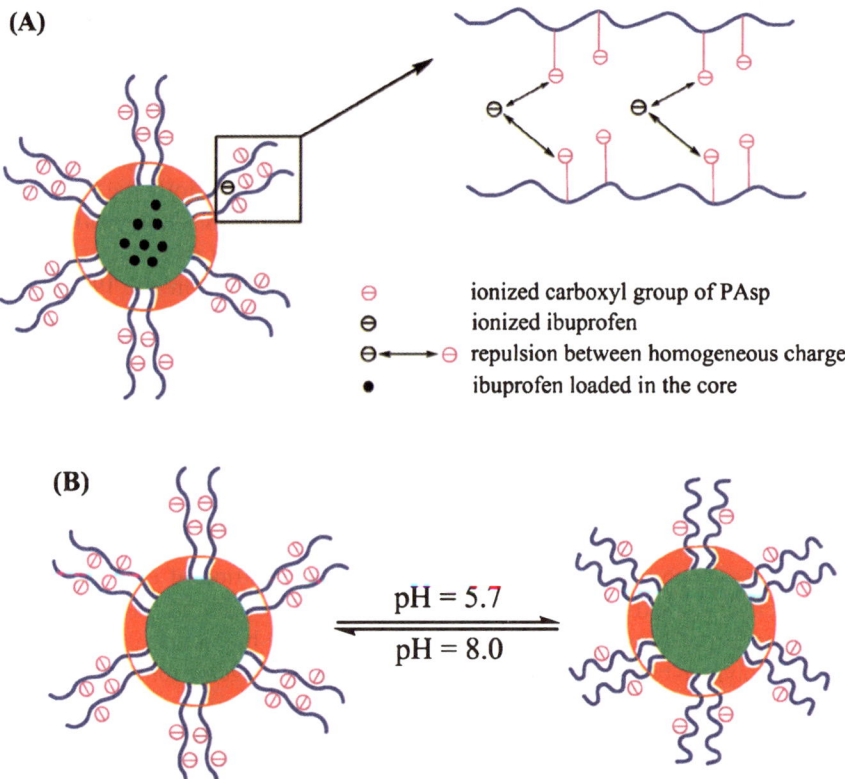

Figure 9.15   (A) Illustration of the release of ibuprofen from PCL-*b*-PAsp/PCL-*b*-PNIPAM complex micelles at 37 °C. (B) Illustration of the contraction of PAsp chains of PCL-*b*-PAsp/PCL-*b*-PNIPAM complex micelles due to the decrease of pH values.[22]

shell. With an increase in temperature, the PNIPAM collapsed and enclosed the PCL core, while the PAsp penetrated through the PNIPAM shell, leading to the formation of negatively charged PAsp channels on the micelle surface. The release behavior of ionic drugs from the complex micelles was remarkably different from that of usual core–shell micelles, where diffusion and solubility of the drugs played a key role. Specifically, it was mainly dependent on the conformation of the PAsp chains and the electrostatic interaction between the PAsp and the drugs, which could partially counteract the influence of the pH-dependent diffusion and solubility of the drugs. As a result, the variation of drug release rate with pH value was suppressed, which was favorable for acquiring a relatively steady plasma drug concentration.

9.4 Polyion Complex Micelles for Drug Delivery

Polyion complex (PIC) micelles generally consist of an electrostatically crosslinked core of two kinds of oppositely charged polyelectrolyte and a hydrophilic shell of PEG. This kind of complex micelle has an advantage in delivery of bioactive substances, such as genes, peptides, *etc*. This section will cover the recent developments of PIC micelles in drug and gene delivery. The excellent properties of PIC micelles for *in vivo* DNA delivery have been confirmed so far: a diameter around 100 nm with a PEG palisade which enables complexes to avoid recognition by reticuloendothelial systems, increased nuclease resistance, increased tolerance under physiological conditions, and excellent gene expression in a serum-containing medium.

PIC micelles were firstly reported by Kataoka *et al.*[23,24] in the 1990s. The PIC micelles are formed when a block copolymer with a neutral hydrophilic block and an ionic block is mixed with counter-charged compounds. The PIC micelles have a core–shell structure with a core consisting of the polyion complexes and a shell comprising the neutral block, as indicated in Figure 9.16.[25] The main driving force for the formation of the PIC micelles is the electrostatic attraction between the ionic block and the counter-charged compounds. Various types of biopharmaceuticals, such as plasmid DNA, oligo-DNA, siRNA, proteins, *etc.*, have been reported as the counter-charged compounds which can form PIC micelles.[26] The application of PIC micelles in drug delivery fields is rapidly increasing due to simple and efficient encapsulation of biopharmaceuticals and outstanding biocompatibility among various polymer-based drug delivery carriers. The change of ionic strength or pH-dependent protonation–deprotonation can be useful for the selective dissociation of PIC micelles. The release of encapsulated biopharmaceuticals of PIC micelles can be effectively controlled by degradation of the chemical bonds in the block copolymer responding to the change of pH or reduction potential.

Kataoka *et al.*[27] have reported PIC micelles of pDNA with acetal-poly(ethylene glycol)-*block*-poly[2-(dimethylamino)ethyl methacrylate] (acetal-PEG-PAMA) as the gene carrier system. The block copolymer effectively

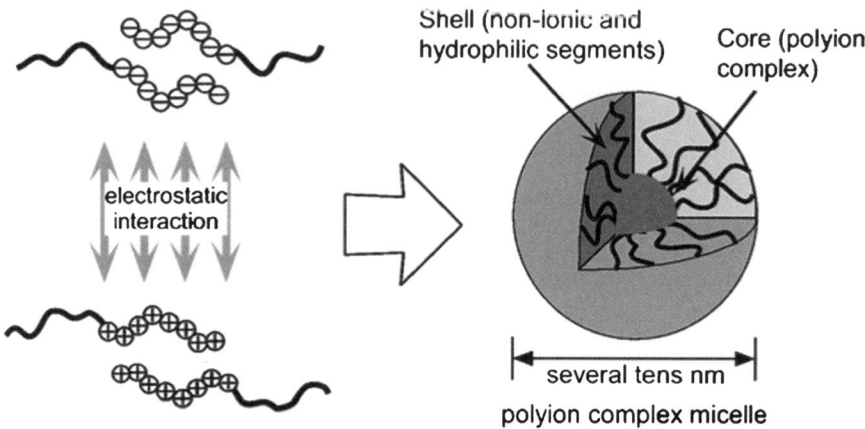

Figure 9.16 PIC micelle formation from a pair of oppositely charged block copolymers.[25]

induced the condensation of pDNA to be incorporated into PIC micelles having approximately 90–100 nm diameter. The PEG segments of the block copolymer are likely to surround the condensed pDNA to ensure the colloidal stability of the PIC micelles. The pDNA in the PIC micelles was adequately protected from DNase I attack. An excess of the counter polyanion potassium poly(vinyl sulfate) (PVSK) solution in 100 mM AcONa/AcOH buffer (pH 5) could dissociate the PIC micelles and thus induced the release of the loaded pDNA. An appreciably higher gene expression was achieved with these PIC micelles. Lactose was also conjugated onto the PIC micelles to enhance the transfection efficiency of the pDNA.[28] α-Lactosyl-poly(ethylene glycol)-poly[2-(dimethylamino)ethyl methacrylate] block copolymer (lactose-PEG-PAMA) was synthesized to construct a PIC micellar-type gene vector potentially useful for selective transfection of hepatic cells. Lactose-PEG-PAMA spontaneously formed a PIC micelle with pDNA encoding luciferase (pGL3-Luc) in aqueous solution without any precipitate formation. The lactosylated PIC micelle thus prepared achieved substantially higher transfection efficiency compared to the control PIC micelle without lactose moieties against HepG2 cells possessing asialoglycoprotein (ASGP) receptors recognizing the h-D-galactose residue. This pronounced transfection efficacy of the lactosylated PIC micelle was inhibited by the addition of excess asialofetuin (ASF), a natural ligand against the ASGP receptor, indicating ASGP receptor-mediated endocytosis to be a major route of the cellular uptake of the lactosylated micelles. Notably, the lactosylated PIC micelle revealed enhanced transfection compared to the control PIC micelle at a lower dose of pDNA, demonstrating the feasibility of using the ligand-conjugated PIC micellar vector for gene delivery to targeted cells.

Mao *et al.*[29] have developed a new block copolymer gene carrier that comprises a PEG segment and a degradable cationic polyphosphoramidate

(PPA) segment. This PEG-*b*-PPA copolymer carrier formed PIC micelles upon condensation with plasmid DNA in aqueous solution. The PEG-*b*-PPA/DNA micelles exhibited uniform and reduced particle size, ranging from 80 to 100 nm, and lowered the surface charge, compared with complexes of DNA with the corresponding cationic PPA carrier. The PIC micelles maintained similar transfection efficiency as PPA/DNA complexes, which was comparable to that of PEI/DNA complexes in HepG2 cells, but yielded about 16-fold lower transgene expression in primary rat hepatocytes than PPA/DNA complexes. Following bile duct infusion in Wistar rats, PEG-*b*-PPA/DNA micelles mediated four-fold higher and more uniform gene expression in the liver than PPA/DNA complexes. Liver function tests and histopathological examination indicated that PEG-*b*-PPA/DNA micelles showed low toxicity and good biocompatibility in the liver. This study demonstrated the potential of PEG-*b*-PPA/DNA micelles as an efficient carrier for liver-targeted gene delivery.

Kataoka *et al.*[30] developed a supramolecular nanocarrier of siRNA from the PEG-based block catiomer PEG-*b*-DPT carrying a diamine side chain with a pK_a directed to enhance intracellular gene silencing (Figure 9.17). The distinctive polymer design managed both a sufficient siRNA complexation and a buffering capacity of the endosomes. The PEG-DPT/siRNA PIC micelles revealed remarkable knockdown of the endogenous gene, even after serum incubation. Different PIC micelles of siRNA with PEG-DPT, PEG-DMAPA, and PEG-PLL showed sufficient knockdown of the GL3 luciferase, while the naked siRNA did not show any knockdown. These results indicated that siRNA was successfully delivered into the cytoplasm. Notably, the gene knockdown abilities of the PEG-DPT/siRNA PIC micelles were superior to those of the other two complexes, especially at higher N/P ratios. The high efficacy of PEG-*b*-DPT was attributed to the existence of additional secondary amines with a lower pK_a to promote the internalization of the siRNA molecules into the cytoplasm through buffering of the endosomal cavity. Similar results were also observed with the polyethylenimine-based polyplex that showed an enhanced transfection efficiency at the higher N/P ratios.

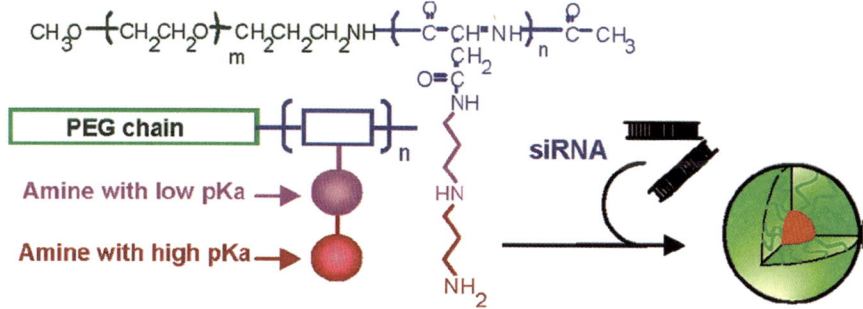

Figure 9.17 Formation of PIC micelles by the block catiomer PEG-*b*-DPT and siRNA.[30]

Proteins can also be incorporated into PIC micelles by complexation with block ionomers. Lysozyme, a positively charged enzyme, was encapsulated into PIC micelles by complexation with negatively charged block copolymers of PEG-*b*-PAsp and PAAm-*b*-PAA through electrostatic interaction in aqueous medium.[31–33] The resultant PIC micelles were stable in mild conditions but could be disintegrated by an enhanced salt concentration, which provided an effective way to trigger the release of the incorporated proteins.[34,35]

Delivery of bioactive antibodies into the cytoplasm of living cells by PIC micelles was achieved by Kataoka *et al.* for controlling cell growth (Figure 9.18).[36] The charge density of a model protein, cytochrome *c*, was temporarily increased by modification of the ε-amines of lysine residues into charge-conversional moieties, citraconic acid amide (Cit) or *cis*-aconitic acid amide (Aco). As the positively charged lysines converted to the negatively charged carboxylic groups by this modification, the modified proteins became strongly anionic and the resulting charge density could be increased significantly to form stable PIC micelles with cationic block copolymers even at physiological salt concentrations. The charge-converted proteins and the cationic block in the copolymer formed the core of the PIC micelle, and the PEG block formed the surface shell. After the PIC micelles were internalized to cells, the Cit and Aco rapidly degraded to reproduce the original lysines at the endosomal pH of 5.5.[37,38] The charge-conversional intracellular antibody delivery was expected to have high potential for the bioimaging of the intracellular structures and functions of living cells, as well as for biotherapeutics to target intracellular antigens. Moreover, charge-conversional PIC micelles could be used in intravenous protein delivery, based on the high

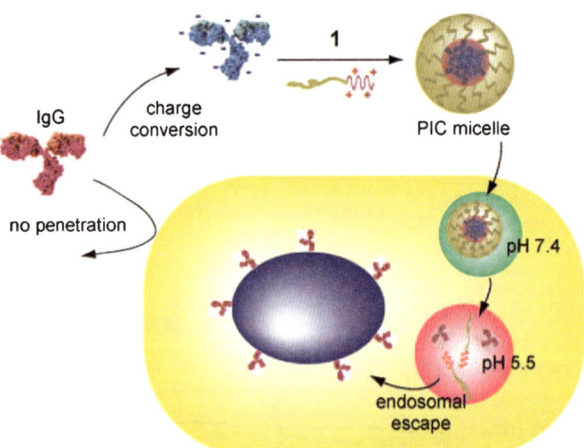

Figure 9.18   Preparation of the charge-conversional PIC micelles between IgG derivatives and PEG-pAsp(DET) and the delivery of IgG into living celles.[36]

biocompatibility and elongated circulation provided by the PEG shell of the PIC micelles.

References

1. U. Kedar, P. Phutane, S. Shidhaye and V. Kadam, *Nanomedicine (Philadelphia, U. S.)*, 2010, **6**, 714.
2. J. F. Gohy, N. Willet, S. Varshney, J. X. Zhang and R. Jerome, *Angew. Chem. Int. Ed.*, 2001, **40**, 3214.
3. W. Q. Zhang, X. W. Jiang, Z. P. He, D. Xiong, P. W. Zheng, Y. L. An and L. Q. Shi, *Polymer*, 2006, **47**, 8203.
4. W. Q. Zhang, L. Q. Shi, Z. J. Miao, K. Wu and Y. L. An, *Macromol. Chem. Phys.*, 2005, **206**, 2354.
5. W. Q. Zhang, L. Q. Shi, L. C. Gao, Y. L. An and K. Wu, *Macromol. Rapid Commun.*, 2005, **26**, 1341.
6. D. A. Xiong, Z. P. He, Y. L. An, Z. Li, H. Wang, X. Chen and L. Q. Shi, *Polymer*, 2008, **49**, 2548.
7. H. Wang, Y. L An, N. Huang, R. J. Ma, J. B. Li and L. Q. Shi, *Macromol. Rapid Commun.*, 2008, **29**, 1410.
8. D. A. Xiong, Y. L. An, Z. Li, R. J. Ma, Y. Liu, C. L. Wu, L. Zou, L. Q. Shi and J. H. Zhang, *Macromol. Rapid Commun.*, 2008, **29**, 1895.
9. M. Elsabahy and K. L. Wooley, *Chem. Soc. Rev.*, 2012, **41**, 2545.
10. K. Zhang, H. F. Fang, Z. H. Wang, J. Taylor and K. L. Wooley, *Biomaterials*, 2009, **30**, 968.
11. R. Shrestha, Y. F. Shen, K. A. Pollack, J. Taylor and K. L. Wooley, *Bioconjugate Chem.*, 2012, **23**, 574.
12. A. N. Koo, K. H. Min, H. J. Lee, S. U. Lee, S. Kim, I. C. Kwon, S. H. Cho, S. Y. Jeong and S. C. Lee, *Biomaterials*, 2012, **33**, 1489.
13. Y. C. Wang, Y. Li, T. M. Sun, M. H. Xiong, J. A. Wu, Y. Y. Yang and J. Wang, *Macromol. Rapid Commun.*, 2010, **31**, 1201.
14. W. W. Wang, D. Cheng, F. M. Gong, X. M. Miao and X. T. Shuai, *Adv. Mater.*, 2012, **24**, 115.
15. Y. Wang, S. J. Gao, W. H. Ye, H. S. Yoon and Y. Y. Yang, *Nat. Mater.*, 2006, **5**, 791.
16. N. Cao, D. Cheng, S. Y. Zou, H. Ai, J. M. Gao and X. T. Shuai, *Biomaterials*, 2011, **32**, 2222.
17. G. Y. Li, L. Q. Shi, R. J. Ma, Y. L. An and N. Huang, *Angew. Chem. Int. Ed.*, 2006, **45**, 4959.
18. J. P. Lin, J. Q. Zhu, T. Chen, S. L. Lin, C. H. Cai, L. S. Zhang, Y. Zhuang and X. S. Wang, *Biomaterials*, 2009, **30**, 108.
19. D. Sutton, N. Nasongkla, E. Blanco and J. M. Gao, *Pharm. Res.*, 2007, **24**, 1029.
20. R. J. Ma and L. Q. Shi, *Macromol. Biosci.*, 2010, **10**, 1397.
21. C. L. Wu, R. J. Ma, H. He, L. Z. Zhao, H. J. Gao, Y. L. An and L. Q. Shi, *Macromol. Biosci.*, 2009, **9**, 1185.

22. X. J. Liu, R. J. Ma, J. Y. Shen, Y. S. Xu, Y. L. An and L. Q. Shi, *Biomacromolecules*, 2012, **13**, 1307.
23. A. Harada and K. Kataoka, *Macromolecules*, 1995, **28**, 5294.
24. K. Kataoka, H. Togawa, A. Harada, K. Yasugi, T. Matsumoto and S. Katayose, *Macromolecules*, 1996, **29**, 8556.
25. A. Harada and K. Kataoka, *Prog. Polym. Sci.*, 2006, **31**, 949.
26. Y. Lee and K. Kataoka, *Soft Matter*, 2009, **5**, 3810.
27. D. Wakebayashi, N. Nishiyama, K. Itaka, K. Miyata, Y. Yamasaki, A. Harada, H. Koyama, Y. Nagasaki and K. Kataoka, *Biomacromolecules*, 2004, **5**, 2128.
28. D. Wakebayashi, N. Nishiyama, Y. Yamasaki, K. Itaka, N. Kanayama, A. Harada, Y. Nagasaki and K. Kataoka, *J. Controlled Release*, 2004, **95**, 653.
29. X. Jiang, H. Dai, C. Y. Ke, X. Mo, M. S. Torbenson, Z. P. Li and H. Q. Mao, *J. Controlled Release*, 2007, **122**, 297.
30. K. Itaka, N. Kanayama, N. Nishiyama, W. D. Jang, Y. Yamasaki, K. Nakamura, H. Kawaguchi and K. Kataoka, *J. Am. Chem. Soc.*, 2004, **126**, 13612.
31. A. Harada and K. Kataoka, *Macromolecules*, 1998, **31**, 288.
32. A. Harada and K. Kataoka, *Langmuir*, 1999, **15**, 4208.
33. S. Lindhoud, R. de Vries, W. Norde and M. A. Cohen Stuart, *Biomacromolecules*, 2007, **8**, 2219.
34. A. Harada and K. Kataoka, *J. Am. Chem. Soc.*, 1999, **121**, 9241.
35. S. Lindhoud, L. Voorhaar, R. de Vries, R. Schweins, M. Stuart and W. Norde, *Langmuir*, 2009, **25**, 11425.
36. Y. Lee, T. Ishii, H. J. Kim, N. Nishiyama, Y. Hayakawa, K. Itaka and K. Kataoka, *Angew. Chem. Int. Ed.*, 2010, **49**, 2552.
37. Y. Lee, S. Fukushima, Y. Bae, S. Hiki, T. Ishii and K. Kataoka, *J. Am. Chem. Soc.*, 2007, **129**, 5362.
38. Y. Lee, T. Ishii, H. Cabral, H. J. Kim, J. H. Seo, N. Nishiyama, H. Oshima, K. Osada and K. Kataoka, *Angew. Chem. Int. Ed.*, 2009, **48**, 5309.

CHAPTER 10

Zwitterionic Polymers for Targeted Drug Delivery

WEIFENG LIN, ZHEN WANG AND SHENGFU CHEN*

State Key Laboratory of Chemical Engineering, Department of Chemical and Biological Engineering, Zhejiang University, Hangzhou 310027, P. R. China
*E-mail: schen@zju.edu.cn

10.1 Introduction

In the past decades, hydrophilic polymer coatings on drug-delivery vehicles have been commonly used both to enhance the water solubility of the hydrophobic drug and to evade recognition by the reticuloendothelial system (RES).[1,2] A poly(ethylene glycol) (PEG) coating is the most frequently used one to prevent nonspecific protein adsorption, lower phagocytosis into macrophage cells, increase circulation time, and thus enhance both passive and active targeting.[3-5] In the authors' view, the resistance to nonspecific protein adsorption of PEG is the key reason for the consequent performance *in vitro* and *in vivo*, while the well-established functionalization methods facilitate the application of PEG. However, it is now recognized that PEG decomposes in the presence of oxygen and transition metal ions found in most biochemically relevant solutions.[6] Moreover, a possible immune reaction to PEGylated protein drug has also been observed.[7] The complex formation between albumin and PEG further suggests a possible unknown mechanism of PEG protection might be involved.[8] Another problem is the accelerated blood clearance (ABC) phenomenon induced upon repeated injections of PEG-coated colloidal carriers.[9,10] All these negative effects

RSC Polymer Chemistry Series No. 3
Functional Polymers for Nanomedicine
Edited by Youqing Shen
© The Royal Society of Chemistry 2013
Published by the Royal Society of Chemistry, www.rsc.org

might limit its long-term applications. Thus, a possible alternative for PEG is desired.

In fact, there are two major classes of materials to resist nonspecific protein adsorption, so-called "nonfouling" materials, namely hydrophilic polymers and zwitterionic polymers. A number of "nonfouling" hydrophilic polymers, including PEG, polysaccharides, and polyamides, have been found to mostly share some common structural and chemical properties: hydrophilic nature, electrical neutrality, and hydrogen-bond acceptors/donors.[11,12] They have been fully investigated for drug delivery due to their significant improvement in protein compatibility and the convenience of preparing copolymers with highly hydrophobic segments. In recent years, zwitterionic polymers, such as poly (2-methacryloyloxyethyl phosphorylcholine) (pMPC),[13–15] poly(sulfobetaine methacrylate) (pSBMA),[16,17] poly(carboxybetaine methacrylate) (pCBMA),[18,19] or simply mixed charge materials,[20,21] have been recognized as superior materials to resist nonspecific protein adsorption, and which can maintain the stability of nanoparticles or micelles in highly complex media such as undiluted serum where PEG coatings fail to stabilize nanoparticles under the same conditions.[22] Furthermore, carboxybetaine (CB)-based zwitterionic polymers are able to conjugate bio-recognizable ligands or antibodies without losing their resistance. Moreover, excellent hemocompatible surfaces resisting the adhesion of blood platelets and highly efficient drug carriers have been achieved by the protection of zwitterionic polymers in recent years, and methods to utilize zwitterionic polymers have been explored. The molecular weight of homopolymers or block copolymers can be well controlled in a narrow range through living radical polymerization.[23] Surface-initiated polymerization could also improve the protection of nanoparticles with fewer drawbacks.[22] In short, zwitterionic polymers could be potential alternatives to PEG for a broad range of applications. Table 10.1 shows the zwitterionic materials found in the literature.

In this chapter, we mainly discuss the unique properties of zwitterionic polymers and their applications in nanomedicine, although this is a rather small amount of work compared with their "nonfouling" hydrophilic polymer-related applications in nanomedicine. The application of zwitterionic polymers on macroscale surfaces, such as blood contact tubules, filters, or stents, will not be reviewed here, although the original motivation for inventing and exploring the zwitterionic polymers coatings was to obtain excellent hemocompatibility of macro scale surfaces.[24,25] Comprehensive reviews about the design principles and biological applications of "nonfouling" materials can be found in previous publications,[20,26] which might provide a broad view through extended reading.

10.2 Principles Toward Protein-Resistant Zwitterionic Polymers

To understand the advantages of zwitterionic polymers for drug delivery systems, the capability in the resistance to nonspecific protein adsorption of polymers should be considered first of all. Although PEG exhibits resistance to

Table 10.1 Overview of zwitterionic "nonfouling" materials.

Materials[a]	Structure	Ref.
Poly(CBAA)		19,62–64
Poly(SBMA)		65,66
Poly(CBMA)		67,68
Poly(MPC)		69–72
Zwitterionic peptide		21

nonspecific protein adsorption, zwitterionic groups are much more efficient than PEG. A single monolayer of phosphorylcholine groups is enough to resist nonspecific protein adsorption,[14] whereas a self-assembled monolayer with six EG units and an optimized loose structure are required to ensure enough surface hydration and thus resistance to protein adsorption.[27] In fact, the ability of resistance to nonspecific protein adsorption of both hydrophilic (PEG and other related polymers) and zwitterionic polymers are tightly correlated with a hydration layer near the surface,[18–20] because a tightly bound water layer forms

Table 10.1 (*Continued*)

Materials[a]	Structure	Ref.
Poly(TM-SA)		73
Poly(METMA-MES)		74,75
Poly(MHist)		76
Poly(SerMA)		77

Table 10.1   (*Continued*)

Materials[a]	Structure	Ref.
Poly(LysMA)		78

[a]Poly(CBAA) = poly(carboxybetaine acrylamide); poly(TM-SA) = poly([2-(Acryloyloxy)ethyl] trimethyl ammonium chloride- carboxylic acid [2-carboxy ethyl acrylate]); poly(METMA-MES) = poly([2-(methacryloyloxy) ethyl]trimethylammonium chloride - -3-sulfopropyl methacrylate potassium salt); poly(MHist) = poly(N-Methacryloyl-L-histidine); poly(SerMA) = poly(o-methacryloyl-L-serine); poly(LysMA) = poly(ε-methacryloyl L-lysine).

a physical and energetic barrier to prevent protein adsorption on the surface. This is an intrinsic cause of the resistance of polymers. Recent results showed that 7–8 water molecules could tightly bind with one sulfobetaine (SB) unit while only ~1 water molecule binds with one EG unit at a lower association constant than with SB.[28] Simulations also reveal different structural and dynamic properties of interfacial waters near the phosphorylcholine self-assembled monolayers (PC-SAM) and oligo(ethylene glycol) self-assembled monolayer (OEG-SAM) surfaces. First, interfacial water molecules near the zwitterionic PC-SAM surface have longer residence times (stay longer) and smaller self-diffusion coefficients (move slower) than those near the hydrophilic OEG-SAM surface, indicating that the PC-SAM surface binds water molecules more strongly than the OEG-SAM surface. Secondly, interfacial water molecules at the PC-SAM surface have much larger reorientational dynamics than those at the OEG-SAM surface, indicating that the PC-SAM binds water molecules *via* ionic solvation while the OEG-SAM binds *via* hydrogen bonding.[14,27,29] After all, the bound water molecules on zwitterionic polymers caused by strongly ionic solvation are more tightly bound than ones on PEG through hydrogen bonding, The superior resistance to nonspecific protein adsorption of zwitterionic polymers is rather unambiguous.

On the other hand, it is necessary to bear in mind the difference between surface resistance and the intrinsic property of polymers to resist nonspecific protein adsorption when unexpected protein adsorption on drug delivery systems occurs. Surface packing (*i.e.* film thickness, packing density, and chain conformation) could also affect the overall surface hydration. Chain flexibility also plays an important role in protein resistance, especially for polymer brush-coated surfaces. When a protein approaches the surface, the compression of the polymer chains causes steric repulsion to resist protein adsorption due to an unfavorable decrease in entropy.[11,12] On the other hand, a long-chain polymer-coated surface might eventually cause protein adsorption after their reconfiguration if the protein could be weakly adsorbed on a long-chain polymer. All these factors will decide the final resistance of the surfaces. In other words, the surface resistance is conditional even if ideal protein-resistant polymers (such as PEG) are used.

Zwitterionic materials all share the same characteristic in a homogenous 1:1 positive to negative charge distribution on a nanometer scale, which also could be used as a simple design rule for new zwitterionic materials to resist protein nonspecific adsorption through mimicking the homogenous charge distribution feature. Following this design rule, peptides with alternative charge distributions by natural amino acids (Glu$^-$, Asp$^-$, and Lys$^+$) have been synthesized and proved to be highly resistant to nonspecific protein adsorption. Also, poly-methacrylates with the same charge feature also show excellent resistance. In fact, the positive charge could be contributed by primary to quaternary amines as well as other monovalent positively charged groups, while the negative charge could be given by $-SO_3^-$, $-COO^-$, monoesters of phosphates, and also other possible monovalent negatively charged groups. Thus, the new zwitterionic polymers with homogenous charge distribution could cover a very wide spectrum of materials.

Moreover, the multiple choice for positive and negative charge providers in zwitterionic polymers facilitates their versatility for conjugation and functionalization, which is essential for drug delivery systems. For example, the CB polymer provides such a solution.[18] Researchers always face a compromise between maximizing the numbers of functional groups for immobilization of bio-recognition molecules and maintaining resistance to nonspecific adsorption. Conventionally, PEGs with reactive end groups, such as COOH and NH_2, are always partially mixed in PEG-based coatings for chemical immobilization of bio-recognition molecules. The unreacted end groups after immobilization become surface defects for nonspecific protein adsorption.[18] However, pCBMA could play dual roles in preventing nonspecific protein adsorption and conjugating with amine groups on bioactive molecules by the large number of carboxyl groups on pCBMA. Furthermore, the conjugation through 1-ethyl-3-(3-dimethylaminopropyl)carbodiimide and N-hydroxysuccinimide (EDC/NHS) chemistry will not degrade the resistance to nonspecific protein adsorption since the remaining ester groups of NHS are hydrolyzed to carboxylate anions, which are paired with cationic quaternary amines to form zwitterionic structures to resist nonspecific protein adsorption. Such an advantage is demonstrated by high reproducibility and sensitivity in undiluted serum of both surface plasmon resonance (SPR) biosensors[19] and superparamagnetic nanosphere-based immunoassay.[30] Thus, this makes poly(carboxybetaine) (PCB), represented by pCBMA, a universal polymer for "theranostics".

Besides a unique capability in conjugation, the zwitterionic related molecules can also provide more choices through the format change response to the environment. For example, the negatively charged group (COO^- or SO_3^-) in a zwitterionic polymer could form pH-responsive hydrolysable esters as positively charged zwitterionic polymer precursors, which can be utilized for a highly efficient degradable zwitterionic polymer-based gene delivery system for DNA delivery,[31] as well as other hydrophobic drugs. The advantage of this method is to enhance DNA unpackaging and lower the toxicity of polymers residing in infected cells upon hydrolysis of the positively charged pCBMA ester to the zwitterionic pCBMA. Moreover, such a charge change could also facilitate a sticky to nonsticky surface change, which might also benefit local drug delivery.[32]

To utilize zwitterionic polymers in drug delivery, various copolymers are synthesized through conventional free radical copolymerization or living radical polymerization. All these copolymers with zwitterionic groups are mainly synthesized in alcoholic solution, usually methanol or ethanol, since the zwitterionic groups are highly hydrophilic. Sometimes, a little water is necessary for these alcoholic solutions in order to ensure the solubility of the final products with a large segment of zwitterionic groups. Also, some less polar solvents, such as THF or DMSO, could be mixed with the alcohol to dissolve hydrophobic monomers. Another route is to prepare copolymers with precursors of zwitterionic groups and convert them to zwitterionic copolymers. Lowe *et al.*[33] prepared a random copolymer of *n*-butyl methacrylate (BMA) with either 10 or 30 mol% 2-(dimethylamino)ethyl methacrylate (DMAEMA) and treated these precursor polymers with propane-1,3-sultone under mild conditions to obtain a zwitterionic copolymer with a narrow molecular weight distribution. Cao *et al.*[18] have developed PLGA-PCB block copolymers through hydrolysis of the ester of PCB. In such a way, a block copolymer with a sharp hydrophilicity/hydrophobicity difference could be prepared, which could be a very useful method to prepare stable nano drug carriers with a very low critical micelle concentration (CMC). On the other hand, Samanta *et al.* developed end-functionalized polyMPC through atom-transfer radical polymerization (ATRP) which could conjugate to amino groups.[34] This MPC polymer provides an alternative to end-functionalized PEG for forming resistance to nonspecific protein adsorption.

10.3 Phosphorylcholine-Based Polymers for Drug Delivery

Phosphorylcholine (PC) is a zwitterionic head group of the major polar groups of phospholipids on cell membranes. 2-Methacryloyloxyethyl phosphorylcholine (MPC), a methacrylate monomer composed of PC, has a cell membrane-like structure and has demonstrated strong non-thrombogenicity.[35] Owing to its excellent biocompatibility, MPC is an ideal candidate for hydrophilic shell-forming blocks and has been widely used for biomedical applications, including drug delivery to solubilize hydrophobic drugs.[36,37]

A representative copolymer for improving the solubility of hydrophobic drugs in aqueous media is poly(2-methacryloyloxyethyl phosphorylcholine-*co-n*-butyl methacrylate) (PMB).[38–40] Wada *et al.*[41] prepared PMB with 70 mol% of the BMA unit (PMB30W) to solubilize paclitaxel (PTX) over 1000 times. The diameter of the PMB aggregate containing 1 mg mL^{-1} of PTX (PTX-PMB30W) was 50 nm in aqueous medium. Antitumor efficacies of PTX-PMB30W and PTX dissolved in polyoxyethylated caster oil were similar, following weekly intraperitoneal administration of 50 mg kg^{-1} PTX in nude mice transplanted with MX-1 cells. However, after the dose was increased to 200 mg kg^{-1} PTX, all animals receiving the PTX-PBM30W survived, whereas all animals died within 1 minute after being injected with the same dose of PTX (Figure 10.1). This result demonstrates that MPC-protected nano-delivery vehicles (NDVs) could lower the immediate toxicity after injection. Furthermore, small hydrophobic drugs incorporated into block copolymer micelles can be much better protected from the bulk phase than those taken up by random copolymer aggregates. The studies by Yusa *et al.* showed that the water solubility of PTX can be enhanced dramatically by use of an aqueous micelle solution of PMB, which demonstrated that block copolymers are much more effective for enhancing solubility because of their associative structure in aqueous media.[42] Chu *et al.*[43] prepared PTX-loaded PMB micelles and investigated the micelles' morphology as well as the *in vitro* drug release kinetics. The TEM studies indicated a small size (less than 30 nm), a narrow size distribution, and a regularly spherical shape for the drug-loaded micelles. An *in vitro* release study showed that the release rate of PTX from the polymeric micelles was slow and sustained. Kano *et al.* demonstrated enhancement of the solubilization of miconazole (MCZ), vidarabine (Ara-

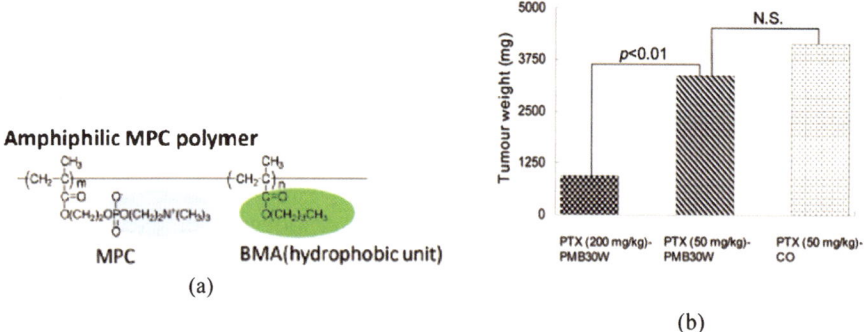

(a)

(b)

Figure 10.1   (a) Chemical structure of poly(MPC-*co-n*-butyl methacrylate) (PMB). (Reproduced from Goda *et al.*[38] with permission from Elsevier.) (b) Growth inhibition of tumor xenografts on day 16. The antitumor efficacy of PTX-PMB30W at 200 mg kg^{-1} PTX was significantly higher than that of PTX-PMB30W at 50 mg kg^{-1} PTX. (Reproduced from Kano *et al.*[41] with permission from The International Institute of Anticancer Research.)

A), and griseofulvin (GRF) by encapsulating in PMBs and the safeness of oral administration to rats.[44]

However, the accumulation of nonbiodegradable polymers, such as PTX-PMB30W, in cells is always a concern in the application of nanomedicine. Biodegradable polymers are much desired. In general, biodegradable polymers represented by poly(L-lactic acid) (PLA) and poly(ε-caprolactone) (PCL) have been used as the materials for making nanoparticles. However, as conventional PLA nanoparticles contact with the blood components after injection into the blood stream, the plasma proteins are immediately adsorbed on the PLA nanoparticles and a conformational change of adsorbed protein is induced.[45,46] Konno *et al.*[47] prepared PMB/PLA nanoparticles by a solvent evaporation technique from oil in water systems using the water-soluble amphiphilic phospholipid polymer PMB as both emulsifier and surface modifier. The phospholipid polymer on the nanoparticles effectively suppressed any unfavorable interactions with the biocomponents and improved the blood compatibility of the conventional PLA nanoparticles. Hsiue *et al.*[48] prepared PMPC-*b*-PLA polymeric nanoparticles, and the low cytotoxicity of the polymer nanoparticles was confirmed by cytotoxicity assay and growth inhibition assays with HFW (human fibroblast cell), indicating good cytocompatibility of the lipid-like diblock copolymer PMPC-*b*-PLA. Liu *et al.*[49] also demonstrated that PMPC-*b*-PLA could form micelles in aqueous solution. The micelles displayed excellent biocompatibility and a high drug-loading efficiency. *In vitro* release profiles indicated that the PMPC-*b*-PLA micelles could be used for the administration of controlled-release hydrophobic anticancer drugs. Tu *et al.*[50] prepared novel star-shaped copolymers having six-arm stars with a zwitterionic PC block copolymer, poly(ε-caprolactone)-*b*-poly(2-methacryloyloxyethyl phosphorylcholine) (6sPCL-*b*-PMPC). The micelles with PC groups expressed at their exterior showed excellent internalization ability into cancer cells. When incorporated with PTX, the 6sPCL-*b*-PMPC micelles show much higher cytotoxicity against HeLa cells than PCL-*b*-PEG micelles, in response to the higher efficiency of cellular uptake. Cooper *et al.*[51] presented the synthesis of water-soluble, biodegradable, zwitterionic aliphatic polyesters using ring-opening polymerization and post click chemistry with PC azide, giving materials with potential applications that benefit from a combination of biodegradability, biocompatibility, and water solubility (Figure 10.2). Iwasaki *et al.*[52] synthesized biodegradable polyphosphate graft copolymers with varying densities of cholesteryl esters and hydrophilic MPC graft chains as novel amphiphilic biomaterials. The graft polymers containing cholesteryl groups effectively enhanced the solubility of PTX in an aqueous solution.

In most cases, the encapsulated chemotherapeutic drugs show a burst release up to 20–30% within several hours post-micelle formation, followed by a slow diffusional drug release lasting for many days. The premature burst release leads to drug loss in micelle storage and blood circulation, which also limits the total applicable dose caused by immediate toxicity after injection. Meanwhile,

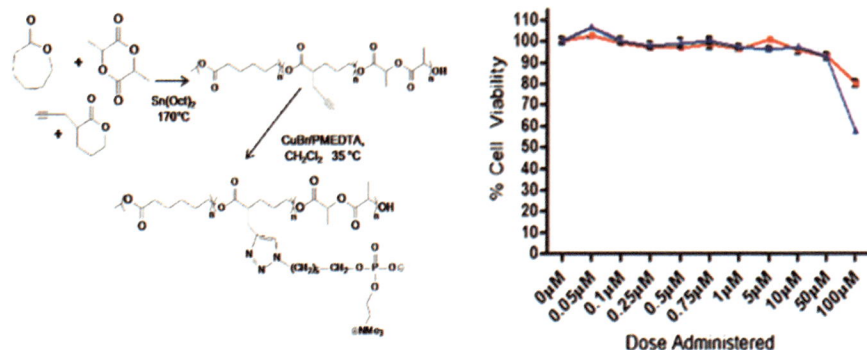

Figure 10.2 *Left*: Synthesis of PC-grafted polyesters. *Right*: Cell titer-glow luminescent cell viability assays of polyester-*graft*-PC after 24 (*red*) and 48 hour (*blue*) incubation in MCF7 cell culture.

the second-stage slow drug release results in low intracellular drug availability, which is insufficient for killing cancer cells. Conjugation of a drug to an amphiphilic polymer will overcome these drawbacks. Chen *et al.*[53] applied click chemistry to camptothecin (CPT) conjugation on the polyMPC backbone, using an acylated and azide-modified form of CPT that gives polyMPC-CPT conjugates with high drug loading, which has potential for future integration into CPT-based injectable cancer therapeutics. Also the drug release rates can be controlled by incorporating different linkers between CPT and the polyMPC backbone (Figure 10.3).

To improve the capability in specific targeting, immobilizable units have to be introduced into a MPC-containing copolymer to conjugate bio-recognizable ligands or antibodies. Miyata *et al.*[54] prepared a PMB polymer with an immobilizable unit, *p*-nitrophenylcarbonyloxyethyl methacrylate (NPMA), which can be modified by the addition of the preS1 domain of hepatitis B surface antigen (HBsAg) (Figure 10.4). The final NDVs with encapsulated PTX (PTX/PMBN-preS1) showed human hepatocyte-specific targeting without serious adverse effects *in vivo*. The 50% inhibitory concentration (IC_{50}) values of PTX/PMBN-preS1 against the human hepatocellular carcinoma cell line (HepG2) were lowered to about one-seventh of PTX/PMBN, the PTX carrier without HBsAg, while the IC_{50} values of PTX/PMBN-preS1 against the human squamous cell carcinoma cell line (A431) were comparable to that of PTX/PMBN. This result could well provide the basis for a human hepatocyte-specific drug delivery system in clinical settings. Shimada *et al.*[55] also use the same method to conjugate epidermal growth factor (EGF) on a MPC polymer. The final EGF-PMBN-PTX micelles represent a more potent targeted therapy for tumors overexpressing EGFR. Iwasaki *et al.*[56] synthesized a water-soluble MPC polymer bearing hydrazide groups functionalized by nonnatural carbohydrates through ketones. The nanoparticles were able to recognize ManLev-treated HeLa cells. Nonspecific

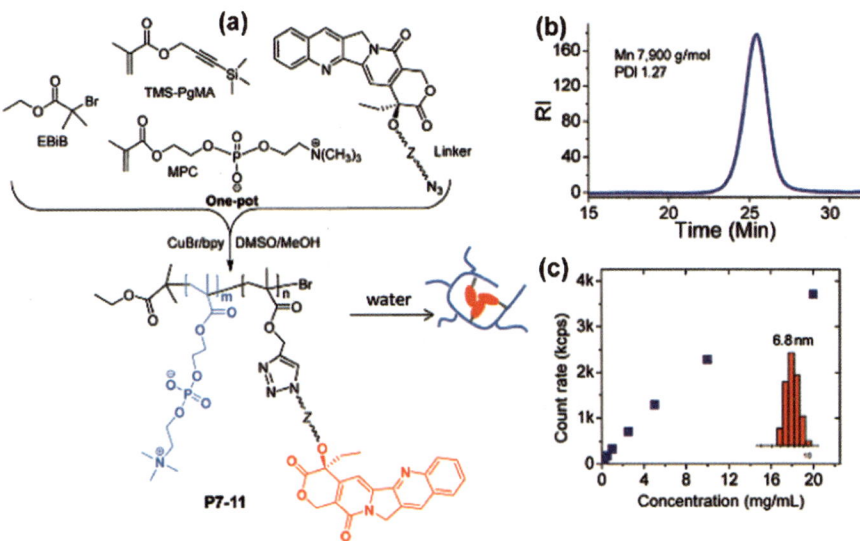

Figure 10.3 (a) One-pot synthesis of polyMPC-*graft*-CPT copolymers. (b) Aqueous GPC trace of copolymer. (c) Light scattering intensity *vs.* concentration of copolymer (*inset* is polymer diameter distribution). (Reproduced from Chen *et al.*[53] with permission from the American Chemical Society.)

delivery of the nanoparticles to the cells was effectively reduced because of the stealth of the MPC units. Licciardi *et al.*[57] developed a successful route involving ATRP of MPC followed by a tertiary amine methacrylate using a 9-fluorenylmethyl chloroformate (Fmoc)-protected ATRP initiator. Folic acid (FA)-functionalized biocompatible block copolymers could be obtained by the deprotection of the Fmoc groups, which could avoid introduction of additional units for conjugation in the final NDVs. FA-MPC-DPA (DPA = diisopropylaminoethyl methacrylate) copolymers have been evaluated as pH-responsive micellar vehicles for the delivery of highly hydrophobic anticancer drugs, namely tamoxifen and paclitaxel.

Besides active targeting, intelligent-responsible properties of the drug vesicles are a very important factor to enhance drug release. Stimuli-sensitive materials can be obtained by the addition of monomers, such as pH-sensitive DPA. Salvage *et al.*[58] prepared nanoparticles using PMPC-*b*-PDPA diblock copolymers. The novel nontoxic biocompatible micelles formed by MPC-DPA diblock copolymers have appropriate size and good colloidal stability, as well as exhibiting pH-modulated drug uptake and release. Lv *et al.*[59] synthesized a series of amphiphilic random copolymers with MPC as the hydrophilic segment, stearyl methacrylate (SMA) as the hydrophobic segment, and glycidyl methacrylate (GMA) as the reactive segment. The polymeric micelles were crosslinked with a difunctional reagent, cystamine. The crosslinked micelles showed improved stability and de-crosslinking of disulfide bonds by

Immobilizable MPC polymer

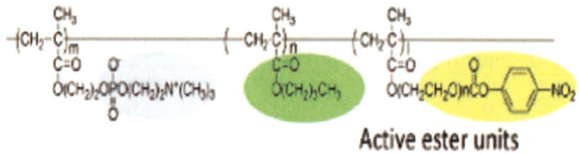

Active ester units

(a)

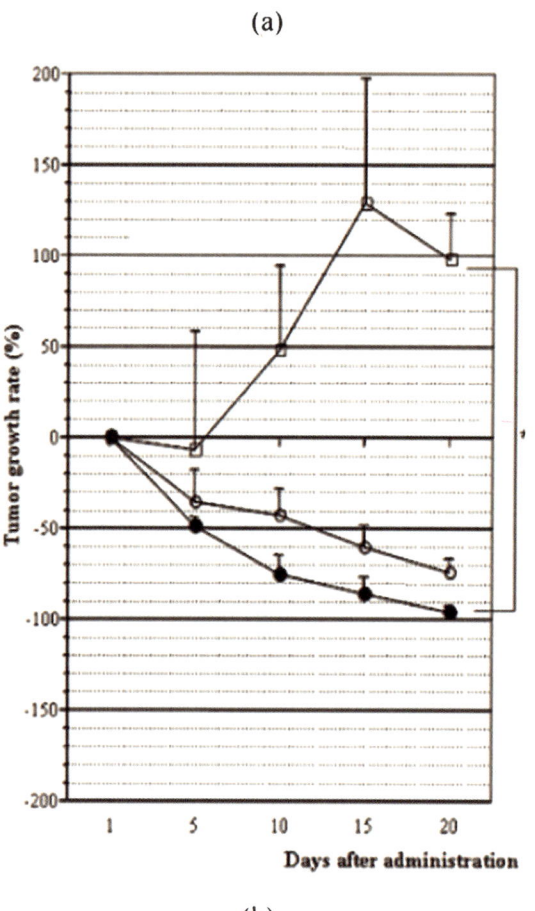

(b)

Figure 10.4   (a) Chemical structure of PMBN. (b) Hepatoma growth inhibition by PTX/PMBN-preS1 *in vivo* on HepG2: PTX (*open rectangles*); PTX/PMBN (*open circles*); PTX/PMBN-preS1 (*filled circles*). Statistical analysis was performed with Student's *t*-test; *$p < 0.05$ *versus* PTX. (Reproduced from Miyata *et al.*[54] with permission from Wiley.)

10 mmol L^{-1} dithiothreitol (DTT), which mimic the reductive intracellular environment after uptake by cells. Cell viability tests indicated that the bioinspired PC-based micelles showed excellent biocompatibility, thus promising great potential for applications in the field of nanocarrier systems.

10.4 CBMA-Based Polymers for Drug Delivery

Little attention has been paid to CB-based polymers in the area of hemocompatible materials in recent years. However, their easy processing, extraordinary stability, and the abundant carboxylate anions of PCBs for the attachment of targeting ligands, therapeutic drugs, and diagnostic labels through conventional NHS/EDC chemistry make PCBs ideal candidates for "theranostics". If the PC-based polymer could be considered as a mimic of cell membranes, CB-based polymers should be a mimic of soluble proteins. The COO$^-$ group is the key negative group on protein surfaces, while NH$_3^+$ is major positive group to balance the negative charge. Although the CBMA-based polymer mainly provides charge by the quaternary ammonium group, it has been proved that the positive charge from other types of amine groups also could form materials to resist nonspecific protein adsorption.[60] Remarkably, the advantage in functionalization of CB-based polymers is superior to other "nonfouling" materials, even in the category of zwitterionic materials. The functional groups, carboxylate anions, are available abundantly and always stay at the top of the protection layer of CBMA-based polymers. This is different from the additional immobilizable units in MPC-based polymers, where the amount and surface abundance of the immobilizable units always need to be considered, as well as the "nonfouling" property altered by the leftover groups.

To demonstrate these advantages, the high stability and targeting efficiency of CB-protected nanoparticles (NPs) has been demonstrated by Cao *et al.*[18] They designed and developed PLGA-PCB block copolymers that self-assemble into NPs with a PLGA core and a PCB shell (Figure 10.5). The strong hydration of zwitterionic CBs and the sharp hydrophilicity/hydrophobicity difference between the PLGA core and the PCB shell give the NP systems high stability in biological media. In their work, no size increase of PLGA-PCB NPs in a phosphate buffered saline (PBS) solution of 10 wt% bovine serum albumin (BSA) or 100% fetal bovine serum (FBS) solution at 37 °C was observed during the 13-hour study, while unmodified PLGA NPs severely aggregated immediately after mixing with these media (Figure 10.5). Furthermore, a long-term study of PLGA-PCB NPs showed that these particles maintain their original size over 5 days in both 10 wt% BSA and 100% FBS media.

Zhang *et al.*[61] prepared a multifunctional and degradable nanogel based on pCBMA. The nanogels encapsulating both a model drug (fluorescently labeled dextran) and an imaging reagent (monodisperse Fe$_3$O$_4$ nanoparticles) have a hydrodynamic size of about 110 nm in saline solution and their size remained

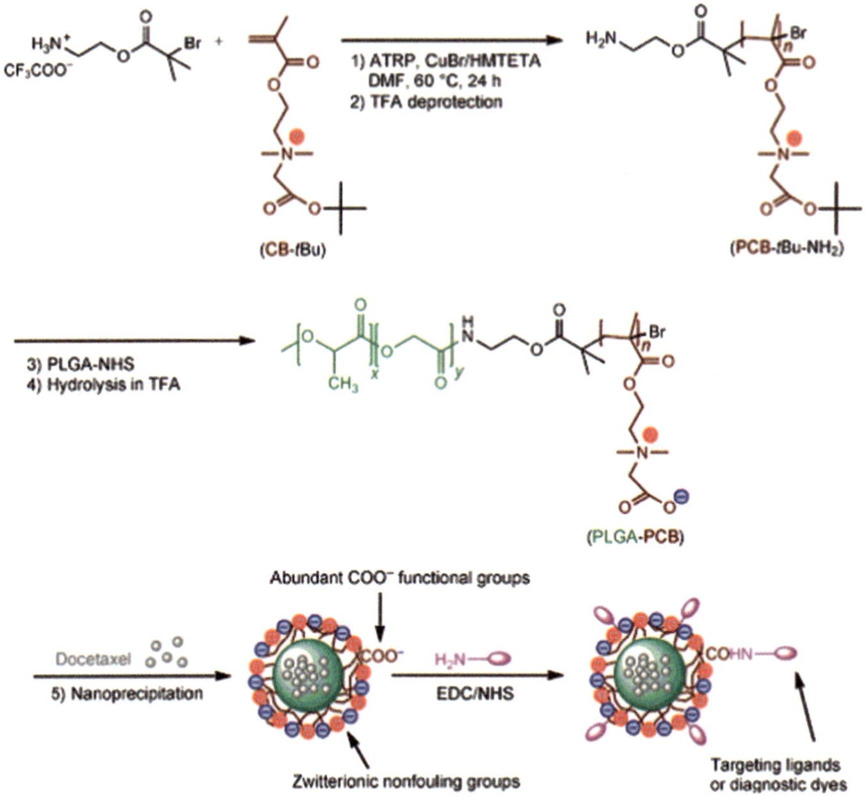

Figure 10.5 Synthesis of PLGA-PCB copolymers, formation of PLGA-PCB/Dtxl NPs, and post-functionalization of NPs with targeting ligands or diagnostic dyes. DMF = *N*,*N*-dimethylformamide; HMTETA = 1,1,4,7,10,10-hexamethyltriethylenetetramine; TFA = trifluoroacetic acid. (Reproduced from Cao *et al.*[18] with permission from Wiley.)

unchanged for over 6 months. The nanogels show low macrophage uptake and significant cellular uptake by human umbilical vein endothelial cells (HUVEC) after being conjugated with a targeting ligand, cyclo[Arg-Gly-Asp-D-Tyr-Lys], implying potential low interaction with the innate immune system and high specific selectivity to targeted cells. The disulfide bonds in the nanogel can facilitate the degradation and accelerate the spontaneous release of the encapsulated model drug and Fe_3O_4 nanoparticles upon exposure to the reducing environment.

Both experiments indicated that nano drug carriers protected by the CB-based "nonfouling" materials could realize compatibility at both the protein level and cellular level. Also the results demonstrated the advantage of the CB "nonfouling" materials in functionalization without compromising the resistance to nonspecific protein adsorption. These results imply that these two systems could potentially be used for *in vivo* drug delivery.

10.5 Conclusion and Perspectives

Zwitterionic materials are very well positioned to play a role in the development of nanomedicine. Zwitterionic polymers have shown excellent stealth and functionalization properties to meet the challenge of long-term stable performance in more complex conditions *in vivo*. They are potential alternatives to PEG for a broad range of applications due to the instability of PEG, especially for CB-based polymers. Their properties, such as excellent stability and the abundant carboxylate anions of CB-based polymers for the attachment of targeting ligands, therapeutic drugs, and diagnostic labels through conventional NHS/EDC chemistry, make pCBs an ideal candidate for "theranostics".

Moreover, since the origin of the repulsive force of zwitterionic materials from ionic hydration is clearly larger than PEG from hydrogen bonding, the thickness of zwitterionic materials to reach "nonfouling" capability on drug carriers could be thinner than a single PEG chain, which might increase the drug loading capability of the drug carrier. It is also possible for the zwitterionic material-protected drug carrier to have a long circulation time and thus enhance both passive and active targeting. The protein surface-like structure of CB zwitterionic materials reminds us that the conventional protein conjugation chemistry could be applicable to them. The possibility of newly designed zwitterionic materials, such as "nonfouling" peptides, might provide a metabolism friendly way to overcome the unnatural degradation of widely used "nonfouling" materials, such as PEG. However, zwitterionic materials are still in the early stage in both proof of conception and practical applications. All these advantages need to be realized through more detailed investigations.

Reference

1. G. Pasut and F. M. Veronese, *Adv. Drug Delivery Rev.*, 2009, **61**, 1177–1188.
2. M. Talelli, C. J. F. Rijcken, C. F. van Nostrum, G. Storm and W. E. Hennink, *Adv. Drug Delivery Rev.*, 2010, **62**, 231–239.
3. S. Kim, Y. Z. Shi, J. Y. Kim, K. Park and J. X. Cheng, *Expert Opin. Drug Delivery*, 2010, **7**, 49–62.
4. J. K. Oh, D. J. Siegwart, H. I. Lee, G. Sherwood, L. Peteanu, J. O. Hollinger, K. Kataoka and K. Matyjaszewski, *J. Am. Chem. Soc.*, 2007, **129**, 5939–5945.
5. Y. Q. Shen, E. L. Jin, B. Zhang, C. J. Murphy, M. H. Sui, J. Zhao, J. Q. Wang, J. B. Tang, M. H. Fan, E. Van Kirk and W. J. Murdoch, *J. Am. Chem. Soc.*, 2010, **132**, 4259–4265.
6. P. Harder, M. Grunze, R. Dahint, G. M. Whitesides and P. E. Laibinis, *J. Phys. Chem. B*, 1998, **102**, 426–436.
7. T. L. Cheng, P. Y. Wu, M. F. Wu, J. W. Chern and S. R. Roffler, *Bioconjugate Chem.*, 1999, **10**, 520–528.
8. J. Xia, P. L. Dubin and E. Kokufuta, *Macromolecules*, 1993, **26**, 6688–6690.

9. H. Xu, K. Q. Wang, Y. H. Deng and D. W. Chen, *Biomaterials*, 2010, **31**, 4757–4763.
10. T. Ishihara, T. Maeda, H. Sakamoto, N. Takasaki, M. Shigyo, T. Ishida, H. Kiwada, Y. Mizushima and T. Mizushima, *Biomacromolecules*, 2010, **11**, 2700–2706.
11. E. Ostuni, R. G. Chapman, R. E. Holmlin, S. Takayama and G. M. Whitesides, *Langmuir*, 2001, **17**, 5605–5620.
12. R. G. Chapman, E. Ostuni, S. Takayama, R. E. Holmlin, L. Yan and G. M. Whitesides, *J. Am. Chem. Soc.*, 2000, **122**, 8303–8304.
13. R. Matsuno and K. Ishihara, *Nano Today*, 2011, **6**, 61–74.
14. S. F. Chen, J. Zheng, L. Y. Li, and S.Y. Jiang, *J. Am. Chem. Soc.*, 2005, **127**, 14473–14478.
15. W. Feng, J. L. Brash and S. P. Zhu, *Biomaterials*, 2006, **27**, 847–855.
16. Z. Zhang, T. Chao, S. F. Chen and S. Y. Jiang, *Langmuir*, 2006, **22**, 10072–10077.
17. Z. Zhang, S. F. Chen, Y. Chang and S. Y. Jiang, *J. Phys. Chem. B*, 2006, **110**, 10799–10804.
18. Z. Q. Cao, Q. M. Yu, H. Xue, G. Cheng and S. Y. Jiang, *Angew. Chem. Int. Ed.*, 2010, **49**, 3771–3776.
19. W. Yang, H. Xue, W. Li, J. L. Zhang and S. Y. Jiang, *Langmuir*, 2009, **25**, 11911–11916.
20. S. F. Chen, L. Y. Li, C. Zhao and J. Zheng, *Polymer*, 2010, **51**, 5283–5293.
21. S. F. Chen, Z. Q. Cao and S. Y. Jiang, *Biomaterials*, 2009, **30**, 5892–5896.
22. W. Yang, L. Zhang, S. L. Wang, A. D. White and S. Y. Jiang, *Biomaterials*, 2009, **30**, 5617–5621.
23. Y. Chang, S. F. Chen, Z. Zhang, and S. Y. Jiang, *Langmuir*, 2006, **22**, 2222–2226.
24. M. Kyomoto, T. Moro, Y. Takatori, H. Kawaguchi, K. Nakamura and K. Ishihara, *Biomaterials*, 2009, **30**, 1017–1024
25. J. H. Seo, R. Matsuno, T. Konno, M. Takai and K. Ishihara, *Biomaterials*, 2008, **29**, 1367–1376.
26. S. Y. Jiang and Z. Q. Cao, *Adv. Mater.*, 2010, **22**, 920–932.
27. L. Y. Li, S. F. Chen, J. Zheng, B. D. Ratner and S. Y. Jiang, *J. Phys. Chem. B*, 2005, **109**, 2934–2941.
28. J. Wu, W. F. Lin, Z. Wang, S. F. Chen and Y. Chang, *Langmuir*, 2012, **28**, 7436–7441.
29. J. C. Hower, M. T. Bernards, S. Chen, H.-K. Tsao, Y.-J. Sheng and S. Jiang, *J. Phys. Chem. B*, 2008, **113**, 197–201.
30. X. A. Zhang, W. F. Lin, S. F. Chen, H. Xu and H. C. Gu, *Langmuir*, 2011, **27**, 13669–13674.
31. L. R. Carr and S. Y. Jiang, *Biomaterials*, 2010, **31**, 4186–4193.
32. Z. Q. Cao, N. Brault, H. Xue, A. Keefe and S. Y. Jiang, *Angew. Chem. Int. Ed.*, 2011, **50**, 6102–6104.
33. A. B. Lowe, M. Vamvakaki, M. A. Wassall, L. Wong, N. C. Billingham, S. P. Armes and A. W. Lloyd, *J. Biomed. Mater. Res.*, 2000, **52**, 88–94.

34. D. Samanta, S. McRae, B. Cooper, Y. Hu and T. Emrick, *Biomacromolecules*, 2008, **9**, 2891–2897.
35. A. L. Lewis, J. Berwick, M. C. Davies, C. J. Roberts, J. H. Wang, S. Small, A. Dunn, V. O'Byrne, R. P. Redman and S. A. Jones, *Biomaterials*, 2004, **25**, 3099–3108.
36. Y. Q. Tang, S. Y. Liu, S. P. Armes and N. C. Billingham, *Biomacromolecules*, 2003, **4**, 1636–1645.
37. J. P. Xu, J. Ji, W. D. Chen and J. C. Shen, *J. Controlled Release*, 2005, **107**, 502–512.
38. T. Goda, Y. Goto and K. Ishihara, *Biomaterials*, 2011, **31**, 2380–2387.
39. D. Soma, J. Kitayama, T. Konno, K. Ishihara, J. Yamada, T. Kamei, H. Ishigami, S. Kaisaki and H. Nagawa, *Cancer Sci.*, 2009, **100**, 1979–1985.
40. T. Kamei, J. Kitayama, H. Yamaguchi, D. Soma, S. Emoto, T. Konno, K. Ishihara, H. Ishigami, S. Kaisaki and H. Nagawa, *Cancer Sci.*, 2010, **102**, 200–205.
41. M. Wada, H. Jinno, M. Ueda, T. Ikeda, M. Kitajima, T. Konno, J. Watanabe and K. Ishihara, *Anticancer Res.*, 2007, **27**, 1431–1435.
42. S. I. Yusa, K. Fukuda, T. Yamamoto, K. Ishihara and Y. Morishima, *Biomacromolecules*, 2005, **6**, 663–670.
43. H. Y. Chu, N. Liu, X. Wang, Z. Jiao and Z. M. Chen, *Int. J. Pharm.*, 2009, **371**, 190–196.
44. T. Kano, C. Kakinuma, S. Wada, K. Morimoto and T. Ogihara, *Drug Metab. Pharmacol.*, 2011, **26**, 79–86.
45. A. G. A. Coombes, S. Tasker, M. Lindblad, J. Holmgren, K. Hoste, V. Toncheva, E. Schacht, M. C. Davies, L. Illum and S. S. Davis, *Biomaterials*, 1997, **18**, 1153–1161
46. M. F. Zambaux, F. Bonneaux, R. Gref, P. Maincent, E. Dellacherie, M. J. Alonso, P. Labrude and C. Vigneron, *J. Controlled Release*, 1998, **50**, 31–40.
47. T. Konno, K. Kurita, Y. Iwasaki, N. Nakabayashi and K. Ishihara, *Biomaterials*, 2001, **22**, 1883–1889.
48. G. H. Hsiue, C. L. Lo, C. H. Cheng, C. P. Lin, C. K. Huang and H. H. Chen, *J. Polym. Sci., Part A: Polym. Chem.*, 2007, **45**, 688–698.
49. G. Y. Liu, L. P. Lv, C. J. Chen, X. F. Hu and J. Ji, *Macromol. Chem. Phys.*, 2011, **212**, 643–651.
50. S. Tu, Y. W. Chen, Y. B. Qiu, K. Zhu and X. L. Luo, *Macromol. Biosci.*, 2011, 1416–1425.
51. B. M. Cooper, D. Chan-Seng, D. Samanta, X. Zhang, S. Parelkar and T. Emrick, *Chem. Commun.*, 2009, 815–817.
52. Y. Iwasaki and K. Akiyoshi, *Biomacromolecules*, 2006, **7**, 1433–1438.
53. X. J. Chen, S. McRae, S. Parelkar and T. Emrick, *Bioconjugate Chem.*, 2009, **20**, 2331–2341.
54. R. Miyata, M. Ueda, H. Jinno, T. Konno, K. Ishihara, N. Ando and Y. Kitagawa, *Int. J. Cancer*, 2009, **124**, 2460–2467.

55. T. Shimada, M. Ueda, H. Jinno, N. Chiba, M. Wada, J. Watanabe, K. Ishihara and Y. Kitagawa, *Anticancer Res.*, 2009, **29**, 1009–1014.
56. Y. Iwasaki, H. Maie and K. Akiyoshi, *Biomacromolecules*, 2007, **8**, 3162–3168.
57. M. Licciardi, G. Giammona, J. Z. Du, S. P. Armes,Y. Q. Tang and A. L. Lewis, *Polymer*, 2006, **47**, 2946–2955.
58. J. P. Salvage, S. F. Rose, G. J. Phillips, G. W. Hanlon, A. W. Lloyd, I. Y. Ma, S. P. Armes, N. C. Billingham and A. L. Lewis, *J. Controlled Release*, 2005, **104**, 259–270.
59. L. P. Lv, J. P. Xu, X. S. Liu, G. Y. Liu, X. A. Yang and J. A. Ji, *Macromol. Chem. Phys.*, 2010, **211**, 2292–2300.
60. S. F. Chen and S. Y. Jiang, *Adv. Mater.*, 2008, **20**, 335–338.
61. L. Zhang, H. Xue, Z. Q. Cao, A. Keefe, J. Wang and S. Y. Jiang, *Biomaterials*, 2011, **32**, 4604–4608.
62. Z. Zhang, H. Vaisocherova, G. Cheng, W. Yang, H. Xue and S. Y. Jiang, *Biomacromolecules*, 2008, **9**, 2686–2692.
63. Z. Zhang, M. Zhang, S. F. Chen, T. A. Horbetta, B. D. Ratner and S. Y. Jiang, *Biomaterials*, 2008, **29**, 4285–4291.
64. Z. Zhang, G. Cheng, L. R. Carr, H. Vaisocherová, S. F. Chen and S. Y. Jiang, *Biomaterials*, 2008, **29**, 4719–4725.
65. Z. Zhang. G. Cheng, S. F. Chen, J. D. Bryers and S. Y. Jiang, *Biomaterials*, 2007, **28**, 4192–4199.
66. G. Z. Li, H. Xue, G. Cheng, S. F. Chen, F. B. Zhang and S. Y. Jiang, *J. Phys. Chem. B*, 2008, **112**, 15269–15274.
67. Z. Zhang, S. F. Chen and S. Jiang, *Biomacromolecules*, 2006, **7**, 3311–3315.
68. C. Gang, X. Hong, Z. Zheng, S. F. Chen and S. Y. Jiang, *Angew. Chem. Int. Ed.*, 2008, **120**, 8963–8966.
69. K. Nam, T. Kimura and A. Kishida, *Biomaterials*, 2007, **28**, 1–8.
70. M. Pitsikalis and N. Hadjichristidis, *Langmuir*, 2007, **23**, 4214–4224.
71. M. Kimura, K. Fukumoto, J. Watanabe, M. Takai and K. Ishihara, *Biomaterials*, 2005, **26**, 6853–6862.
72. F. Goto, K. Ishihara, Y. Iwasaki, K. Katayama, R. Enomoto and S. Yusa, *Polymer*, 2011, **52**, 2810–2818.
73. L. Mi, M. T. Bernards, G. Cheng, Q. Yu and S. Jiang, *Biomaterials*, 2010, **31**, 2919–2925.
74. G. Z. Li, H. Xue, C. L. Gao, F. B. Zhang and S. Y. Jiang, *Macromolecules*, 2010, **43**, 14–16.
75. M. T. Bernards, G. Cheng, Z. Zhang, S. F. Chen and S. Y. Jiang, *Macromolecules*, 2008, **41**, 4216–4219.
76. T. Ishi, A. Wada, S. Tsuzuki, M. Casolaro and Y. Ito, *Biomacromolecules*, 2007, **8**, 3340–3344.
77. K. Shiraishi, T. Ohnishi and K. Sugiyama, *Macromol. Chem. Phys.*, 1998, **199**, 2023–2028.
78. S. Nagaoka, A. Shundo, T. Satoh, K. Nagira, R. Kishi, K. Ueno, K. Iio and H. Ihara, *Synth. Commun.*, 2005, **35**, 2529–2534.

CHAPTER 11

Polymer-Based Prodrugs for Cancer Chemotherapy

QIHANG SUN[b], JINQIANG WANG[a], MACIEJ RADOSZ[b]
AND YOUQING SHEN*[a]

[a] Center for Bionanoengineering and State Key Laboratory of Chemical Engineering, Department of Chemical and Biological Engineering, Zhejiang University, Hangzhou 310027, P. R. China; [b] Department of Chemical and Petroleum Engineering, Soft Materials Laboratory, University of Wyoming, Laramie, WY 82071, USA
*E-mail: shenyq@zju.edu.cn

11.1 Introduction

Chemotherapy has achieved great success in cancer treatment during recent past decades, but it is still challenged by poor solubility, low tumor selectivity, and associated toxicity of most anticancer drugs.[1] The prodrug strategy is one of the most commonly used chemical/biochemical strategies towards improving the therapeutic index of anticancer drugs.[2]

A prodrug is defined as a chemically modified drug derivative that is inactive or less active but metabolized *in vivo* to release the parent active component in the pharmacological environment[3] to improve the drug's desirability, including water solubility, patient acceptability (*e.g.* decreasing pain on injection), and pharmacokinetics (absorption, biodistribution, metabolism, and elimination; ADME). A typical prodrug normally consists of three parts (Figure 11.1): (1) the parent drug exerting therapeutic effects; (2) the chemical linker bridging the parent drug and the modifier; and (3) the modifier endowing the prodrug with

RSC Polymer Chemistry Series No. 3
Functional Polymers for Nanomedicine
Edited by Youqing Shen
© The Royal Society of Chemistry 2013
Published by the Royal Society of Chemistry, www.rsc.org

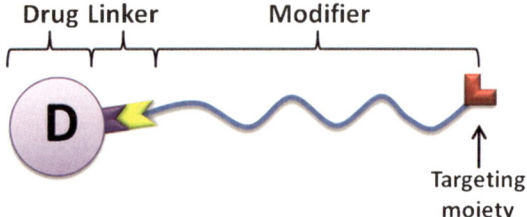

Figure 11.1   Sketch of a typical polymer prodrug.

various properties and functionalities. In this chapter, we focus on the low molecular weight drugs chemically modified with polymers for cancer drug delivery, some of which are generally called polymer–drug conjugates.[4]

   The ultimate goal of cancer drug delivery is to increase the tumor selectivity or targeting ability to enhance the therapeutic efficacy and reduce side effects.[5] According to the three stages of the cancer drug-delivery process discussed in Chapter 3, such an ideal anticancer prodrug should be able to sustainably circulate in the blood compartments, efficiently extravasate into the tumor and penetrate through the tumor tissue, and finally get into tumor cells and release the parent drug.[6] Most polymer-based prodrugs endow the parent drugs with adequate aqueous solubility and improved tumor targeting. However, current polymer prodrugs have some inherent drawbacks, *e.g.* low drug content and low tumor specificity, that limit their further translation from the benchtop to the bedside. Thus, design of novel prodrugs with desirable properties is needed. In this chapter, we briefly review the methods for preparing conventional polymer prodrugs and their associated problems, and summarize new strategies showing great promise and the remaining challenges in translational prodrugs.

11.2   Design of Polymer-Based Prodrugs

As shown in Figure 11.1, a polymer-based prodrug consists of three parts: the drug, the linker, and the modifier. The drug determines the potency, while the linker and modifier determine where the drug goes to exert the potency (targeting). Anticancer drugs good for making prodrugs must have at least one reactive site to be anchored to the polymer modifier *via* a linker. The parent drugs must have high potency to avoid using too much excipient(s). The most investigated anticancer drugs are doxorubicin (DOX), paclitaxel (PTX), camptothecin (CPT), and its derivatives (*e.g.* SN38).

11.2.1   Linkers

Common chemical linkers for the synthesis of prodrugs for cancer therapy, such as ester, amide, hydrazone, and disulfide bonds, have already been summarized by Mahato *et al.*[7] An ideal linker rendering the prodrug maximal therapeutic efficacy and minimal toxicity would be the one that is stable in the blood compartments but labile in cancer cells.

The linker should first be stable in blood circulation to ensure low toxicity.[8] For example, the PK1 conjugate in which DOX was covalently bound to poly[*N*-(2-hydroxypropyl)methacrylamide] (PHPMA) *via* a blood-stable but lysosome-labile peptidyl linker had very low dose-limiting toxicity.[9] In phase I clinical and pharmacokinetic studies, PK1 had a maximum tolerated dose (MTD) of 320 mg m^{-2}, and no congestive cardiac failure despite individual cumulative doses up to 1,680 mg m^{-2}.[9] However, both the PHPMA-PTX prodrug named PNU166945[10] and the poly(methacryloylglycinamide) (PMAG)-CPT prodrug[11] experienced dose-limiting toxicity in phase I clinical study because their easily hydrolysable ester linkers released the drugs while in circulation.[12]

Upon reaching the intracellular target, the prodrug must efficiently release the parent drug to exert its pharmaceutical action because only the liberated drug becomes active.[13] Stable prodrugs, *e.g.* drugs bonded to poly(lactic-*co*-glycolic acid) (PLGA)[14] or poly(L-aspartic acid) [P(Asp)],[15] showed low or even no anticancer activity. It is preferable that the linker be cleavable in the tumor microenvironment. This is achieved by using labile linkers responsive to the tumor's extracellular or intracellular stimuli (Figure 11.2).[16] Lysosomal pH-labile linkers (*e.g.*, amide, hydrazone, or *cis*-aconityl bonds) ensure the intracellular drug release in lysosomes and are widely used.[17] For instance, DOX was conjugated to the P(Asp) block of its block copolymer poly(ethylene glycol) (PEG)-*b*-P(Asp) *via* acid-labile amide[18] or hydrazone linkers.[19] DOX

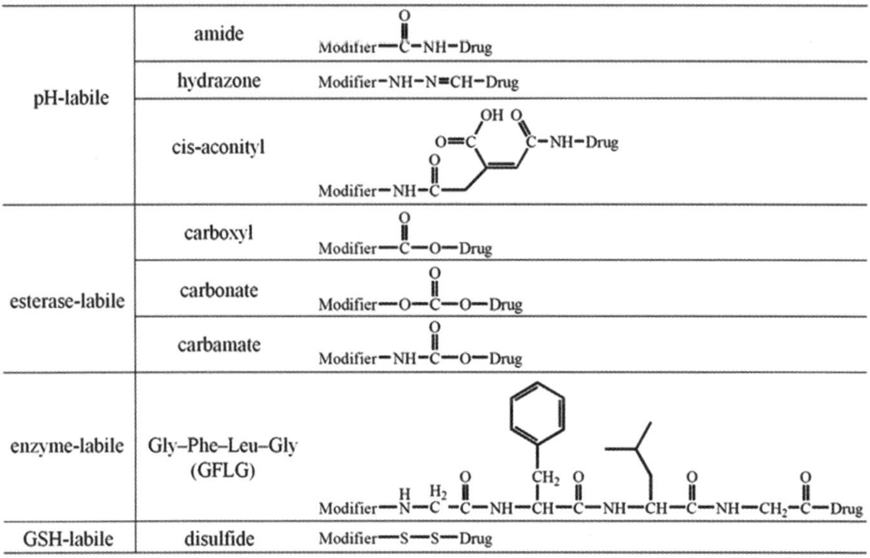

pH-labile	amide	Modifier—C(=O)—NH—Drug
	hydrazone	Modifier—NH—N=CH—Drug
	cis-aconityl	O=C(OH)—...—C(=O)—NH—Drug, Modifier—NH—C(=O)—
esterase-labile	carboxyl	Modifier—C(=O)—O—Drug
	carbonate	Modifier—O—C(=O)—O—Drug
	carbamate	Modifier—NH—C(=O)—O—Drug
enzyme-labile	Gly–Phe–Leu–Gly (GFLG)	Modifier—N(H)—CH$_2$—C(=O)—NH—CH(CH$_2$Ph)—C(=O)—NH—CH(CH(CH$_3$)$_2$)—C(=O)—NH—CH$_2$—C(=O)—Drug
GSH-labile	disulfide	Modifier—S—S—Drug

Figure 11.2 Commonly used extracellular and intracellular stimuli-labile linkers in prodrugs.

conjugated to PHPMA and biodegradable star polymers *via* such linkers demonstrated a fast DOX release at pH 4.[20] An interesting example is the dual pH-responsive prodrug PPC-Hyd-DOX-DA developed by Du and co-workers.[21] At tumor extracellular pH ($\sim$6.8), the prodrug reversed its surface charge from negative to positive *via* acid-labile amide bonds for fast internalization; at endosomal or lysosomal pH ($\sim$5.0), the drug was fast released due to the breakable hydrazone bonds. Esterases are ubiquitously distributed in the body and can readily hydrolyze ester bonds in prodrugs; thereby esterase-labile linkers, such as carboxyl, carbonate, and carbamate esters, are also employed.[22] For instance, in IT-101, CPT was conjugated to the linear β-cyclodextrin (β-CD) and PEG copolymer *via* the ester bond.[23] The conformation of released CPT was not changed compared to the parent CPT and the release rate of the conjugated CPT can be tuned.[24] Lysosomal degradable peptides [*e.g.*, glycylphenylalanylleucylglycine (GFLG)], which are cleavable by lysosomal enzymes, are also useful linkers in prodrugs.[25] Recently, the disulfide bond,[26] which can be cleaved by intracellular glutathione (GSH), has attracted increasing attention as linkers for intracellularly triggered drug release. The underlying rationale is the elevated intracellular GSH concentration, particularly in cancer cells, but low in the blood.[27]

11.2.2 Modifiers

The modifier may consist of a polymer chain and targeting moieties and others such as a tracer moiety. The main roles of the polymer chain are to endow the prodrug with water solubility and a long blood-circulation time for altered ADME and tumor targeting capability. The targeting moiety facilitates the prodrug's active targeting, as discussed in Chapter 2.

Most anticancer drugs are water insoluble, giving them poor bioavailability.[28] Anchoring drugs to water-soluble polymer chains makes them water soluble. For instance, PTX is an extremely water-insoluble anticancer drug ($<$0.01 mg mL^{-1}).[29] Conjugating PTX to a PEG[30] or poly(L-glutamic acid) (PGA)[31] resulted in highly water-soluble prodrugs with antitumor effects superior to PTX itself.[31,32]

It is generally assumed that by prolonging the blood circulation time, a polymer prodrug has more opportunity to pass through the hyperpermeable tumor blood vessels and extravasate into tumor tissue *via* the EPR effect.[1b] Prodrugs with a stealth property may evade the reticuloendothelial system (RES) screening[33] and thus circulate for a long time in the blood compartments, resulting in greatly increased tumor drug concentrations (10-fold or higher) and MTD relative to administration of the free drug.[34] The stealth character of a polymer prodrug is mainly determined by the polymer's properties. The polymer must be water soluble and, very importantly, not immunogenic. A very interesting example is the natural biopolymer dextran. It is usually assumed to be non-immunogenic, but drugs conjugated to dextran and carboxymethyldextran can be captured by the RES, causing dose-limiting toxicity.[35] PHPMA[36] and PEG[37] are the most studied bio-inert polymers able to render prodrugs stealthy. For example, NKTR-102, a PEG prodrug of irinotecan (a chemotherapy drug that is metabolized to its active metabolite SN38 in the body), increased the half-life of SN38 90-fold (from 4 h to 15 days) and 10-fold in colorectal and lung cancer treatment, respectively.[38] Other PEG-based prodrugs of docetaxel (NKTR-105), SN38 (EZN-2208), and CPT (Pegamotecan) also showed longer half-life times and increased accumulation at tumor sites and thereby tumor growth suppression.[39]

Incorporating targeting ligands in the polymer modifier, such as antibodies, peptides, aptamers, and folic acid, may promote tumor targeting (building on the EPR effect) by receptor-mediated delivery,[40] as discussed in Chapter 2. For example, PK1 conjugated with galactosamine, named PK2,[41] delivered 3.3 ± 5.6% of the dose to the tumor and 16.9 ± 3.9% to the liver, whereas PK1 showed no obvious targeting.[42] Zhu *et al.*[43] and Borgman *et al.*[44] also proved that incorporating targeting ligands in polymer modifiers increased the prodrug's tumor accumulation. More examples have been reviewed elsewhere.[45]

11.2.3 Drawbacks of Current Polymer-Based Prodrugs

An inherent dilemma in polymer prodrugs is their drug-loading content *versus* the water solubility and thereby the stealth capability. A high drug-loading content in a nanocarrier reduces use of excipients and minimizes the related side effects, and should be considered as a demanding criterion to judge a prodrug's quality.[46] As most parent drugs are highly hydrophobic, the modifier or the water-soluble polymer chain must be sufficiently long to make the prodrug water soluble. Therefore, in most polymer prodrugs the drug content is less than 10 wt%.[47] Typical examples are PEG-DOX prodrugs using PEG with different molecular weights from 5,000 to 20,000 g mol^{-1}, which contain 2.7–8.0 wt% DOX.[48] A four-armed PEG-SN38 prodrug named EZN-2208, prepared by coupling SN38 to multiarm PEGs (Figure 11.3A), had an increased drug content, but only of 3.7 wt%, compared to the CPT-PEG-diol prodrug Pegamotecan.[39] As for PHPMA–drug conjugates, for instance PK1, the drug contents are also as low as 8 wt%.[49] Some prodrugs have high drug-loading contents. For instance, the prodrug CT-2103, PTX conjugated to PGA through its 2′-hydroxyl group (Figure 11.3B),[50] has a drug content of approximately 37 wt% and a PGA-20(S)-CPT prodrug has a 30–35 wt% drug content.[51] However, increasing the drug content not only lowers the prodrug water solubility but may also cause opsonization,[37] resulting in rapid blood clearance.

Another complication is also related to the polymer chain's molecular weight. To have a long blood-circulation time for passive tumor targeting, the molecular weight of a prodrug must be higher than the polymer's renal threshold, *e.g.* 40 kDa for PEG[52] and 45 kDa for PHPMA,[53] but the safety

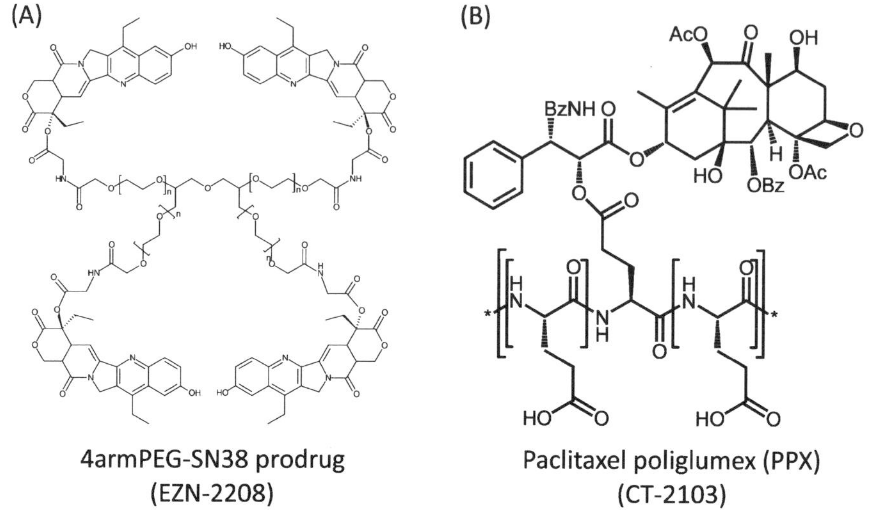

(A)   (B)

4armPEG-SN38 prodrug   Paclitaxel poliglumex (PPX)
(EZN-2208)   (CT-2103)

Figure 11.3   Chemical structures of EZN-2208 and CT-2103 polymer prodrugs.
(Reproduced from Pasut and Veronese[39] and Singer[50b] with permission
from Elsevier.)

prerequisite for *in vivo* applications is that the prodrug and the modifier must
be smaller than the renal threshold if their excretion pathway is through the
kidneys. The current strategy to reconcile these conflicting requirements is to
choose a polymer molecular weight close to the threshold, for example 30 kDa
for PHPMA, but this way inevitably shortens the blood circulation time of the
prodrug. Therefore, new strategies resolving these dilemmas are required to
design novel polymer prodrugs with higher drug content and better therapeutic
indices.

11.3   New Strategies for Polymer Prodrugs

11.3.1   Self-Assembling Prodrugs

Given that prodrugs with a high hydrophobic drug content are not water
soluble, we may take advantage of the hydrophobicity and directly use the
drugs as the hydrophobic part and the modifier as the hydrophilic segment to
make amphiphilic prodrugs that can self-assemble into vesicles or micelles as
nanocarriers. An important advantage of this way is that even though a single
prodrug molecule is small (several kDa) and far below the renal threshold, the
formed vesicles or micelles are generally larger (tens of nanometers in
diameter) than the kidney threshold (~ 5.5 nm) and thus can be retained for
long blood circulation.[54] The second advantage of this strategy is that because
the drug molecules are used as a part of the nanocarrier and replace some of
the inert carrier materials, the nanocarrier's drug content is high. For instance,
the anticancer drug CPT is very hydrophobic, with a water solubility of

2.5 µg mL^{-1}.[55] Conjugating one or two molecules to short oligomer chains of ethylene glycol (OEG) *via* β-thioester bonds produces an amphiphilic phospholipid-mimicking prodrug OEG-CPT or OEG-DiCPT (Figure 11.4). These prodrugs can self-assemble in aqueous solution into stable liposome-like nanocapsules of around 100 nm in diameter (Figure 11.4A). As shown in Figure 11.4, the hydrophobic wall of the vesicle contains no aliphatic chains like normal liposomes but is solely made of hydrophobic CPT drug molecules. Thus, the drug content is as high as 40 or 58 wt%. Another important characteristic is that since the nanocapsules or vesicles are made of the prodrug with well-defined structures, their drug content is fixed independently of the

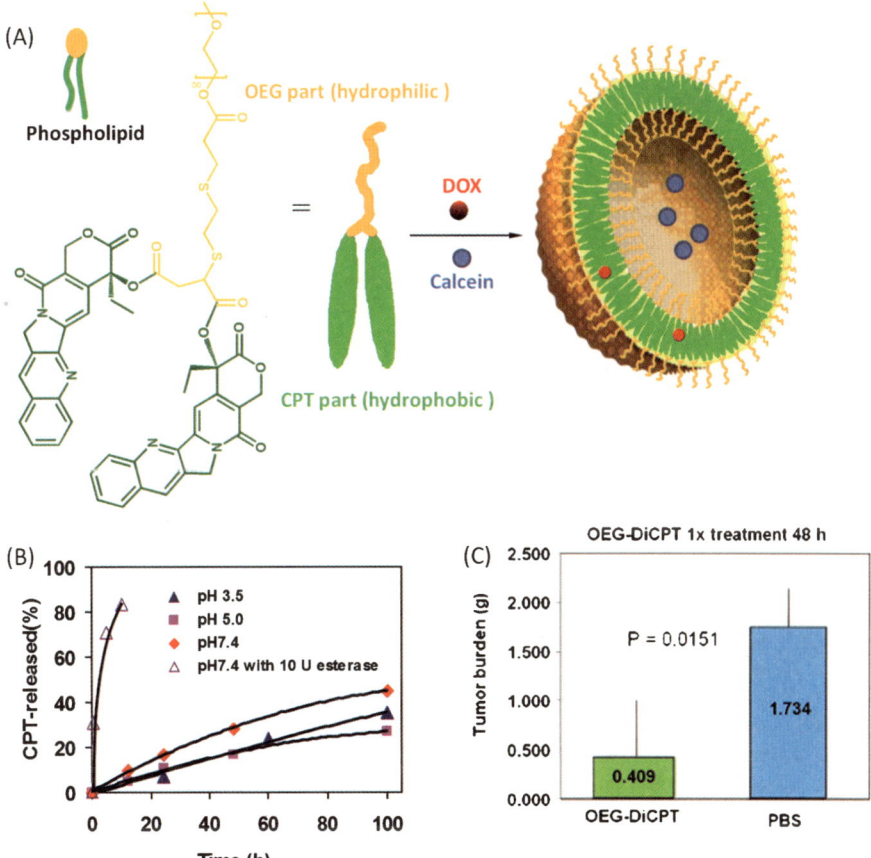

Figure 11.4 (A) Scheme of phospholipid-mimicking prodrugs OEG-DiCPT self-assembling into stable liposome-like nanocapsules to load other drugs. (B) CPT release kinetics from the OEG-DiCPT nanocapsules. (C) *In vivo* antitumor activity of OEG-DiCPT to xenografted intraperitoneal SKOV-3 ovarian tumors in BALB/c strain nu/nu mice. (Adapted from Shen *et al.*[56] with permission from the American Chemical Society.)

nanocarrier size and preparation methods. This character makes it easy to scale-up for translation, as discussed in Chapter 3. The nanocapsules have no burst release but can release CPT quickly once inside the cells (Figure 11.4B). *In vivo* tests showed that the nanocapsules had strong antitumor activity (Figure 11.4C).[56]

Another example is a self-assembling curcumin prodrug. Curcumin, a substance in turmeric, has been shown to have high cytotoxicity towards various cancer cell lines,[57] but its extremely low water solubility and stability make its bioavailability exceedingly low and generally inactive in *in vivo* anticancer tests.[58] Using the same concept, we conjugated curcumin with two short OEG chains (Curc-OEG) *via* β-thioester bonds that are labile in the presence of intracellular glutathione and esterases and obtained an intracellular-labile amphiphilic surfactant-like curcumin prodrug: curcumin-OEG (Curc-OEG) (Figure 11.5).[59] Curc-OEG formed stable nanoparticles in aqueous conditions and served two roles: an anticancer prodrug and a drug

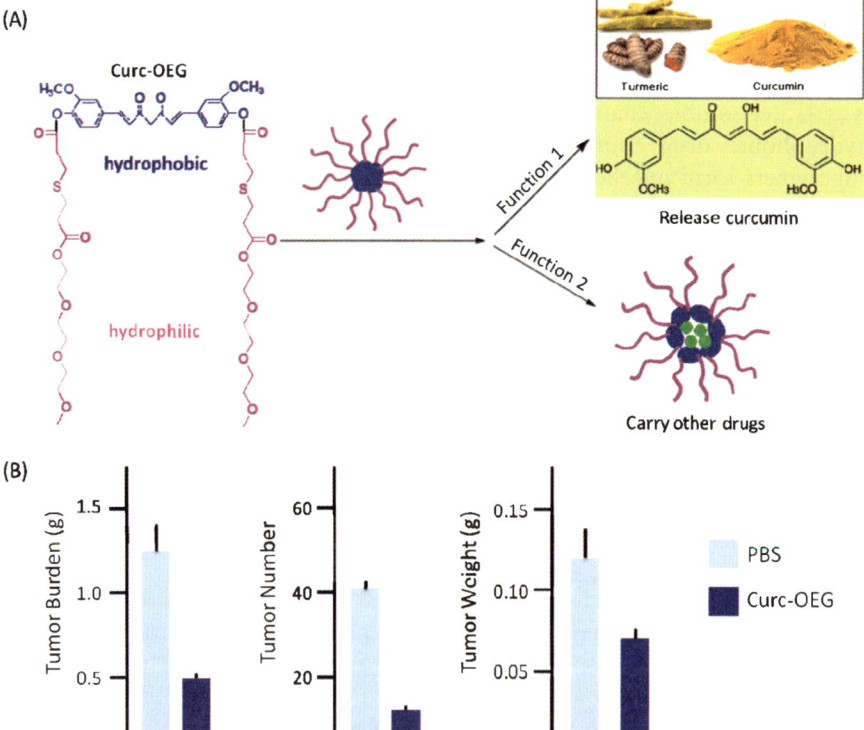

Figure 11.5   (A) Structure of Curc-PEG and its self-assembly into micelles to perform two functions: release curcumin and carry other drugs. (B) *In vivo* antitumor activity of Curc-OEG against SKOV-3 (*left* and *middle*) and MDA-MB-468 (*right*) carcinoma xenograft models. (Adapted from Tang *et al.*[59] with permission from Future Medicine.)

carrier (Figure 11.5A). As an anticancer prodrug, the formed nanoparticles had a high and fixed curcumin-loading content of 25.3 wt%, and released active curcumin in the intracellular environment. Curc-OEG had high inhibition ability for several cancer cell lines due to apoptosis. Intravenously injected Curc-OEG had a bioavailability (the area under curve value) 250 times of that of the parent curcumin, and thus a high overall tumor concentration. As a result, it significantly reduced tumor weights and tumor numbers in the athymic mice xenografted with intraperitoneal SKOV-3 tumors and subcutaneous (mammary fat pad) MDA-MB-468 tumors (Figure 11.5B). Preliminary systemic toxicity studies found that Curc-OEG did not cause acute or subchronic toxicities in mouse visceral organs at high doses. As a drug carrier, Curc-OEG nanoparticles could carry other anticancer drugs, such as DOX, and ship them into drug-resistant cells, greatly enhancing the cytotoxicity of the loaded drug. Thus, Curc-OEG is a promising prototype that merits further study for cancer therapy.

11.3.2 Prodrug Micelles

Besides the phospholipid-mimicking amphiphilic prodrugs, another type of prodrug capable of self-assembly is the diblock copolymer in which one block is a water-soluble chain such as PEG and the other is conjugated with hydrophobic drug molecules as the water-insoluble block. Such block copolymers form micelles in aqueous solution with drug molecules anchored in the micelle core. For instance, the Matsumura group conjugated SN38 to the PGA block of a PEG-PGA copolymer.[60] The resulting conjugate PEG-PGA(SN38) containing approximately 20 wt% drug, named NK012, self-assembled into micelles with a diameter of 20 nm (Figure 11.6). NK012 was reported to eradicate liver metastases and achieve a significantly longer survival rate than CPT-11 ($P = 0.0006$). More recently, our group conjugated SN38 as the hydrophobic group to short poly(methacrylic acid) and obtained amphiphilic PEG-polySN38.[61] This prodrug formed nanoparticles with a high drug-loading content (~ 20 wt%) and tailorable sizes. Sub-100 nm nanoparticles were also self-assembled from a PEG-poly(L-lactide) (PEG-PLA)-based polymer–cisplatin prodrug in which cisplatin was anchored to the PLA terminal *via* a hydrazone bond.[62]

Another interesting self-assembly prodrug for continuous slow release of PTX was developed by Kwon and co-workers.[63] PTX was conjugated to PEG-p(Asp) through hydrazone linkers made from levulinic acid (LEV) or 4-acetylbenzoic acid (4AB) to give amphiphilic PEG-p(Asp-Hyd-LEV-PTX) and PEG-p(Asp-Hyd-4AB-PTX), which assembled into polymeric micelles with diameters of 42 and 137 nm, respectively. Mixing the two prodrugs produced micelles with diameters of 85 and 113 nm, respectively, having pH-dependent release. Detailed reviews can be found elsewhere.[64]

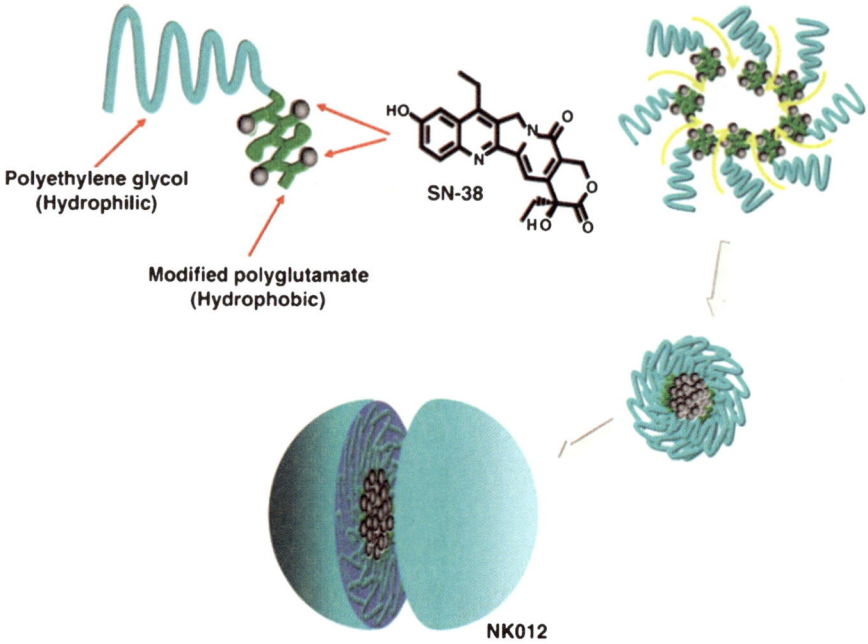

Polyethylene glycol
(Hydrophilic)

SN-38

Modified polyglutamate
(Hydrophobic)

NK012

Figure 11.6   Synthesis and assembly process of NK012. (Reproduced from Matsumura[60a] with permission from Elsevier.)

11.3.3   Drug Polymers

An alternative to using the drug molecules themselves as a part of the nanocarrier is to directly use biofunctional drug molecules as monomers to prepare backbone-type polymer–drug conjugates, or *drug polymers*. One such example is polycurcumin (PCurc; Figure 11.7).[65] Inspired by the bihydroxyl functionality of the curcumin molecule, curcumin was used as a co-monomer to make the curcumin-containing PCurc by polycondensation polymerization. A series of water-soluble PCurcs were prepared with different covalent bonds (*e.g.*, ester, disulfide, acetal) with high and fixed curcumin-loading content and efficiency in comparison with previously developed curcumin-loading nano-medicines.[66] For instance, curcumin was condensed with divinyl ether to produce linear polycurcumin linked *via* acetal bonds. To make the polymer water soluble, PEG-diol was used as a co-monomer. Water-soluble poly-curcumin, PCurc 8, could contain up to 25 wt% curcumin. The polymers were stable at neutral conditions, but hydrolyzed rapidly at lysosomal pH to release curcumin (Figure 11.7A), causing high toxicity to three cancer cell lines (even more cytotoxic than the parent curcumin itself) (Figure 11.7B) and remarkable antitumor activity in the SKOV-3 intraperitoneal xenograft tumor model (Figure 11.7C).

Another drug polymer we developed is the platinum(IV)-coordinated polymer using the platinum(IV) prodrugs DHP and DSP as co-monomers

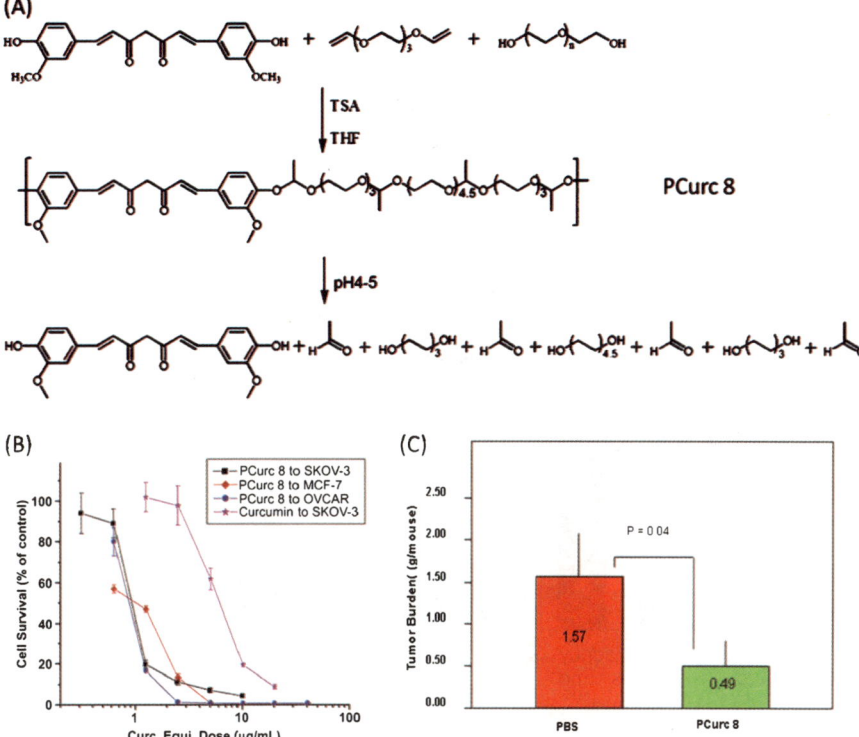

Figure 11.7 (A) Synthesis and acid-catalyzed hydrolysis of PCurc 8. (B) Cytotoxicity of PCurc 8 to SKOV-3, OVCAR-3, and MCF-7 cancer cell lines. (C) *In vivo* antitumor activity of Curc-OEG against SKOV-3 carcinoma xenografts. (Adapted from Tang *et al.*[65])

(Figure 11.8A).[67] The polymers had good water solubility and high and fixed platinum-loading contents, 27.7% for P(DSP-EDA) and 29.6% for P(DSP-PA), respectively, which were not achievable by any conventional conjugation methods. Additionally, the biodistribution results suggested that the conjugated platinum polymer might greatly minimize the liver and renal toxicity (Figure 11.8B).[67]

11.4 Future Challenges

Although polymer–drug conjugates or prodrugs have been investigated for several decades, the number of approved polymer-based prodrugs is still small.[68] Thus, the design of polymer-based prodrugs with more efficient drug delivery to achieve higher therapeutic efficacy is still a pressing need. Three aspects should be taken into consideration when designing a new prodrug:

(A)

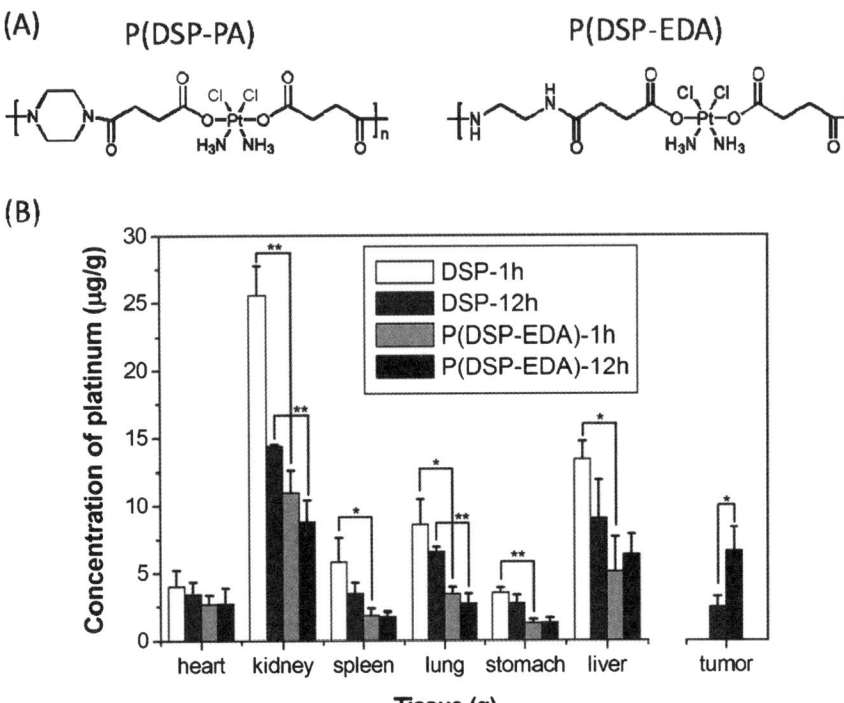

(B)

Figure 11.8 (A) Chemical structures of backbone-type platinum(IV) prodrugs. (B) The biodistribution of platinum in various organs and SKOV-3 xenograft tumors after a single i.v. administration of free DSP or P(DSP-EDA) at a DSP-equivalent dose of 20 mg kg^{-1}. (Adapted from Yang et al.[67] with permission from Elsevier.)

(1) the drug's configuration or structure should not be changed and its therapeutic efficacy should not be impaired; (2) the drug should remain tightly conjugated in circulation in order to reduce the drug-associated toxicity; and (3) the drug should be intracellularly released from conjugates efficiently once inside the tumor cells. The first aspect is the prerequisite of a conjugate, while the latter two aspects are the elementary 2R (*R*etention in circulation *versus* *R*elease in cells) rules[6] discussed in Chapter 3. Certainly, elementary 2S (*S*tealthy in circulation *versus* *S*ticky in tumor) rules[6] also should be addressed as further requirements of a superior prodrug.

Challenges remain even for those proved polymer–drug conjugates, including safety issues, good manufacturing process, and the cost-effective scale-up. Most of the proven conjugates use PHPMA or PEG as the modifier.[69] Although PHPMA and PEG are well tolerated clinically, their non-biodegradability is a significant disadvantage, especially in cases of chronic parenteral administration and high required doses.[69] Compared to other polymeric nanocarriers, the self-assembling prodrugs can be easily formulated with a fixed composition and a relatively good manufacturing process, providing a novel design concept for translation of polymer-based prodrugs from benchtop to the bedside.

References

1. (*a*) R. Tong, D. A. Christian, L. Tang, H. Cabral, J. R. Baker, Jr., K. Kataoka, D. E. Discher and J. Cheng, *MRS Bull.*, 2009, **34**, 422–431; (*b*) F. Danhier, O. Feron and V. Preat, *J. Controlled Release*, 2010, **148**, 135–146.
2. J. S. Sohn, J. I. Jin, M. Hess and B. W. Jo, *Polym. Chem.*, 2010, **1**, 778–792.
3. J. Rautio, H. Kumpulainen, T. Heimbach, R. Oliyai, D. Oh, T. Jarvinen and J. Savolainen, *Nat. Rev. Drug Discovery*, 2008, **7**, 255–270.
4. F. Greco and M. J. Vicent, *Adv. Drug Delivery Rev.*, 2009, **61**, 1203–1213.
5. I. Brigger, C. Dubernet and P. Couvreur, *Adv. Drug Delivery Rev.*, 2002, **54**, 631–651.
6. Q. Sun, M. Radosz and Y. Shen, *J. Controlled Release*, 2012, **164**, 156–169.
7. R. Mahato, W. Tai and K. Cheng, *Adv. Drug Delivery Rev.*, 2011, **63**, 659–670.
8. (*a*) J. W. Singer, S. Shaffer, B. Baker, A. Bernareggi, S. Stromatt, D. Nienstedt and M. Besman, *Anti-Cancer Drugs*, 2005, **16**, 243–254; (*b*) S. Jaracz, J. Chen, L. V. Kuznetsova and L. Ojima, *Bioorg. Med. Chem.*, 2005, **13**, 5043–5054.
9. P. A. Vasey, S. B. Kaye, R. Morrison, C. Twelves, P. Wilson, R. Duncan, A. H. Thomson, L. S. Murray, T. E. Hilditch, T. Murray, S. Burtles, D. Fraier, E. Frigerio and J. Cassidy, *Clin. Cancer Res.*, 1999, **5**, 83–94.
10. J. M. M. Terwogt, W. W. T. Huinink, J. H. M. Schellens, M. Schot, I. A. M. Mandjes, M. G. Zurlo, M. Rocchetti, H. Rosing, F. J. Koopman and J. H. Beijnen, *Anti-Cancer Drugs*, 2001, **12**, 315–323.
11. N. E. Schoemaker, C. van Kesteren, H. Rosing, S. Jansen, M. Swart, J. Lieverst, D. Fraier, M. Breda, C. Pellizzoni, R. Spinelli, M. G. Porro, J. H. Beijnen, J. H. M. Schellens and W. W. T. Huinink, *Br. J. Cancer.*, 2002, **87**, 608–614.
12. R. Duncan, *Nat. Rev. Cancer*, 2006, **6**, 688–701.
13. (*a*) J. Kopecek, P. Kopeckova, T. Minko, Z. R. Lu and C. M. Peterson, *J. Controlled Release*, 2001, **74**, 147–158; (*b*) A. Malugin, P. Kopeckova and J. Kopecek, *J. Controlled Release*, 2007, **124**, 6–10.
14. H. S. Yoo, K. H. Lee, J. E. Oh and T. G. Park, *J. Controlled Release*, 2000, **68**, 419–431.
15. M. Yokoyama, S. Fukushima, R. Uehara, K. Okamoto, K. Kataoka, Y. Sakurai and T. Okano, *J. Controlled Release*, 1998, **50**, 79–92.
16. (*a*) C. Wei, J. Guo and C. Wang, *Macromol. Rapid Commun.*, 2011, **32**, 451–455; (*b*) L. Wong, M. Kavallaris and V. Bulmus, *Polym. Chem.*, 2011, **2**, 385–393.

17. (*a*) B. Rihova, T. Etrych, M. Sirova, L. Kovar, O. Hovorka, M. Kovar, A. Benda and K. Ulbrich, *Mol. Pharmaceutics*, 2010, **7**, 1027–1040; (*b*) T. Etrych, L. Kovar, J. Strohalm, P. Chytil, B. Rihova and K. Ulbrich, *J. Controlled Release*, 2011, **154**, 241–248.

18. M. Yokoyama, G. S. Kwon, T. Okano, Y. Sakurai, T. Seto and K. Kataoka, *Bioconjugate Chem.*, 1992, **3**, 295–301.

19. (*a*) M. Harada, I. Bobe, H. Saito, N. Shibata, R. Tanaka, T. Hayashi and Y. Kato, *Cancer Sci.*, 2011, **102**, 192–199; (*b*) A. Ponta and Y. Bae, *Pharm. Res.*, 2010, **27**, 2330–2342; (*c*) Y. Bae, A. W. G. Alani, N. C. Rockich, T. S. Z. C. Lai and G. S. Kwon, *Pharm. Res.*, 2010, **27**, 2421–2432.

20. (*a*) K. Ulbrich, T. Etrych, P. Chytil, M. Pechar, M. Jelinkova and B. Rihova, *Int. J. Pharm.*, 2004, **277**, 63–72; (*b*) K. Ulbrich, T. Etrych, P. Chytil, M. Jelinkova and B. Rihova, *J. Controlled Release*, 2003, **87**, 33–47.

21. J.-Z. Du, X.-J. Du, C.-Q. Mao and J. Wang, *J. Am. Chem. Soc.*, 2011, **133**, 17560–17563.

22. (*a*) B. M. Liederer and R. T. Borchardt, *J. Pharm. Sci.*, 2006, **95**, 1177–1195; (*b*) R. B. Greenwald, Y. H. Choe, J. McGuire and C. D. Conover, *Adv. Drug Delivery Rev.*, 2003, **55**, 217–250.

23. M. E. Davis, *Adv. Drug Delivery Rev.*, 2009, **61**, 1189–1192.

24. J. J. Cheng, K. T. Khin, G. S. Jensen, A. J. Liu and M. E. Davis, *Bioconjugate Chem.*, 2003, **14**, 1007–1017.

25. (*a*) Y. Shiose, H. Kuga, H. Ohki, M. Ikeda, F. Yamashita and M. Hashida, *Bioconjugate Chem.*, 2009, **20**, 60–70; (*b*) V. Subr, J. Strohalm, K. Ulbrich, R. Duncan and I. C. Hume, *J. Controlled Release*, 1992, **18**, 123–132.

26. (*a*) G. Saito, J. A. Swanson and K. D. Lee, *Adv. Drug Delivery Rev.*, 2003, **55**, 199–215; (*b*) Y. E. Kurtoglu, R. S. Navath, B. Wang, S. Kannan, R. Romero and R. M. Kannan, *Biomaterials*, 2009, **30**, 2112–2121.

27. D. P. Jones, J. L. Carlson, P. S. Samiec, P. Sternberg, V. C. Mody, R. L. Reed and L. A. S. Brown, *Clin. Chim. Acta*, 1998, **275**, 175–184.

28. R. Duncan, *Nat. Rev. Drug Discovery*, 2003, **2**, 347–360.

29. D. M. Vyas, *Pharmacochem. Libr.*, 1995, **22**, 103–130.

30. C. Li, D. F. Yu, T. Inoue, D. J. Yang, L. Milas, N. R. Hunter, E. E. Kim and S. Wallace, *Anti-Cancer Drugs*, 1996, **7**, 642–648.

31. C. Li, D. F. Yu, R. A. Newman, F. Cabral, L. C. Stephens, N. Hunter, L. Milas and S. Wallace, *Cancer Res.*, 1998, **58**, 2404–2409.

32. Y. Y. Zou, H. Fu, S. Ghosh, D. Farquhar and J. Klostergaard, *Clin. Cancer Res.*, 2004, **10**, 7382–7391.

33. S. D. Li and L. Huang, *Mol. Pharmaceutics*, 2008, **5**, 496–504.

34. D. W. Northfelt, F. J. Martin, P. Working, P. A. Volberding, J. Russell, M. Newman, M. A. Amantea and L. D. Kaplan, *J. Clin. Pharm.*, 1996, **36**, 55–63.

35. S. A. Veltkamp, E. O. Witteveen, A. Capriati, A. Crea, F. Animati, M. Voogel-Fuchs, I. J. G. M. van den Heuvel, J. H. Beijnen, E. E. Voest and J. H. M. Schellens, *Clin. Cancer Res.*, 2008, **14**, 7535–7544.
36. K. Ulbrich and V. Subr, *Adv. Drug Delivery Rev.*, 2010, **62**, 150–166.
37. K. Knop, R. Hoogenboom, D. Fischer and U. S. Schubert, *Angew. Chem. Int. Ed.*, 2010, **49**, 6288–6308.
38. M. A. Eldon, C.-M. Staschen, T. X. Viegas and M. D. Bentley, *Mol. Cancer Ther.*, 2007, **6**, 3577S–3578S.
39. G. Pasut and F. M. Veronese, *Adv. Drug Delivery Rev.*, 2009, **61**, 1177–1188.
40. (*a*) T. M. Allen, *Nat. Rev. Cancer*, 2002, **2**, 750–763; (*b*) R. A. Petros and J. M. DeSimone, *Nat. Rev. Drug Discovery*, 2010, **9**, 615–627.
41. P. J. Julyan, L. W. Seymour, D. R. Ferry, S. Daryani, C. M. Boivin, J. Doran, M. David, D. Anderson, C. Christodoulou, A. M. Young, S. Hesslewood and D. J. Kerr, *J. Controlled Release*, 1999, **57**, 281–290.
42. L. W. Seymour, D. R. Ferry, D. Anderson, S. Hesslewood, P. J. Julyan, R. Poyner, J. Doran, A. M. Young, S. Burtles and D. J. Kerr, *J. Clin. Oncol.*, 2002, **20**, 1668–1676.
43. S. Zhu, L. Qian, M. Hong, L. Zhang, Y. Pei and Y. Jiang, *Adv. Mater.*, 2011, **23**, H84–H89.
44. M. P. Borgman, O. Aras, S. Geyser-Stoops, E. A. Sausville and H. Ghandehari, *Mol. Pharmaceutics*, 2009, **6**, 1836–1847.
45. (*a*) G. Trapani, N. Denora, A. Trapani and V. Laquintana, *J. Drug Targeting*, 2012, **20**, 1–22; (*b*) M. Wang and M. Thanou, *Pharmacol. Res.*, 2010, **62**, 90–99.
46. J. Szebeni, *Toxicology*, 2005, **216**, 106–121.
47. H. S. Yoo and T. G. Park, *J. Controlled Release*, 2004, **100**, 247–256.
48. F. M. Veronese, O. Schiavon, G. Pasut, R. Mendichi, L. Andersson, A. Tsirk, J. Ford, G. F. Wu, S. Kneller, J. Davies and R. Duncan, *Bioconjugate Chem.*, 2005, **16**, 775–784.
49. R. Duncan, J. K. Coatsworth and S. Burtles, *Hum. Exp. Toxicol.*, 1998, **17**, 93–104.
50. (*a*) J. W. Singer, B. Baker, P. De Vries, A. Kumar, S. Shaffer, E. Vawter, M. Bolton and P. Garzone, in *Polymer Drugs in the Clinical Stage: Advantages and Prospects (Advances in Experimental Medicine and Biology*, vol. 519), ed. H. Maeda, A. Kabanov, K. Kataoka and T. Okano, Kluwer/ Plenum, New York, 2003, pp. 81–99; (*b*) J. W. Singer, *J. Controlled Release*, 2005, **109**, 120–126.
51. R. L. Bhatt, P. de Vries, J. Tulinsky, G. Bellamy, B. Baker, J. W. Singer and P. Klein, *J. Med. Chem.*, 2003, **46**, 190–193.
52. K. D. Jensen, A. Nori, M. Tijerina, P. Kopeckova and J. Kopecek, *J. Controlled Release*, 2003, **87**, 89–105.
53. L. W. Seymour, R. Duncan, J. Strohalm and J. Kopecek, *J. Biomed. Mater. Res.*, 1987, **21**, 1341–1358.

54. (*a*) V. P. Torchilin, *Eur. J. Pharm. Sci.*, 2000, **11**, S81–S91; (*b*) H. Maeda, J. Wu, T. Sawa, Y. Matsumura and K. Hori, *J. Controlled Release*, 2000, **65**, 271–284.

55. W. D. Kingsbury, J. C. Boehm, D. R. Jakas, K. G. Holden, S. M. Hecht, G. Gallagher, M. J. Caranfa, F. L. McCabe, L. F. Faucette, R. K. Johnson and R. P. Hertzberg, *J. Med. Chem.*, 1991, **34**, 98–107.

56. Y. Shen, E. Jin, B. Zhang, C. J. Murphy, M. Sui, J. Zhao, J. Wang, J. Tang, M. Fan, E. Van Kirk and W. J. Murdoch, *J. Am. Chem. Soc.*, 2010, **132**, 4259–4265.

57. G. Bar-Sela, R. Epelbaum and M. Schaffer, *Curr. Med. Chem.*, 2010, **17**, 190–197.

58. P. Anand, A. B. Kunnumakkara, R. A. Newman and B. B. Aggarwal, *Mol. Pharmaceutics*, 2007, **4**, 807–818.

59. H. D. Tang, C. J. Murphy, B. Zhang, Y. Q. Shen, M. H. Sui, E. A. Van Kirk, X. W. Feng and W. J. Murdoch, *Nanomedicine (London, U. K.)*, 2010, **5**, 855–865.

60. (*a*) Y. Matsumura, *Adv. Drug Delivery Rev.*, 2011, **63**, 184–192; (*b*) A. Takahashi, N. Ohkohchi, M. Yasunaga, J.-I. Kuroda, Y. Koga, H. Kenmotsu, T. Kinoshita and Y. Matsumura, *Clin. Cancer Res.*, 2010, **16**, 4822–4831.

61. Y. Shen, *Polymer preprints American Chemical Society, Division of Polymer Chemistry*, 2012, **53** (1), 439.

62. S. Aryal, C.-M. J. Hu and L. Zhang, *ACS Nano*, 2010, **4**, 251–258.

63. A. W. G. Alani, Y. Bae, D. A. Rao and G. S. Kwon, *Biomaterials*, 2010, **31**, 1765–1772.

64. (*a*) M. J. Vicent and R. Duncan, *Trends Biotechnol.*, 2006, **24**, 39–47; (*b*) R. Satchi-Fainaro, R. Duncan and C. M. Barnes, *Adv. Polym. Sci.*, 2006, **193**, 1–65; (*c*) F. Greco and M. J. Vicent, *Front. Biosci.*, 2008, **13**, 2744–2756; (*d*) R. Haag and F. Kratz, *Angew. Chem. Int. Ed.*, 2006, **45**, 1198–1215.

65. H. D. Tang, C. J. Murphy, B. Zhang, Y. Q. Shen, E. A. Van Kirk, W. J. Murdoch and M. Radosz, *Biomaterials*, 2010, **31**, 7139–7149.

66. (*a*) M. S. Cartiera, E. C. Ferreira, C. Caputo, M. E. Egan, M. J. Caplan and W. M. Saltzman, *Mol. Pharmaceutics*, 2010, **7**, 86–93; (*b*) W. Shi, S. Dolai, S. Rizk, A. Hussain, H. Tariq, S. Averick, W. L'Amoreaux, A. El Ldrissi, P. Banerjee and K. Raja, *Org. Lett.*, 2007, **9**, 5461–5464.

67. J. Yang, W. Liu, M. Sui, J. Tang and Y. Shen, *Biomaterials*, 2011, **32**, 9136–9143.

68. C. Li and S. Wallace, *Adv. Drug Delivery Rev.*, 2008, **60**, 886–898.

69. R. Duncan, *Curr. Opin. Biotechnol.*, 2011, **22**, 492–501.

CHAPTER 12

Nonviral Vector Recombinant Mesenchymal Stem Cells: A Promising Targeted-Delivery Vehicle in Cancer Gene Therapy

YU-LAN HU[a], YING-HUA FU[b], YASUHIKO TABATA[c] AND JIAN-QING GAO*[a]

[a] Institute of Pharmaceutics, College of Pharmaceutical Sciences, Zhejiang University, P. R. China; [b] Department of Pharmaceutics, School of Medicine, Jiaxing University, P. R. China; [c] Department of Biomaterials, Institute for Frontier Medical Sciences, Kyoto University, Japan
*E-mail: gaojianqing1029@yahoo.com.cn

12.1 Introduction

With the development of molecular biology, cancer gene therapy becomes a promising field in the treatment of malignant tumors. Cytokines, such as IL-2, IL-12, and IFN-β, can stimulate antitumor responses through the activation of T cells, which mediates immune response to eliminate tumors. Nevertheless, the therapeutic application of exogenously administered cytokines is limited by their short half-lives and poor accessibility to tumor sites.[1] These cytokines exhibit rapid blood clearance and poor retention times in the target, which results in the necessity for frequent administration of such agents. Therefore, therapeutic utility of these cytokines *in vivo* is limited by its excessive toxicity when administered systemically at high doses and with high frequency.[1]

RSC Polymer Chemistry Series No. 3
Functional Polymers for Nanomedicine
Edited by Youqing Shen
© The Royal Society of Chemistry 2013
Published by the Royal Society of Chemistry, www.rsc.org

Table 12.1 Current targeting drug delivery systems for cancer therapy.

Targeting drug delivery system		Advantages	Disadvantages	Ref.
Nanocarriers	Liposome	1. Can carry both hydrophilic and hydrophobic agents. 2. Are composed of natural materials and are essentially nontoxic.	1. Rapidly cleared from the bloodstream by the reticuloendothelial defense mechanism. 2. Instability of the carrier and burst drug release. 3. Nonspecific uptake by the mononuclear phagocytic system (MPS).	86–88
	Polymeric nanoparticles	1. Could reduce dosage and minimize side-effects. 2. Protects drugs against degradation and enhances drug stability. 3. Potential for tumor targeting and could improve cancer therapy.	1. Inherent structural heterogeneity of polymers. 2. Low drug loading capacities. 3. Prone to form agglomerates and may lead to an occlusion of capillaries.	89–91
	Microbubbles	1. Used for site-specific delivery of the bioactive agents. 2. Could reduce the dosage.	1. Burst drug release	92
	Nano-emulsions	1. Are relatively safe after entering into the body. 2. Have sustained-release properties.	1. Physiochemical stability problems.	93
	Inorganic nanoparticles	1. Inert and nontoxic. 2. Ease of synthesis. 3. Ready functionalization. 4. Have photophysical properties which could trigger drug release at remote place.	1. Not biodegradable or small enough to be cleared easily. 2. Potential accumulation in the body and may cause long-term toxicity.	94,95

Table 12.1 (*Continued*)

Targeting drug delivery system		Advantages	Disadvantages	Ref.
Surface-modified delivery system	Antibody-conjugated delivery system	1. Has high selectivity and specificity for tumor site.	1. The receptor may be expressed on normal cells and may cause side effects. 2. Low ligand conjugation efficiency. 3. The different expression level of antigen/receptors on different tumor cells may decrease the site-specific targeting.	96–99
	Ligand-conjugated delivery system			100–103
Other carriers	Magnetically targeted drug carrier	1. Magnetic fields are focused over the target site and allow the particles to be captured at the target sites.	1. Only effective for tumor sites close to the body's surface; sites deeper within the body are difficult to target.	104–108
Cell-based drug carriers	Immune cells [macrophages, T cells, natural killer (NK) cells]	1. Could enhance the tumor site target efficiency and improve the therapeutic effect. 2. Have tumoricidal activity.	1. Difficulty in obtaining sufficient numbers. 2. Not easy to transduce these cells with the plasmids. 3. Suffer from the immunosuppressive nature of many tumors.	83,109
	Mesenchymal stem cells (MSCs)	1. Easily available and to expand by simple propagation procedures. 2. Have high migration potential to the tumor sites. 3. Have a low immunogenicity and intrinsic mutation rate. 4. Do not possess neurotoxicity or tumorigenicity.	1. The vectors and cytokines used are limited 2. Little is understood about the fate of MSCs *in vivo*.	9–11

Additionally, previous studies have largely relied on viral vectors to deliver these therapeutic genes, which are associated with safety concerns[2] and limit the clinical application of these cytokines.

Targeted delivery of anticancer agents is one of the promising fields in anticancer therapy. A major disadvantage of anticancer agents is their lack of selectivity for tumor tissue, which causes severe side effects and results in low therapeutic efficiency. Therefore, tumor-targeting approaches have been developed for improved efficiency and minimizing systematic toxicity by altering the biodistribution profiles of anticancer agents. In recent years, targeting drug delivery systems (TDDS) have attracted the extensive attention of researchers. More and more drug/gene targeted delivery carriers, such as stealth liposomes,[3] magnetic nanoparticles,[4] ligand-conjugated nanoparticles,[5] and ultrasound microbubbles,[6] have been developed and are under investigation for their tumor target efficiency and effectiveness for cancer treatment (Table 12.1).[7] However, these drug/gene delivery vehicles are limited by their several disadvantages. For example, the magnetic nanoparticles have low drug-loading capacities, non-uniform particle size distributions, and are prone to form agglomerates that may lead to an occlusion of capillaries.[4] Also, the rapid recognition and clearance of liposomes themselves from the bloodstream by the reticuloendothelial system (RES) limits the usefulness of liposomes as drug carriers (Table 12.1).

Cell-based therapies are emerging as promising therapeutic options for cancer treatment. However, the clinical application of differentiated cells is hindered by the difficulty in obtaining a large quantity of cell numbers and their lack of ability to expand *in vitro*, as well as the poor engraftment efficiency to targeted tumor sites. Mesenchymal stem cells (MSCs) have been attractive cell therapy vehicles for the delivery of agents into tumor cells because of their capability of self-renewal, relative ease of isolation and expansion *in vitro*, and homing capacity, allowing them to migrate toward and engraft into tumor sites.[8] Several studies have provided evidence supporting the rationale for genetically modified MSCs to deliver therapeutic cytokines directly into the tumor microenvironment to produce high concentrations of antitumor proteins at the tumor sites, which have been shown to inhibit tumor growth in experimental animal models. The antitumor effects of intravenous injections of gene-modified MSCs have been demonstrated in lung, brain, and subcutaneous tumors.[9–12]

12.2   Gene Recombination of MSCs

To develop MSCs as therapeutic agents, efficient gene transfer to the cells is a prerequisite. The strategies for gene delivery into MSCs include using viral vectors, nonviral vectors, and three-dimensional/reverse transfection systems.

12.2.1 Viral Vectors

The vectors of retroviruses, lentiviruses, adenoviruses, and adeno-associated viruses have been widely used for gene transfer. Viral methods using viral vectors for gene transfer to tumor tissue have proved to be very effective approaches, which allow the achievement of high transfection efficiency. In series of studies, fiber-mutant adenovirus vectors were developed and used for cancer gene therapy and induced significant antitumor activity when cytokines and/or chemokines were employed as therapeutic genes.[13–17] Furthermore, viral vectors modified with PEG induced evasion ability against the neutralizing antibody and the accumulation of vectors in tumor tissue as well as effective tumor-targeted gene transfer.[13,18] A variety of studies has been reported to use different viral vector systems to transduce MSCs.[19–22] However, many clinical trials in which viral vectors were used have been terminated, since the application of these vectors had induced unexpected adverse effects such as toxicities, immunogenicity, and oncogenicity.[23–25]

12.2.2 Nonviral Vectors

An increasing number of nonviral vectors are being developed for the purpose of gene delivery. These vectors have several advantages, such as ease of synthesis, cell/tissue targeting, low immune response, and unrestricted plasmid size.[10,26] Until now, a variety of nonviral delivery carriers, including calcium phosphate,[27] liposomes,[28] and niosomes,[29] as well as nanoparticles,[30] have been under development.[31] More recently, a novel transfection vector, spermine-pullulan, was synthesized by cationization of pullulan with the chemical introduction of spermine.[32] We compared its transfection efficiency on rat MSCs with other vectors, including peptide TAT-modified polyethylenimine (PEI)-β-cyclodextrin (Tat-cyd), PEI-cyclodextrin (Cyd), and Lipofectamine 2000 (Lipofectamine 2k, Invitrogen) (Figure 12.1). It was proved that the transfection efficiency of spermine-pullulan was equal to that of Cyd and was higher than that of Tat-cyd and Lipofectamine 2000, indicating that spermine-pullulan could be used as an effective nonviral gene transfection vector.[33] Although nonviral vectors hold promise in delivering therapeutic genes to MSCs, most current studies on these vectors are still limited to the *in vitro* evaluation of their transfection efficiency. As viral vectors can integrate into the host genome, they could result in a stable and long-term gene expression *in vivo*.[34] However, the gene expression by the nonviral system is transient, so that this system is not suitable to carry out studies involving cells for the treatment of diseases for which gene expression is required over long periods of time, as the nonviral vectors have a shorter expression time for the transgenes and a relatively lower transfection efficiency. It was reported that nonviral gene carriers can also prepare cells stably expressed through antibiotic selection.[35] However, this procedure is not practical, as it takes a long time to be performed and to optimize the conditions. Therefore, to tackle this issue the controlled release of anticancer

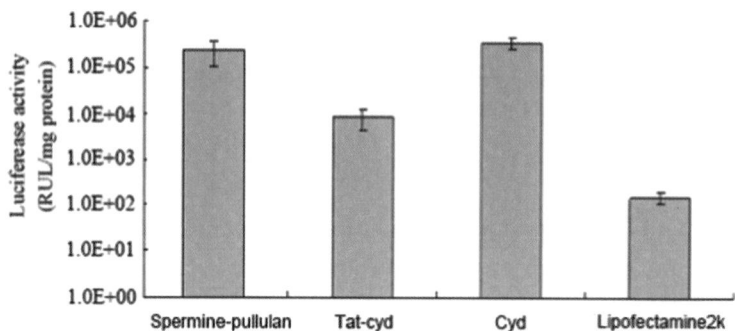

Figure 12.1   Transfection efficiency of different nonviral vectors, including spermine-pullulan, peptide TAT-modified PEI-β-cyclodextrin (Tat-cyd), PEI-cyclodextrin (Cyd), and Lipofectamine 2000 (Lipofectamine 2k). For the transfection assay, known amounts of MSCs were seeded into the 24-well plate and cultured for 18 h before transfection. The vector/DNA complex was prepared by mixing 1 μg of pGL3 plasmid with appropriate amounts of vector. The vector/DNA complexes were added to the 24-well plate and incubated for 6 h at 37 °C under 5% CO_2 atmosphere in serum-free DMEM. Then the DMEM was replaced with fresh DMEM with 10% serum. After 96 h, a luciferase assay was carried out according to the manufacture's instructions (Promega, USA). Light units (LUs) due to luciferase activity were measured with chemilumin-ometer (Autolumat LB953, EG&G Derthold, Germany). All the experiments were carried out in triplicate to ascertain the reproducibility of the results.

agents inside the cells or in the tissue will be more promising. Hence, a targeting drug delivery system could be practically used to modify and regulate the level and time period of gene expression.[32] In this case, MSCs could be used for the targeted delivery of tumor therapeutic genes transfected by nonviral vectors for their engraftment efficiency to tumor sites.

12.2.3   Three-Dimensional and Reverse Transfection Systems

Although a great improvement in the transfection efficiency has been made using nonviral vectors, in most cases, nonviral systems could not reach the high transfection efficiency and allow long-term transfection as viral vectors do. To enhance the transfection efficiency, several other transfection approaches, such as using a reverse transfection method and/or three dimensional (3D) systems, were developed. By optimizing the transfection system, such as changing the order of adding the gene complexes and cells, the cell culture environment can be greatly improved, not only in the increased transfection efficiency of the gene carrier, but also in a longer gene expression time.[31]

The technology and methodology of gene transfection have become more and more important to enhance the efficiency of gene therapy for several diseases. The transfection system can greatly affect the growth status of cells

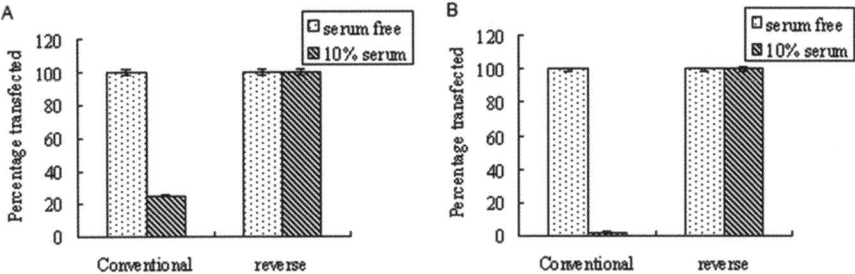

Figure 12.2 The transfection efficiency of nonviral vectors using different transfection methods: (A) PEI; (B) spermine-pullulan. For the reverse transfection, the reverse gene transfection system was composed of anionic gelatin, Pronectin, and gene complexes, all of which were coated on the surface of culture substrate before the introduction of the MSCs. The level of gene expression was compared with that of cells transfected in the medium containing the gene complexes (conventional transfection method) either in the presence or absence of serum.

and the sensitivity of the carriers to gene uptake, as well as the condition of carriers in the serum.[32] Up to now, the reverse transfection approach, which is different from the conventional approach, has been demonstrated to be able to protect the carriers from the inhibiting effects of the serum. In our studies, it was shown that using the reverse method, a high transfection efficiency of MSCs was observed even in the presence of serum. For the conventional method, in which the complex is added to the culture medium, the presence of serum reduced the transfection level to $25 \pm 1.5\%$ and $1.25 \pm 0.5\%$ using PEI and spermine-pullulan as the vectors, respectively (Figure 12.2). Also, the reverse method employing biodegradable or nonbiodegradable 3D scaffolds has been reported.[31,36,37]

12.3 MSCs as a Promising Targeted-Delivery Vehicle in Cancer Gene Therapy

12.3.1 Rationale for Using MSCs as a Vehicle for Gene Delivery

In 1987, Friedenstein *et al.*[38] found that the bone marrow single-cell can differentiate into bone and cartilage-forming adipocyte cells under certain conditions. These cells retain the ability of forming bone and cartilage after being transplanted in diffusion chambers after 20–30 cell doublings *in vitro*, and were called mesenchymal stem cells (MSCs) or bone marrow stromal cells. MSCs could also be isolated from other tissues, such as adipose tissue[39] and placenta.[40] It was reported[38] that human MSCs derived from adult marrow had been thought to be multipotent cells, which have the potential to differentiate to lineages of mesenchymal tissues, including bone, cartilage, fat, tendon, muscle, and marrow stroma. Stem cells, which have the characteristics of human MSCs, were isolated from marrow aspirates of volunteer donors. It was shown[41] that these cells displayed a stable phenotype and remained as a monolayer *in vitro*, and could be induced to differentiate into several mesenchymal lineages, including osteoblasts, adypocytes, and chondrocytes. It was also found by Pittenger *et al.*[41] that individual stem cells retained their multilineage potential when they were expanded to colonies.

Several properties of MSCs favor the development of engineered MSCs as a vehicle to deliver therapeutic genes aimed at cancer therapy (Table 12.1). (1) MSCs are easily obtained using less invasive methods and easy to expand by simple propagation procedures *in vitro.*[42] It was reported that they can be expanded for more than 50 population doublings in culture without the loss of their phenotype and multilineage potential.[40] (2) MSCs have high migration potential to the injury or tumor sites.[43] Using a human–sheep *in utero* xenotransplantation model, Mackenzie *et al.*[44] demonstrated that human MSCs have the ability to engraft, undergo site-specific differentiation into multiple cell types, and survive for more than 1 year in fetal lamb recipients. Also, MSC-derived cells appear to be present in increased numbers in wounded or regenerating tissues. Tumor microenvironments are pathologically altered tissues that resemble unresolved wounds,[45–47] which favor the homing of MSCs to the tumor sites. (3) MSCs have a low immunogenicity and intrinsic mutation rate. The low immunogenicity of MSCs is associated with the lack or low expression levels of the expression of MHC class I and class II molecules as well as co-stimulatory molecules.[48] No immunosuppression was observed when MSCs were engrafted into the bone marrow cavities of the recipient rats for at least 13 weeks after transplantation. Therefore, the MSCs are adult stem cells with unique immunologic tolerance allowing their engraftment into a xenogeneic environment, while preserving their ability to be recruited to an injured myocardium through the blood circulation.[49] (4) MSCs do not possess neurotoxicity or tumorigenicity. Sato *et al.*[50] injected EGFR-MSCs into the C57BL/6 background athymic mice and no tumors were found during 100 days observation, supporting the lack of tumorigenicity of these cells. (5) It is possible to genetically modify MSCs for the purpose of expressing therapeutic proteins and secreting these proteins into the tumor microenvironment.[51,52] These features have led to the suggestion that MSCs may be useful as cellular vehicles for gene delivery to multiple tumor sites.

12.3.2 Targeting of MSCs to Tumor Cells

Recent studies have shown the ability of MSCs to migrate to and incorporate within the connective tissue stroma of tumors.[11,53] This property of MSCs can be used to achieve targeting antitumor agents to tumor cells and their micro-metastases with an improvement in murine tumor models of glioma,[11,54] melanoma,[43] and breast[9] and colon[55] cancers.

The ability of MSCs to migrate toward gliomas has been assessed both *in vitro* and *in vivo.*[11,50] *In vitro* Matrigel invasion assays demonstrated that the MSCs derived from human bone marrow (hMSCs) have the capacity of migration towards gliomas. Furthermore, the tropism of these hMSCs for gliomas may be mediated by specific growth factors/chemokines. It was also observed that murine MSCs transfected with epidermal growth factor receptor (EGFR) could enhance migratory responses toward glioma-conditioned media in comparison to primary MSCs *in vitro*. Enhanced migration of EGFR-MSC may be partially dependent on EGF-EGFR, PI 3-kinase, MAP kinase, protein kinase C, and actin polymerization.[50] Another *in vivo* test indicated that MSCs can localize to human gliomas either after regional intra-arterial delivery or after local intracranial delivery.[11] It was also reported that intravenous injection of MSC-IFN-β cells into mice with established MDA 231 or A375SM pulmonary metastases led to the incorporation of MSCs in the tumor architecture.[9] However, in the healthy organs examined, no engraftment of intravenously administered MSCs was observed,[9] indicating MSCs themselves may not cause side effects in the healthy organs.

Compared with the tumor-targeted nanocarrier systems, which simply involves the ligand–receptor interaction, more factors are implicated in the homing of MSCs to tumor sites, and therefore a higher tumor target efficiency of MSCs would be expected (Figure 12.3). However, the processes and factors

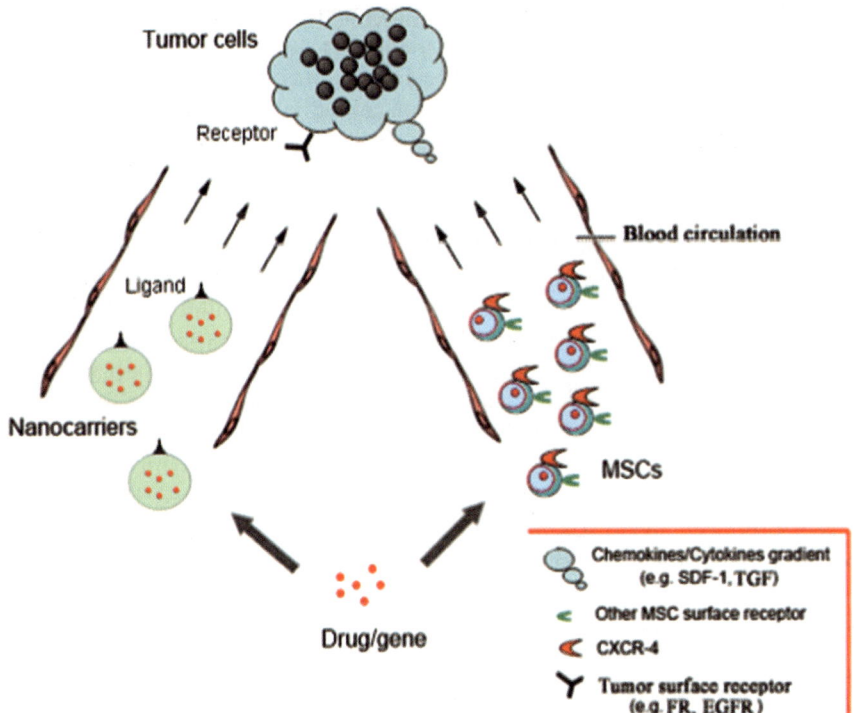

Figure 12.3 Schematic diagram of nanocarrier systems and MSCs for site-targeted drug/gene delivery (modified from Tanaka *et al.*[5] and Chen *et al.*[67]). FR: folate receptor; EGFR: epidermal growth factor receptor.

underling the migration of MSCs to tumors sites have not been well characterized. Until now, two possible mechanisms have been proposed (Figure 12.3). (1) Secretion of chemokines/cytokines from tumor tissues increases the migration of MSCs. The tumor tropism of MSCs might be mediated by several receptor–ligand combinations.[56] Cytokines, such as vascular endothelial cell growth factors, transforming growth factors (TGFs), fibroblast growth factors (FGFs), platelet-derived growth factors, monocyte chemo-attractants, protein-1, and IL-8 released from the neoplasm or inflammatory tissue are possible factors that mediate the activation of MSC migration.[10,57] It is already known that these factors released from cancer cells promote the migration of endothelial cell and stromal cell progenitors from the bone marrow towards the cancer bed[58,59] or tissues surrounding the tumor, therefore enhancing the formation of tumor-stroma.[60] Similar mechanisms would be anticipated for tumor-stromal formation in glioma, and the migration of implanted MSCs. Additionally, adhesion molecules, such as b1- and b2-integrins and L-selectin, may also play a significant role in the mobilization and homing of MSCs to gliomas.[61,62] MSCs injected intratumorally are mostly distributed at the border zone between tumor and normal parenchyma. They develop a capsule-like structure, and also infiltrate into the tumor bed relatively uniformly.[10] Tissue repair is a balance between damage and repair. When the balance is broken, the injured vessel requires the recruitment of more progenitor cells which contribute to lesion formation. MSCs exhibit multipotent differentiation potential, and have been shown to give rise to different mesodermal cell lineages, including osteoblasts, chondroblasts, and adipocytes under proper experimental conditions both *in vitro* and *in vivo*.[63] Therefore, MSCs could migrate towards the tumor site and participate in the formation of tumor stroma, which provides a new strategy for tumor therapy. Using MSCs as a tumor-targeted vehicle for the delivery of tumor therapeutic genes may decrease the side effects of these genes. For example, Studeny *et al.*[43] demonstrated that bone marrow-derived MSCs transducted with an adenoviral vector carrying the human IFN-β gene can produce biological agents locally at tumor sites. They also showed that MSCs with enhanced expression of IFN-β inhibited the growth of malignant cells *in vivo*. Importantly, this effect required the integration of MSCs into the tumors, and therefore could not be achieved by systemically delivered IFN-β or by IFN-β produced by MSCs at a site distant from the tumors. These results indicated that MSCs may serve as a platform for delivering biological agents into tumors. The successful engraftment of MSC in tissues would most likely be triggered by tissue damage or tumor growth, which makes MSCs excellent candidates for the cell-based delivery of therapeutics to tumor sites.[47] (2) The interaction of the cytokines or chemokines with their corresponding receptors would induce the migration of MSCs towards the tumor microenvironment. These receptors, such as CXCR4, CX₃CRl, CXCR6, CCRI, CCR7, *etc.*, were expressed on MSCs, and could interact with their respective ligands, namely CXCLl2, CX3CLl, CXCLl6, CCL3, or CCLl9.[64] Currently, CXCR4 and its receptor, stromal cell-derived factor-1 (SDF-1), are thought to be the most important pair of cytokines in attracting MSCs to migrate to the tumor, as CXCR4/SDF-1α interaction plays an important role in inflammation, tumor tropism of stem cells, and the pathology of gliomas.[65] Therefore, the microenvironment of the tumor site plays an important role in the migration of MSCs.[66] Also, a better understanding of the signaling transduction pathways associated with the tropism of these MSCs to gliomas will help to elucidate the role of MSCs in tumor growth and may permit more efficient targeted delivery of MSCs to the desired sites for therapeutic purposes.[67]

12.3.3 MSCs as Tumor Target Vehicles for Gene Delivery

The evidence that the tumor microenvironment favors the homing of exogenous MSCs has supported the rationale for developing engineered MSCs as a tool to track tumor sites and deliver anticancer agents within these areas. Several reports have proven the efficiency of MSCs as cell carriers for *in vivo* delivery of various clinically relevant anticancer agents following engraftment within tumor sites (Table 12.2). The observation that unmodified MSCs could have an anti-tumorigenic activity further supports the use of MSCs in cancer immunotherapy.[68] Besides, MSCs have proven to be effective in delivering oncolytic adenovirus for the treatment of cancer with low systemic toxicity,[69,70] suggesting MSCs are promising cell vehicles for cancer therapy.

Several research groups have reported promising results following the injection of genetically engineered MSCs into animal models bearing different tumors.[9,43,71,72] For example, Nakamura *et al.*[10] used gene-modified MSCs as a new tool for gene therapy of malignant brain neoplasms. Gene modification of MSCs by infection with an adenoviral vector encoding human IL-2 clearly enhanced the antitumor effect and further prolonged the survival of tumor-bearing rats, indicating that gene therapy employing MSCs as a targeting vehicle is promising as a new therapeutic approach for brain tumors. Studeny *et al.*[9] injected MSCs transduced with an adenoviral expression vector carrying the human IFN-β gene into mice with MDA231 or A375SM pulmonary metastases, resulting in suppression of the pulmonary metastases and prolonged survival. This suggests MSCs may be an effective platform for the targeted delivery of therapeutic proteins to cancer sites. Loebinger *et al.*[73] also demonstrated that MSCs could accumulate in tumor tissues (Figure 12.4) and the tumor necrosis factor-related apoptosis-inducing ligand (TRAIL)-expressing MSCs were able to inhibit tumor growths, suggesting that the TRAIL-expressing MSCs have a wide potential therapeutic application, which includes the treatment of both primary tumors and their metastases. Also, MSCs were transduced by the rAAV-IFN-β or green fluorescent protein *ex vivo* and used as cellular vehicles to target lung metastasis of TRAMP-C2 prostate cancer cells in a therapy model. The results indicated a significant reduction of tumor volume in lungs following IFN-β expressing MSC therapy.

Table 12.2   MSCs used for tumor delivery of anticancer agents.

Cytokines and chemokines	Animal model	Results	Ref.
IL-2	Rat glioma model	Gene modification of MSCs by infection with an adenoviral vector encoding human interleukin-2 (IL-2) augmented the antitumor effect and prolonged the survival of tumor-bearing rats.	10
IL-12	Mouse xenograft model of renal cell carcinoma (RCC)	Systemic administration of MSC/IL-12 reduced the growth of 786-0 RCC and significantly prolonged survival of tumor-bearing mouse.	110
IFN-β	Mouse prostate cancer lung metastasis model	A significant reduction in tumor volume in lungs following IFN-β expressing MSC therapy.	74
TRAIL	Mice metastatic lung tumors (MDAMB231)	TRAIL-expressing MSCs were able to reduce tumor growth and clear the metastatic disease.	73
CX₃CL1	Mice metastatic lung tumors (LLC)	Intratracheal administration of MSCs expressing CX_3CL1 suppressed the growth of multiple lung metastases and prolonged survival in mice.	55
CCL5	Mice primary pancreatic tumor	The active homing of MSCs into primary pancreatic tumor stroma and activation of the CCL5 promoter was verified using eGFP- and RFP-reporter genes. In the presence of ganciclovir, HSV-Tk transfected MSCs led to a significant reduction of primary pancreatic tumor growth and incidence of metastases.	111

Immunohistochemical examination of the lung tissue demonstrated an increase in tumor cell apoptosis, and a decrease in both tumor cell proliferation and blood vessel counts. A significant increase in the natural kill cell activity was observed following IFN-β therapy, correlating the antitumor effect. The systemic level of IFN-β was not significantly elevated from this targeted cell therapy. These data demonstrate the potential of MSC-based IFN-β therapy for prostate cancer lung metastasis.[74]

Currently, most studies on gene-modified MSCs for the treatment of tumors use viral vectors for gene transfer.[75] However, they may be associated with safety problems. Recently, we employed the nonviral vector polyethylenimine (PEI) to transfer IL-12 into MSCs for the treatment of lung metastases in C57BL/6 mice, and found that MSC-IL12 injected mice showed significantly less lung metastases numbers compared with the direct injection of IL-12 and

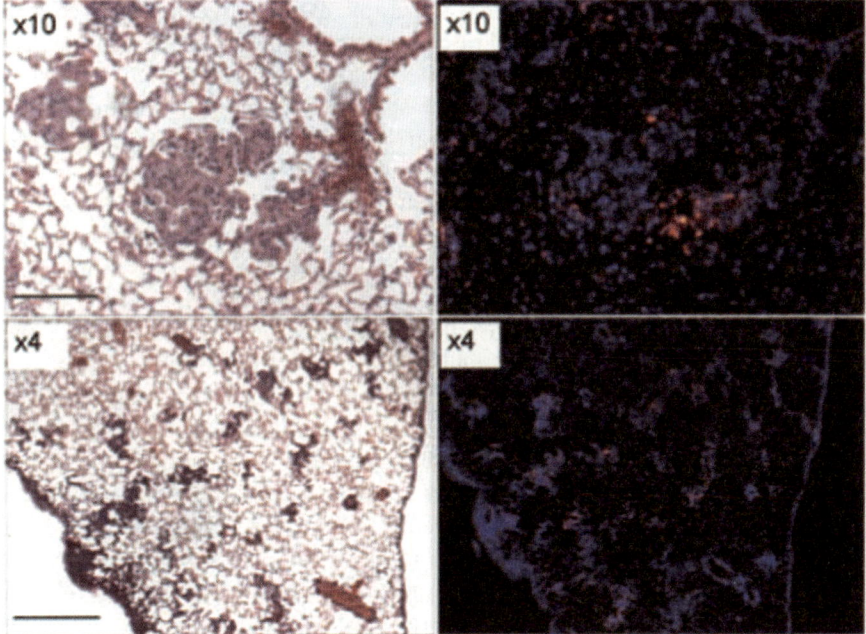

Figure 12.4   DiI-labeled MSCFLTs (*red*) injected intravenously at day 10 and shown to localize to lung metastases on fluorescent microscopy with DAPI nuclear counterstain with H&E contiguous sections from day 30 harvested lungs (magnification ×10 μm, bar 20 μm; magnification ×4, bar 60 μm). (Reproduced from Loebinger *et al.*[73] with permission from the American Association for Cancer Research.)

vector (data not shown).Our study supports the effectiveness of nonviral vectors in transferring the therapeutic gene to MSCs.

12.4   Future Perspectives

Targeted delivery of anticancer drugs/genes to tumor cells/tissues can improve the therapeutic index of drugs by minimizing their toxic effects. Currently, a variety of delivery systems have been employed for developing TDDS for anticancer agents to enhance their therapeutic values.[76] For instance, nanocarrier drug delivery systems were designed to reach target cells and tissues or respond to stimuli in a well-controlled manner to induce desired physiological responses.[7] Also, cell- or tissue-specific receptor/antigens can provide a useful target for targeted delivery of anticancer drugs (see Figure 12.3). Clinical developments of anticancer agents utilizing TDDS are now only at their initial stages by pharmaceutical companies. The application of MSCs to targeted gene delivery has been the subject of current experimental studies and may bring new hope to cancer patients. After transfecting MSCs with a tumor therapeutic gene, the MSCs would deliver the gene to tumor sites, which could be widely used in tumor therapy. Besides, recent evidence also suggests that MSCs selectively migrate to areas of injury/inflammation, where they are involved in tissue repair. This effect makes MSCs attractive candidates for widely used delivery vehicles for site-specific therapy. In the future, these strategies could be extended by using various genes to achieve a general cell-based gene therapy platform for the treatment of different diseases.

MSCs have attracted considerable attention as novel tools for the delivery of therapeutic proteins *in vivo*, and the ability to efficiently transfer genes of interest into these cells would create a number of therapeutic opportunities.[77] However, the potential concerns in using MSCs as delivery vehicles are that we understand little about the fate of this cell population *in vivo*[78,79] and the possibility that MSCs themselves might enhance or initiate tumor growth.[80–82] Although more and more studies are using gene-engineered MSCs for the treatment of tumor disease and they are considered as a powerful therapeutic weapon for the targeting of advanced malignancies,[75] until now the cytokines and the vectors used are still limited.

The use of MSCs for cancer gene therapy usually provides some degree of tumor selectivity; however, strategies to improve tumor homing might greatly increase their applicability as tumor-targeted cell delivery vehicles.[83] The cell homing ability may be enhanced by using specific culture agents or medium treatments to alter the expression profile of cell-surface receptors.[84,85] It might be possible to manipulate cell targeting or homing properties by pre-programming the MSCs to alter their properties during the gene delivery process to make them most effective in targeting the tumor site. This approach represents another option of investigation to enhance the efficiency of cellular vehicle strategies.

Still, much work remains to be done before the clinical application of MSC-based therapy. The development of optimally targeted cellular vectors will require both advances in molecular and cell biology and the improvement in the methods for gene transduction to cells. A lot of effort still needs to be made to develop delivery systems capable of effectively transporting various therapeutic cytokines. Together, the further development of MSC-based therapy will benefit from a better understanding of biology, tissue stem cell engineering, and pharmaceutics, as well as gene therapy.

Acknowledgements

This work was financially supported by the National Natural Science Foundation of China (30873173, 30973648, 81273441, 81001410), the Zhejiang Provincial Natural Science Foundation of China (R2090176), and the China–Japan Scientific Cooperation Program (81011140077) supported by both the NSFC, China, and the JSPS, Japan. We would like to thank Dr. Guping Tang (Institute of Chemical Biology and Pharmaceutical Chemistry, Zhejiang University) for providing Cyd and TAT-cyd and thank Ms. Cai-Xia He for technical assistance.

References

1. S. Einhorn and D. Grander, *J. Interferon Cytokine Res.*, 1996, **16**, 275–281.
2. H. Okada and I. F. Pollack, *Expert Opin. Biol. Ther.*, 2004, **4**, 1609–1620.
3. X. Li, L. Ding, Y. Xu, Y. Wang and Q. Ping, *Int. J. Pharm.*, 2009, **373**, 116–123.
4. S. Moritake, S. Taira, Y. Ichiyanagi, N. Morone, S. Y. Song, T. Hatanaka, S. Yuasa and M. Setou, *J. Nanosci. Nanotechnol.*, 2007, **7**, 937–944.
5. T. Tanaka, S. Shiramoto, M. Miyashita, Y. Fujishima and Y. Kaneo, *Int. J. Pharm.*, 2004, **277**, 39–61.
6. Y. Liu, H. Miyoshi and M. Nakamura, *J. Controlled Release*, 2006, **114**, 89–99.
7. D. Peer, J. M. Karp, S. Hong, O. C. Farokhzad, R. Margalit and R. Langer, *Nat. Nanotechnol.*, 2007, **2**, 751–760.
8. J. E. Dennis, N. Cohen, V. M. Goldberg and A. I. Caplan, *J. Orthop. Res.*, 2004, **22**, 735–741.
9. M. Studeny, F. C. Marini, J. L. Dembinski, C. Zompetta, M. Cabreira-Hansen, B. N. Bekele, R. E. Champlin and M. Andreeff, *J. Natl. Cancer Inst.*, 2004, **96**, 1593–1603.
10. K. Nakamura, Y. Ito, Y. Kawano, K. Kurozumi, M. Kobune, H. Tsuda, A. Bizen, O. Honmou, Y. Niitsu and H. Hamada, *Gene Ther.*, 2004, **11**, 1155–1164.
11. A. Nakamizo, F. Marini, T. Amano, A. Khan, M. Studeny, J. Gumin, J. Chen, S. Hentschel, G. Vecil, J. Dembinski, M. Andreeff and F. F. Lang, *Cancer Res.*, 2005, **65**, 3307–3318
12. L. Kucerova, V. Altanerova, M. Matuskova, S. Tyciakova and C. Altaner, *Cancer Res.*, 2007, **67**, 6304–6313.
13. J. Q. Gao, Y. Eto, Y. Yoshioka, F. Sekiguchi, S. Kurachi, T. Morishige, X. Yao, H. Watanabe, R. Asavatanabodee, F. Sakurai, H. Mizuguchi, Y. Okada, Y. Mukai, Y. Tsutsumi, T. Mayumi, N. Okada and S. Nakagawa, *J. Controlled Release*, 2007, **122**, 102–110.
14. J. Q. Gao, N. Kanagawa, D. H. Xu, M. Han, T. Sugita, Y. Hatanaka, Y. Tani, H. Mizuguchi, Y. Tsutsumi, T. Mayumi, N. Okada and S. Nakagawa, *Cancer Immunol. Immunother.*, 2008, **57**, 1657–1664.
15. J. Q. Gao, T. Sugita, N. Kanagawa, K. Iida, Y. Eto, Y. Motomura, H. Mizuguchi, Y. Tsutsumi, T. Hayakawa, T. Mayumi and S. Nakagawa, *Biochem. Biophys. Res. Commun.*, 2005, **328**, 1043–1050.
16. J. Q. Gao, Y. Tsuda, K. Katayama, T. Nakayama, Y. Hatanaka, Y. Tani, H. Mizuguchi, T. Hayakawa, O. Yoshie, Y. Tsutsumi, T. Mayumi and S. Nakagawa, *Cancer Res.*, 2003, **63**, 4420–4425.
17. J. Q. Gao, N. Kanagawa, Y. Motomura, T. Yanagawa, T. Sugita, Y. Hatanaka, Y. Tani, H. Mizuguchi, Y. Tsutsumi, T. Mayumi, N. Okada and S. Nakagawa, *Gene Ther.*, 2007, **14**, 491–502.

18. Y. Eto, J. Q. Gao, F. Sekiguchi, S. Kurachi, K. Katayama, H. Mizuguchi, T. Hayakawa, Y. Tsutsumi, T. Mayumi and S. Nakagawa, *Biol. Pharm. Bull.*, 2004, **27**, 936–938.

19. P. A. Conget and J. J. Minguell, *Exp. Hematol.*, 2000, **28**, 382–390.

20. L. Ding, S. Lu, R. Batchu, R. S. Iii and N. Munshi, *Gene Ther.*, 1999, **6**, 1611–1616.

21. E. A. Frolova-Jones, A. Ensser, A. J. Stevenson, S. E. Kinsey and D. M. Meredith, *J. Hematother. Stem Cell Res.*, 2000, **9**, 573–581.

22. K. Lee, M. K. Majumdar, D. Buyaner, J. K. Hendricks, M. F. Pittenger and J. D. Mosca, *Mol. Ther.*, 2001, **3**, 857–866.

23. W. Walther and U. Stein, *Drugs*, 2000, **60**, 249–271.

24. E. Marshall, *Science*, 1999, **286**, 2244–2245.

25. S. Hacein-Bey-Abina, C. von Kalle, M. Schmidt, F. Le Deist, N. Wulffraat, E. McIntyre, I. Radford, J. L. Villeval, C. C. Fraser, M. Cavazzana-Calvo and A. Fischer, *N. Engl. J. Med.*, 2003, **348**, 255–256.

26. C. W. Cho, Y. S. Cho, B. T. Kang, J. S. Hwang, S. N. Park and D. Y. Yoon, *Cancer Lett.*, 2001, **162**, 75–85.

27. H. Fu, Y. Hu, T. McNelis and J. O. Hollinger, *J. Biomed. Mater. Res., A*, 2005, **74**, 40–48 .

28. C. R. Dass and P. F. Choong, *J. Controlled Release*, 2006, **113**, 155–163 .

29. Y. Huang, J. Chen, X. Chen, J. Gao and W. Liang, *J. Mater. Sci. Mater. Med.*, 2008, **19**, 607–614.

30. N. Pimpha, P. Sunintaboon, S. Inphonlek and Y. Tabata, *J. Biomater. Sci., Polym. Ed.*, 2010, **21**, 205–223.

31. C. X. He, Y. Tabata and J. Q. Gao, *Int. J. Pharm.*, 2010, **386**, 232–242.

32. J. Jo and Y. Tabata, *Eur. J. Pharm. Biopharm.*, 2008, **68**, 90–104.

33. F. Yang, S. W. Cho, S. M. Son, S. R. Bogatyrev, D. Singh, J. J. Green, Y. Mei, S. Park, S. H. Bhang, B. S. Kim, R. Langer and D. G. Anderson, *Proc. Natl. Acad. Sci. U. S. A.*, 2010, **107**, 3317–3322 .

34. G. D. Trobridge, D. G. Miller, M. A. Jacobs, J. M. Allen, H. P. Kiem, R. Kaul and D. W. Russell, *Proc. Natl. Acad. Sci. U. S. A.*, 2006, **103**, 1498–1503.

35. A. Jimenez and J. Davies, *Nature*, 1980, **287**, 869–871.

36. H. Hosseinkhani, T. Azzam, H. Kobayashi, Y. Hiraoka, H. Shimokawa, A. J. Domb and Y. Tabata, *Biomaterials*, 2006, **27**, 4269–4278.

37. A. Okazaki, J. Jo and Y. Tabata, *Tissue Eng.*, 2007, **13**, 245–251.

38. A. J. Friedenstein, R. K. Chailakhyan and U. V. Gerasimov, *Cell Tissue Kinet.*, 1987, **20**, 263–272.

39. P. A. Zuk, M. Zhu, H. Mizuno, J. Huang, J. W. Futrell, A. J. Katz, P. Benhaim, H. P. Lorenz and M. H. Hedrick, *Tissue Eng.*, 2001, **7**, 211–228.

40. P. S. In't Anker, S. A. Scherjon, C. Kleijburg-van der Keur, G. M. de Groot-Swings, F. H. Claas, W. E. Fibbe and H. H. Kanhai, *Stem Cells*, 2004, **22**, 1338–1345.

41. M. F. Pittenger, A. M. Mackay, S. C. Beck, R. K. Jaiswal, R. Douglas, J. D. Mosca, M. A. Moorman, D. W. Simonetti, S. Craig and D. R. Marshak, *Science*, 1999, **284**, 143–147.

42. M. Compte, A. M. Cuesta, D. Sanchez-Martin, V. Alonso-Camino, J. L. Vicario, L. Sanz and L. Alvarez-Vallina, *Stem Cells*, 2009, **27**, 753–760.

43. M. Studeny, F. C. Marini, R. E. Champlin, C. Zompetta, I. J. Fidler and M. Andreeff, *Cancer Res.*, 2002, **62**, 3603–3608.

44. T. C. Mackenzie and A. W. Flake, *Blood Cells Mol. Dis.*, 2001, **27**, 601–604.

45. H. F. Dvorak, *N. Engl. J. Med.*, 1986, **315**, 1650–1659.

46. J. Riss, C. Khanna, S. Koo, G. V. Chandramouli, H. H. Yang, Y. Hu, D. E. Kleiner, A. Rosenwald, C. F. Schaefer, S. A. Ben-Sasson, L. Yang, J. Powell, D. W. Kane, R. A. Star, O. Aprelikova, K. Bauer, J. R. Vasselli, J. K. Maranchie, K. W. Kohn, K. H. Buetow, W. M. Linehan, J. N. Weinstein, M. P. Lee, R. D. Klausner and J. C. Barrett, *Cancer Res.*, 2006, **66**, 7216–7224.

47. B. Hall, M. Andreeff and F. Marini, *Handb. Exp. Pharmacol.*, 2007, **180**, 263–283.

48. P. A. Sotiropoulou, S. A. Perez, M. Salagianni, C. N. Baxevanis and M. Papamichail, *Stem Cells*, 2006, **24**, 462–471.

49. T. Saito, J. Q. Kuang, B. Bittira, A. Al-Khaldi and R. C. Chiu, *Ann. Thorac. Surg.*, 2002, **74**, 19–24.

50. H. Sato, N. Kuwashima, T. Sakaida, M. Hatano, J. E. Dusak, W. K. Fellows-Mayle, G. D. Papworth, S. C. Watkins, A. Gambotto, I. F. Pollack and H. Okada, *Cancer Gene Ther.*, 2005, **12**, 757–768.

51. J. Chan, K. O'Donoghue, J. de la Fuente, I. A. Roberts, S. Kumar, J. E. Morgan and N. M. Fisk, *Stem Cells*, 2005, **23**, 93–102.

52. S. Yip, R. Sabetrasekh, R. L. Sidman and E. Y. Snyder, *Eur. J. Cancer*, 2006, **42**, 1298–1308.

53. L. G. Menon, S. Picinich, R. Koneru, H. Gao, S. Y. Lin, M. Koneru, P. Mayer-Kuckuk, J. Glod and D. Banerjee, *Stem Cells*, 2007, **25**, 520–528.

54. S. M. Kim, J. Y. Lim, S. I. Park, C. H. Jeong, J. H. Oh, M. Jeong, W. Oh, S. H. Park, Y. C. Sung and S. S. Jeun, *Cancer Res.*, 2008, **68**, 9614–9623.

55. H. Xin, M. Kanehira, H. Mizuguchi, T. Hayakawa, T. Kikuchi, T. Nukiwa and Y. Saijo, *Stem Cells*, 2007, **25**, 1618–1626.

56. T. Kosztowski, H. A. Zaidi and A. Quinones-Hinojosa, *Expert Rev. Anticancer Ther.*, 2009, **9**, 597–612.

57. D. Hanahan and R. A. Weinberg, *Cell*, 2000, **100**, 57–70.

58. M. De Palma, M. A. Venneri, C. Roca and L. Naldini, *Nat. Med.*, 2003, **9**, 789–795.

59. L. M. Coussens and Z. Werb, *Nature*, 2002, **420**, 860–867.

60. V. M. Weaver, A. H. Fischer, O. W. Peterson and M. J. Bissell, *Biochem. Cell Biol.*, 1996, **74**, 833–851.

61. S. Rafii and D. Lyden, *Nat. Med.*, 2003, **9**, 702–712.
62. B. R. Son, L. A. Marquez-Curtis, M. Kucia, M. Wysoczynski, A. R. Turner, J. Ratajczak, M. Z. Ratajczak and A. Janowska-Wieczorek, *Stem Cells*, 2006, **24**, 1254–1264.
63. K. Suzuki, M. Oyama, L. Faulcon, P. D. Robbins and C. Niyibizi, *Cell Transplant.*, 2000, **9**, 319–327.
64. S. A. Azizi, D. Stokes, B. J. Augelli, C. DiGirolamo and D. J. Prockop, *Proc. Natl. Acad. Sci. U. S. A.*, 1998, **95**, 3908–3913.
65. E. J. Schwarz, G. M. Alexander, D. J. Prockop and S. A. Azizi, *Hum. Gene Ther.*, 1999, **10**, 2539–2549.
66. I. Sekiya, D. C. Colter and D. J. Prockop, *Biochem. Biophys. Res. Commun.*, 2001, **284**, 411–418.
67. Y. Chen, J. Z. Shao, L. X. Xiang, X. J. Dong and G. R. Zhang, *Int. J. Biochem. Cell Biol.*, 2008, **40**, 815–820.
68. V. Fritz and C. Jorgensen, *Curr. Stem Cell Res. Ther.*, 2008, **3**, 32–42.
69. A. M. Sonabend, I. V. Ulasov, M. A. Tyler, A. A. Rivera, J. M. Mathis and M. S. Lesniak, *Stem Cells*, 2008, **26**, 831–841.
70. M. A. Stoff-Khalili, A. A. Rivera, J. M. Mathis, N. S. Banerjee, A. S. Moon, A. Hess, R. P. Rocconi, T. M. Numnum, M. Everts, L. T. Chow, J. T. Douglas, G. P. Siegal, Z. B. Zhu, H. G. Bender, P. Dall, A. Stoff, L. Pereboeva and D. T. Curiel, *Breast Cancer Res. Treat.*, 2007, **105**, 157–167.
71. M. Kassem, M. Kristiansen and B. M. Abdallah, *Basic Clin. Pharmacol. Toxicol.*, 2004, **95**, 209–214.
72. J. Xiang, J. Tang, C. Song, Z. Yang, D. G. Hirst, Q. J. Zheng and G. Li, *Cytotherapy*, 2009, **11**, 516–526.
73. M. R. Loebinger, A. Eddaoudi, D. Davies and S. M. Janes, *Cancer Res.*, 2009, **69**, 4134–4142.
74. C. Ren, S. Kumar, D. Chanda, L. Kallman, J. Chen, J. D. Mountz and S. Ponnazhagan, *Gene Ther.*, 2008, **15**, 1446–1453.
75. X. Chen, X. Lin, J. Zhao, W. Shi, H. Zhang, Y. Wang, B. Kan, L. Du, B. Wang, Y. Wei, Y. Liu and X. Zhao, *Mol. Ther.*, 2008, **16**, 749–756.
76. C. K. Kim and S. J. Lim, *Arch. Pharm. Res.*, 2002, **25**, 229–239.
77. X. Y. Zhang, V. F. La Russa and J. Reiser, *J. Virol.*, 2004, **78**, 1219–1229.
78. D. Baksh, L. Song and R. S. Tuan, *J. Cell. Mol. Med.*, 2004, **8**, 301–316.
79. R. E. Feldmann, Jr., K. Bieback, M. H. Maurer, A. Kalenka, H. F. Burgers, B. Gross, C. Hunzinger, H. Kluter, W. Kuschinsky and H. Eichler, *Electrophoresis*, 2005, **26**, 2749–2758.
80. C. A. Iacobuzio-Donahue, P. Argani, P. M. Hempen, J. Jones and S. E. Kern, *Cancer Res.*, 2002, **62**, 5351–5357.
81. D. M. Parham, *Mod. Pathol.*, 2001, **14**, 506–514.
82. A. E. Karnoub, A. B. Dash, A. P. Vo, A. Sullivan, M. W. Brooks, G. W. Bell, A. L. Richardson, K. Polyak, R. Tubo and R. A. Weinberg, *Nature*, 2007, **449**, 557–563.

83. K. Harrington, L. Alvarez-Vallina, M. Crittenden, M. Gough, H. Chong, R. M. Diaz, G. Vassaux, N. Lemoine and R. Vile, *Hum. Gene Ther.*, 2002, **13**, 1263–1280.

84. J. Fiedler, N. Etzel and R. E. Brenner, *J. Cell Biochem.*, 2004, **93**, 990–998.

85. D. L. Kraitchman, M. Tatsumi, W. D. Gilson, T. Ishimori, D. Kedziorek, P. Walczak, W. P. Segars, H. H. Chen, D. Fritzges, I. Izbudak, R. G. Young, M. Marcelino, M. F. Pittenger, M. Solaiyappan, R. C. Boston, B. M. Tsui, R. L. Wahl and J. W. Bulte, *Circulation*, 2005, **112**, 1451–1461.

86. K. Yachi, H. Kikuchi, N. Suzuki, R. Atsumi, M. Aonuma and Y. Kawato, *Biopharm. Drug Dispos.*, 1995, **16**, 653–667.

87. A. Samad, Y. Sultana and M. Aqil, *Curr. Drug Delivery*, 2007, **4**, 297–305.

88. O. P. Medina, Y. Zhu and K. Kairemo, *Curr. Pharm. Des.*, 2004, **10**, 2981–2989.

89. J. K. Vasir and V. Labhasetwar, *Technol. Cancer Res. Treat.*, 2005, **4**, 363–374.

90. S. Mitra, U. Gaur, P. C. Ghosh and A. N. Maitra, *J. Controlled Release*, 2001, **74**, 317–323.

91. Y. Liu, H. Miyoshi and M. Nakamura, *Int. J. Cancer*, 2007, **120**, 2527–2537.

92. J. M. Tsutsui, F. Xie and R. T. Porter, *Cardiovasc. Ultrasound*, 2004, **2**, 23.

93. A. C. Lo Prete, D. A. Maria, D. G. Rodrigues, C. J. Valduga, O. C. Ibanez and R. C. Maranhao, *J. Pharm. Pharmacol.*, 2006, **58**, 801–810.

94. G. Han, P. Ghosh and V. M. Rotello, *Nanomedicine (London, U. K.)*, 2007, **2**, 113–123.

95. P. Ghosh, G. Han, M. De, C. K. Kim and V. M. Rotello, *Adv. Drug Delivery Rev.*, 2008, **60**, 1307–1315.

96. L. G. Remsen, P. A. Trail, I. Hellstrom, K. E. Hellstrom and E. A. Neuwelt, *Neurosurgery*, 2000, **46**, 704–709.

97. V. Guillemard and H. U. Saragovi, *Cancer Res.*, 2001, **61**, 694–699.

98. A. Muvaffak, I. Gurhan, U. Gunduz and N. Hasirci, *J. Drug Targeting*, 2005, **13**, 151–159.

99. R. Suzuki, N. Utoguchi, K. Kawamura, N. Kadowaki, N. Okada, T. Takizawa, T. Uchiyama and K. Maruyama, *Yakugaku Zasshi*, 2007, **127**, 301–306.

100. S. Ni, S. M. Stephenson and R. J. Lee, *Anticancer Res.*, 2002, **22**, 2131–2135.

101. X. Q. Pan and R. J. Lee, *Anticancer Res.*, 2005, **25**, 343–346.

102. X. Q. Pan, H. Wang and R. J. Lee, *Pharm. Res.*, 2003, **20**, 417–422.

103. Y. H. Bae, *J. Controlled Release*, 2009, **133**, 2–3.

104. C. Alexiou, R. Jurgons, C. Seliger and H. Iro, *J. Nanosci. Nanotechnol.*, 2006, **6**, 2762–2768.

105. C. Alexiou, R. J. Schmid, R. Jurgons, M. Kremer, G. Wanner, C. Bergemann, E. Huenges, T. Nawroth, W. Arnold and F. G. Parak, *Eur. Biophys. J.*, 2006, **35**, 446–450.

106. S. C. Goodwin, C. A. Bittner, C. L. Peterson and G. Wong, *Toxicol. Sci.*, 2001, **60**, 177–183.

107. M. Babincova and P. Babinec, *Biomed. Pap. Med. Fac. Univ. Palacky Olomouc Czech Repub.*, 2009, **153**, 243–250.

108. S. C. McBain, H. H. Yiu and J. Dobson, *Int. J. Nanomed.*, 2008, **3**, 169–180.

109. S. H. Thorne and C. H. Contag, *Cell. Mol. Life Sci.*, 2007, **64**, 1449–1451.

110. P. Gao, Q. Ding, Z. Wu, H. Jiang and Z. Fang, *Cancer Lett.*, 2010, **290**, 157–166.

111. C. Zischek, H. Niess, I. Ischenko, C. Conrad, R. Huss, K. W. Jauch, P. J. Nelson and C. Bruns, *Ann. Surg.*, 2009, **250**, 747–753.

CHAPTER 13

Near-Critical Micellization for Nanomedicine: Enhanced Drug Loading, Reduced Burst Release

JADE GREEN[a], MACIEJ RADOSZ*[a] AND YOUQING SHEN[b]

[a] Department of Chemical and Petroleum Engineering, Soft Materials Laboratory, University of Wyoming, Laramie, WY 82071, USA; [b] Center for Bionanoengineering and State Key Laboratory of Chemical Engineering, Department of Chemical and Biological Engineering, Zhejiang University, Hangzhou 310027, P. R. China
*E-mail: Radosz@uwyo.edu

13.1 Introduction

Micelles, a self-assembled formation of diblock copolymer chains, have been shown to have utility for delivery of highly hydrophobic drugs.[1-5] This is due to their ability to solubilize these drugs into the core of the micelle, thereby increasing the effective dose of drug available for delivery, and to target cancer cells *via* the enhanced permeation and retention (EPR) effect.[6-8] In most conventional methods, such diblock micelles are formed by first dissolving the copolymer in a nonselective solvent, such as acetone. Then, micelle formation is induced by the slow addition of a selective solvent, such as water. Finally, the micelles are removed and purified by freeze drying, solvent evaporation, or dialysis.[8-11] However, there are problems associated with such methods,

RSC Polymer Chemistry Series No. 3
Functional Polymers for Nanomedicine
Edited by Youqing Shen
© The Royal Society of Chemistry 2013
Published by the Royal Society of Chemistry, www.rsc.org

including trace amounts of potentially toxic organic solvents left in the micelle, low drug loading, and burst release of the drug, to name a few.

We recently developed a new technology that can address these problems by using a near-critical solvent, which is a compressed gas, and simply reducing the pressure to induce micellization, referred to as the near-critical micellization (NCM) method. In addition to low viscosity and high diffusivity, which promote mass transfer, near-critical solvents are easy to remove by spontaneous and complete evaporation. More importantly, solvent capacity and selectivity can be tuned precisely with its density *via* pressure, temperature, or both, which, in contrast to liquid solvents, allows for high-resolution control of the NCM process.[8,12]

13.2   Early Feasibility Studies on Model Systems

The first step in developing the NCM process was to understand its physics on simple, well-defined model systems, such as polystyrene and polyisoprene. Towards this end, Winoto *et al.*[13] used an experimental set-up shown in Figure 13.1, in which a small high-pressure, variable-volume (~ 1 cm^3) cell is used to detect cloud pressure (CP) and micellization pressure (MP) transitions. It is equipped with a floating piston, so that the pressure can be changed

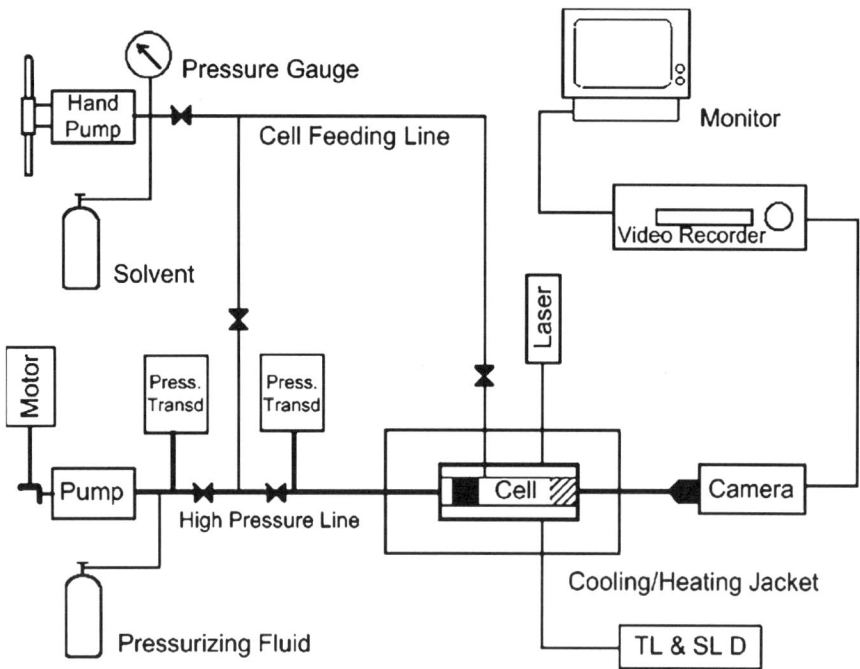

Figure 13.1   A simplified apparatus schematic diagram. (Reproduced from Winoto *et al.*[13] with permission from the American Chemical Society.)

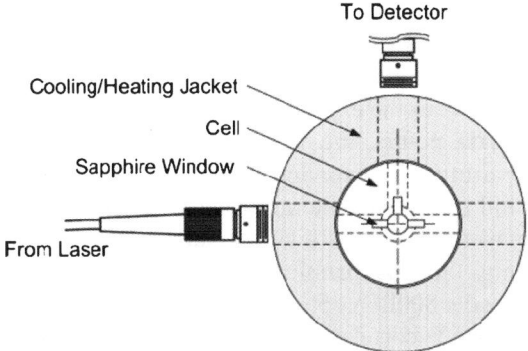

Figure 13.2 The optical fiber interface with the cell. (Reproduced from Winoto *et al.*[13] with permission from the American Chemical Society.)

without changing the mixture composition. It also has a borescope for visual observation of the phase transitions, and transmitted- and scattered-light intensity probes, arranged as shown in Figure 13.2. The apparatus is also equipped with a data-acquisition system that allows for measurements at constant pressure and temperature and for increasing and decreasing the pressure and temperature at constant rate. The CP is detected by a transmitted-light intensity (TLI) probe, which measures light at 180° from the laser. The MP is detected using a scattered-light intensity (SLI) probe, which is measured at 90° from the laser.

The CP at a given temperature is the specific pressure at which the capacity of the solvent is just enough to allow complete dissolution of the polymer. Above the CP, the polymer is dissolved. Below the CP, the polymer forms a distinct separate phase, at the onset of which the solution becomes opaque, or "cloudy." Similarly, for block copolymers the MP at a given temperature is the specific pressure at which solvent selectivity forces the less-soluble block to aggregate as the micelle cores. In the micellar region, below the MP but above the CP, a micellar solution exists as a result of nano, rather than bulk, phase separation. At pressures above the MP, the solvent is no longer selective enough to induce solvent–phobic block aggregation, so the micelles cease to exist.

In a typical experiment, a known amount of copolymer is loaded into the cell and pressurized with a known amount of solvent, both obtained by weighing the cell. Following weighing, the cell is brought to and maintained at a desired pressure and temperature to dissolve the copolymer completely and left, stirred, in the one-phase region for about 90 min to reach equilibrium.

To obtain a CP data point, the pressure is lowered slowly until the solution becomes cloudy at the onset of bulk phase separation, which is indicated by a drop in TLI. Pressure drop rates as low as 15 bar min^{-1} are used to obtain a reproducible CP, to within 3 bar. The bulk phase boundary can also be approached from the two-phase side upon compression, which leads to an increase in TLI, but produces roughly the same result, with no significant hysteresis. The solution is allowed to equilibrate in the one-phase region at a much higher pressure than the expected CP for 15 min before a new data point is taken. For all data points, the TLI results are stored and analyzed as a function of temperature, pressure, and time.

Since the SLI and hydrodynamic radius increase sharply at the MP, micelle formation can be measured using high-pressure dynamic light scattering. For these measurements, the high-pressure cell is coupled with an Ar^+ ion laser (National Laser, model 800BL) operating at a $\lambda = 488$ nm wavelength with a Brookhaven BI-9000AT correlator. The coherence area is controlled by a pinhole placed before the detector. The laser and detector are linked to the high-pressure cell by optical fibers produced by Thorlabs. The high-pressure optical interface design is different from, but inspired by, the approach described by Koga *et al.*[3] The scattered-light intensity is recorded and measured for isothermal measurements upon increasing or decreasing pressures at a rate of 30–60 bar min^{-1}. More detailed descriptions are available elsewhere.[13–15]

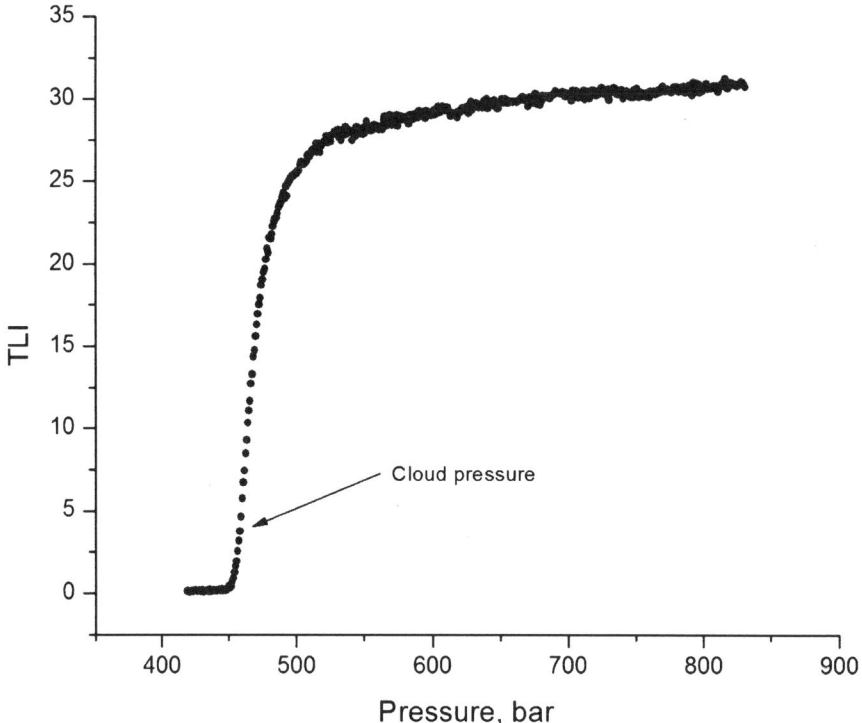

Figure 13.3   Transmitted light intensity plot of PEG-b-PCL (5k-b-11k) in 33% trifluoromethane and 67% dimethyl ether at 40 °C and 1 wt%.

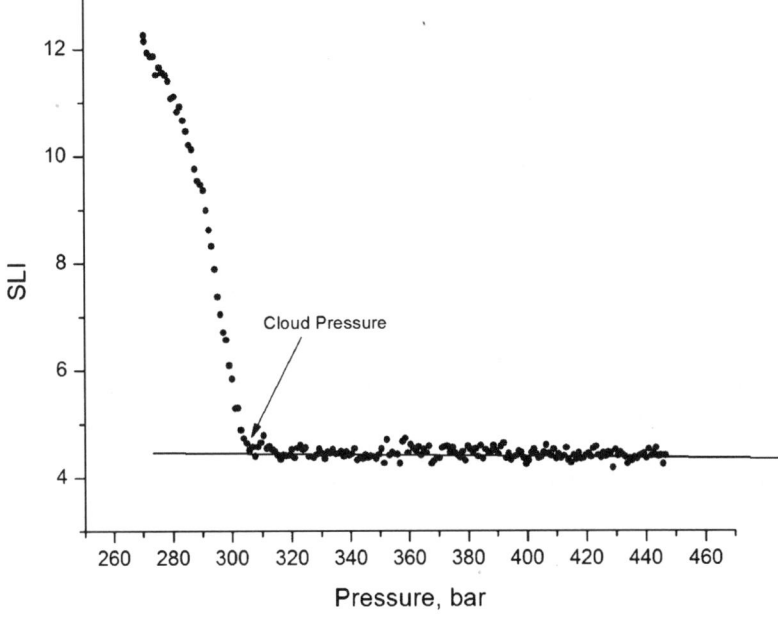

Figure 13.4 Scattered light intensity plot (without micellization) of PEG-*b*-PCL (5k-*b*-11k) in dimethyl ether at 100 °C and 1 wt%.

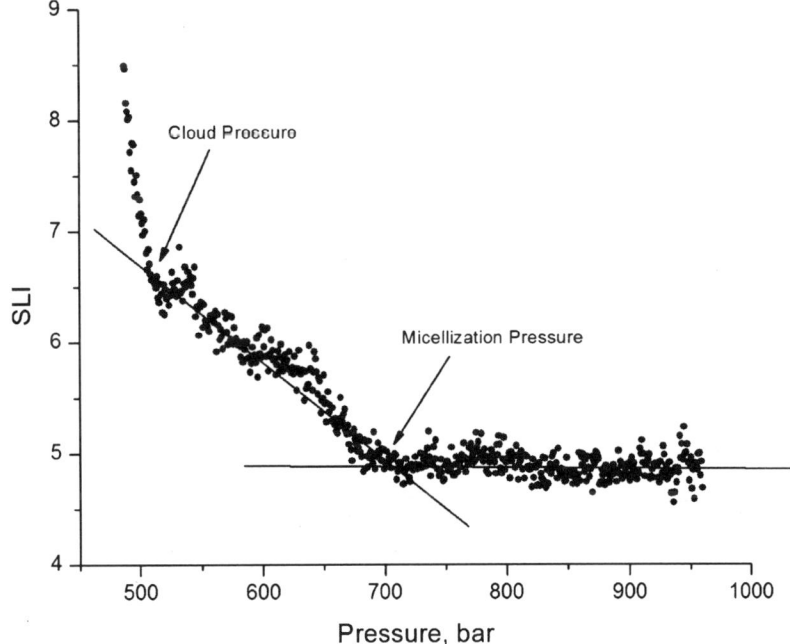

Figure 13.5 Scattered light intensity plot (with micellization) of PEG-*b*-PCL (5k-*b*-11k) in 33% trifluoromethane and 67% dimethyl ether at 60 °C and 1 wt%.

Figure 13.3 shows a sample graph of the TLI as a function of pressure. The CP is determined to be the inflection point of the TLI curve, which corresponds to a peak in its first derivative. Sample graphs of two typical cases of SLI curves are shown in Figures 13.4 and 13.5. Figure 13.4 illustrates an example with no micellization; the SLI increases sharply at the CP, but remains constant until then. By contrast, Figure 13.5 illustrates an example with micellization, where upon decreasing pressure, the SLI starts increasing at the MP, and then increases sharply at the CP.

A complete example of those early feasibility studies by Winoto *et al.*[13] is shown in Figure 13.6 for 0.5 wt% polystyrene-*block*-polyisoprene (MW 11.5k-*b*-10.5k) in propane. As explored by Tan *et al.*,[16] the CP and MP can be estimated from statistical associating fluid theory (SAFT1) that captures the behavior of the corresponding homopolymers. Such mean-field theories cannot account for detailed micelle structure, but can capture the onset of micelle formation, for example on the basis of a nanophase separation hypothesis. While all these results showed in general that that NCM can be used to form micelles, the results were for model copolymers that are not directly applicable to drug delivery.

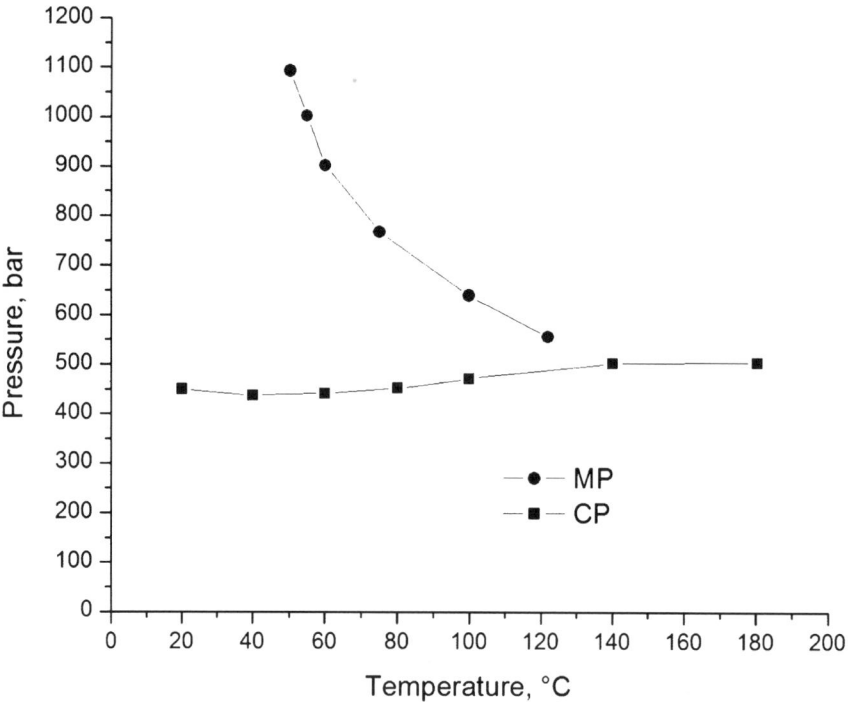

Figure 13.6   Pressure–temperature phase diagram of polystyrene-*block*-polyisoprene (11.5k-*b*-10.5k) in near-critical propane. The concentration of polymer in solution was 0.5 wt%. Re-plotted on the basis of data taken from Winoto *et al.*[13]

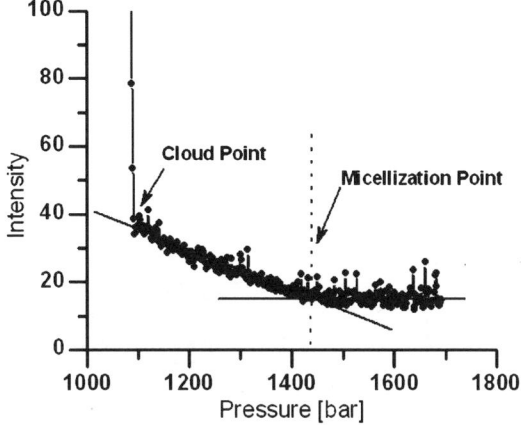

Figure 13.7 Scattered light intensity graph showing micelle formation. This is for PEG-*b*-PCL (5k-*b*-2.3k) in near-critical trifluoromethane at 100 °C and 5 wt%. (Reproduced from Tyrrell *et al.*[18] with permission from the American Chemical Society.)

13.3 Extension to PEG-*b*-PCL

Therefore, the next logical step in developing NCM was to extend it to a biomedically relevant polymer, such as poly(ethylene glycol)-*block*-poly(ε-caprolactone),[17] selected for NCM studies by Tyrrell *et al.*[18] Poly(ethylene glycol) (PEG) is a common hydrophilic corona-forming block while poly(ε-caprolactone) (PCL) is a common hydrophobic core-forming block; both are biodegradable and FDA approved.

Tyrrell *et al.*[18] chose chlorodifluoromethane and trifluoromethane as possible solvents for NCM, based on polarity and critical properties. While chlorodifluoromethane showed no micellization (due to narrow difference in block CPs), trifluoromethane exhibited evidence of micelle formation, as shown in Figure 13.7. A compilation of the CP and MP results shown in Figure 13.8 confirms a robust micellar region for trifluoromethane.

However, such micelles formed at high pressure may in principle undergo a structural rearrangement or decomposition upon decompression. So, the next step was to confirm that micelles still exist upon dispersion of the copolymer precipitate in water. Towards this end, the solvent was removed from the NCM system depicted in Figure 13.8 by depressurization, and the precipitate was re-dispersed in water and then filtered through a 200 μm filter. The filtered solution was then characterized for particle size. As shown in Figure 13.9, the particle size distribution is narrow and centered around 100 nm, reminiscent of micelle structure in the NCM solution. This was in sharp contrast to a control solution (copolymer dissolved in water without NCM treatment), which could not be dissolved at all.

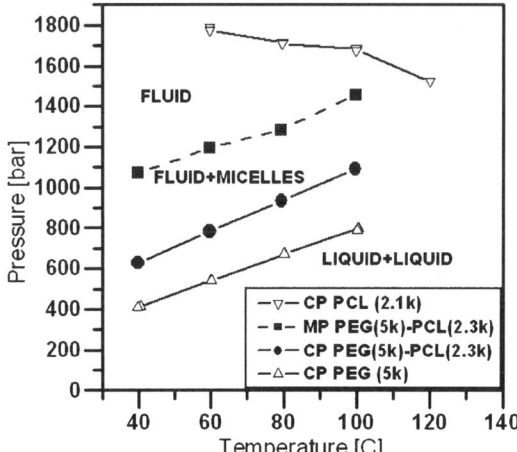

Figure 13.8   Pressure–temperature phase diagram of PEG (5k), PCL (2.3k), and the corresponding block copolymer in near-critical trifluoromethane. The concentrations of polymer were 1 wt% for the homopolymers and 5 wt% for the copolymer. (Reproduced from Tyrrell *et al.*[18] with permission from the American Chemical Society.)

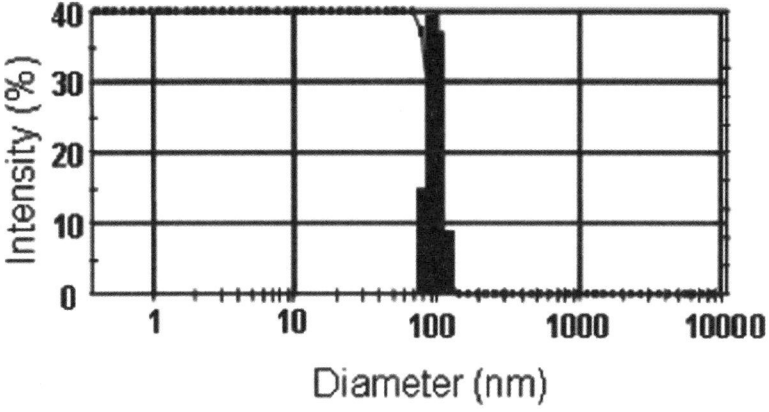

Figure 13.9   Particle size results of PEG-*b*-PCL (5k-*b*-2.3k) in water after near-critical trifluoromethane processing ($P = 650$ bar, $T = 35$ °C). The approximate concentration of polymer in water was ∼0.1 wt%. The nominal average diameter over three runs was 101 ± 5 nm. (Reproduced from Tyrrell *et al.*[18] with permission from the American Chemical Society.)

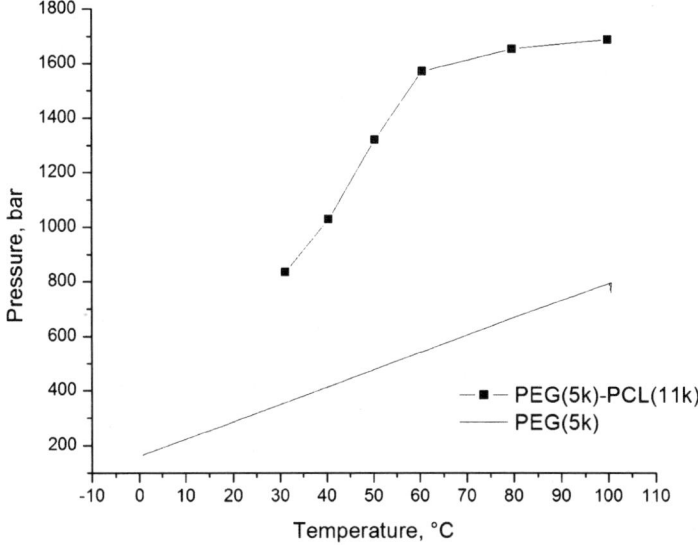

Figure 13.10 Cloud pressures for PEG (5k) and the corresponding PCL block copolymer (5k-*b*-11k) in near-critical trifluoromethane. The concentration of polymer in solution was 1 wt%. Note: micelles are present above the copolymer CP to over 1800 bar (the actual MP was beyond the experimental pressure range). Re-plotted on the basis of data from Green *et al.*[19]

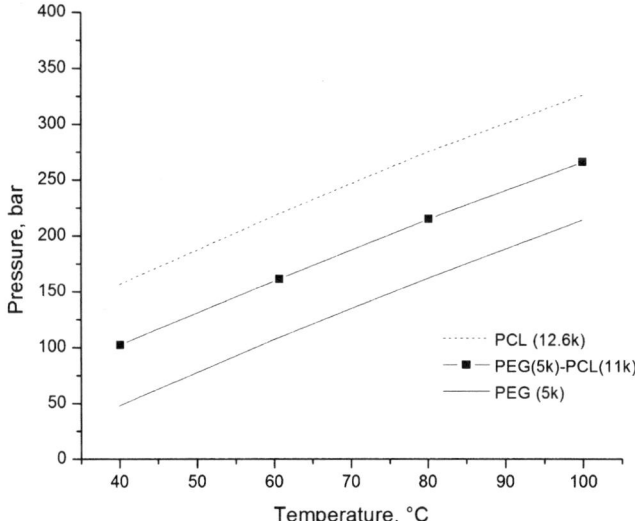

Figure 13.11 Cloud pressures for PEG (5k), PCL (12.6k), and the corresponding block copolymer (5k-*b*-11k) in near-critical dimethyl ether. The concentration of polymer in solution was 1 wt%. Note: micelles were formed at a concentration of 2 wt% for a small region at lower temperatures. Re-plotted on the basis of data from Green *et al.*[19]

Figures 13.7–13.9 illustrate that NCM can produce polymeric micelles that are recoverable in water and hence can be potentially useful for drug delivery. However, the phase behavior of that early prototype system was not very practical due to relatively high pressures. Thus, the next objective was to understand how to reduce the NCM pressure, for example, by optimizing the solvent composition effects.

13.4 Optimizing the NCM Solvent

To address this challenge, Green *et al.*[19] chose trifluoromethane, 1,1,1,2-tetrafluoroethane, hexafluoroethane, and dimethyl ether as model solvents, and PEG-*b*-PCL as a model solute, in which the PCL block was substantially larger than that used in the previous figures. Figure 13.10 illustrates sample results for trifluoromethane, in which micelles are found above the CP curve (the actual MP is beyond the pressure limit). Figure 13.11 illustrates that a new solvent, dimethyl ether, is not selective enough for PEG-*b*-PCL to form

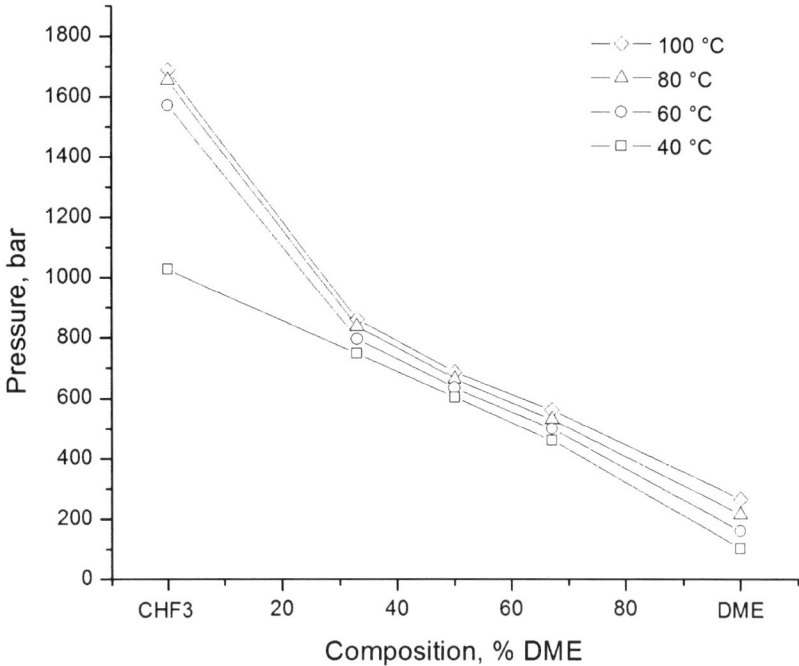

Figure 13.12   Pressure–composition phase diagram for PEG-*b*-PCL (5k-*b*-11k) in near-critical mixtures of trifluoromethane and dimethyl ether at various temperatures. The concentration of polymer in solution was 1 wt%. (Reproduced from Green *et al.*[19] with permission from the American Chemical Society.)

micelles (at this concentration), but the CPs in dimethyl ether are substantially lower than those in trifluoromethane. Hexafluoroethane was found too weak to dissolve the copolymer. While tetrafluoroethane dissolved the copolymer, it did not produce micelles.

These two figures are useful to approximate the solvent capacity and selectivity, which are usually defined in terms of mole or weight fractions. Capacity is a measure of the affinity of the solvent for the polymer. For a given solvent, the capacity roughly scales with density and hence increases with increasing pressure. Therefore, in comparing two different solvents, the one with a lower CP for a given polymer is deemed to have a higher capacity. For example, considering the difference in CP, dimethyl ether should have a higher capacity for the diblock than trifluoromethane due to a much lower CP. Selectivity is a measure of the affinity of a given solvent for one polymer over another. So, for the same solvent, a large difference in CP for two polymers means a high selectivity. Therefore, when the polymers are joined in a

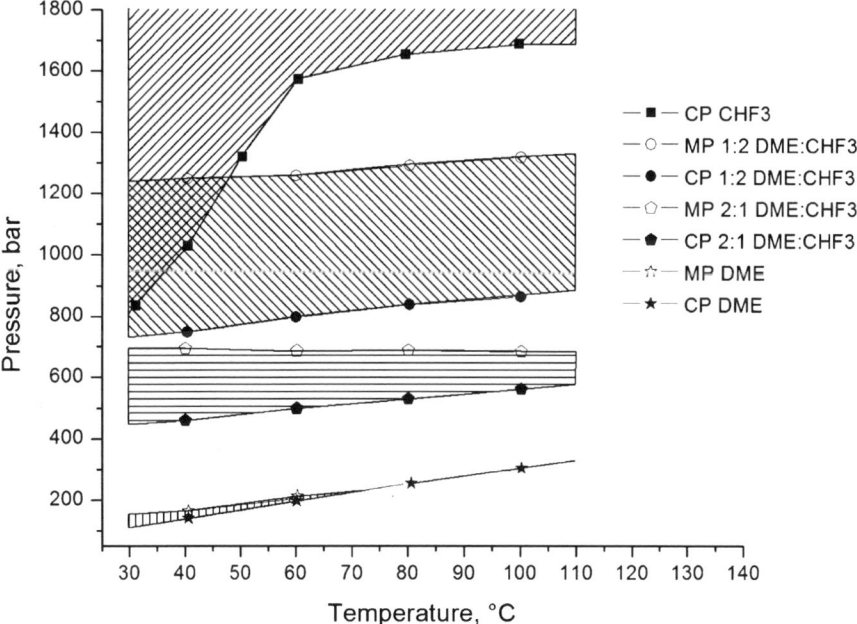

Figure 13.13 Pressure–temperature phase diagram of PEG-*b*-PCL (5k-*b*-11k) in near-critical trifluoromethane, dimethyl ether, and mixtures. Micellar regions are the shaded regions between the CP and MP curves. Note: the upper limit of the shaded region for pure trifluoromethane is the experimental boundary, not the MP. The concentration of polymer in solution was 1 wt% for all but the pure dimethyl ether, which was 2 wt%. (Reproduced from Green *et al.*[19] with permission from the American Chemical Society.)

diblock, high selectivity means a high likelihood of micelle formation. If the solvent is selective enough, the less soluble blocks (high-CP) are induced to precipitate, and form the core of micelles, while the more soluble blocks (low-CP) form the corona. As shown in Figures 13.10 and 13.11, PEG has a low CP in trifluoromethane, while PCL has a very high one (above the experimental range). This means trifluoromethane has high selectivity and high likelihood of micelle formation with PEG-*b*-PCL, which is indeed the case. However, in dimethyl ether, the difference between the two homopolymer CPs is small, which means low selectivity and low probability of micelle formation.

In fact, dimethyl ether did not produce micelles at 1 wt%, but it did at 2 wt% in a small low-temperature region. More important, dimethyl ether exhibited much lower CPs, which suggests a high capacity for PEG-*b*-PCL. This suggested combining the high selectivity of trifluoromethane with the high capacity of dimethyl ether simply by mixing the two pure solvents. Figure 13.12 shows that mixing the two solvents produced a nearly linear

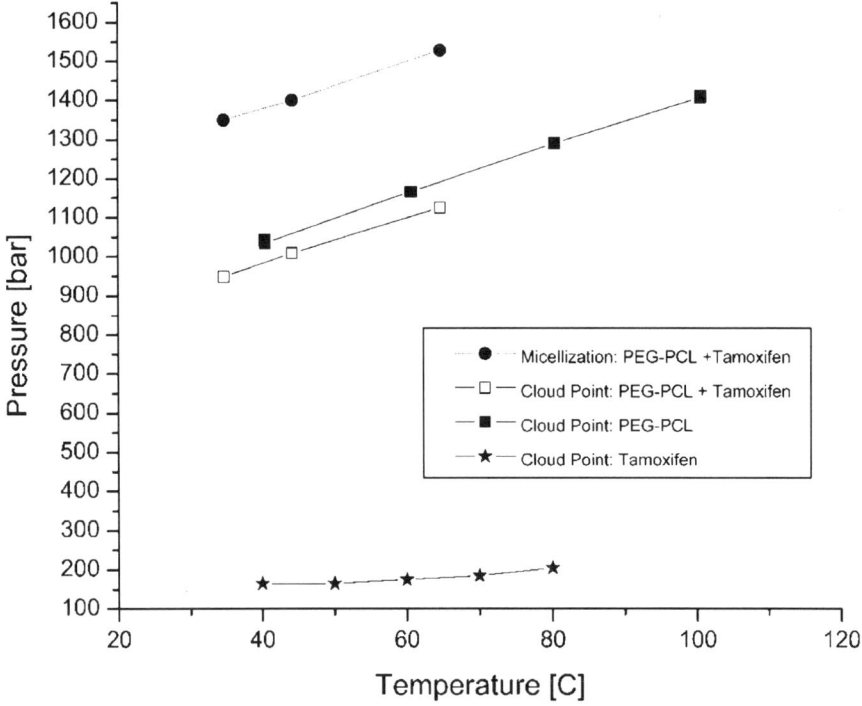

Figure 13.14   Cloud pressures of tamoxifen, PEG-*b*-PCL (5k-*b*-5k), and mixture and micellization pressure of PEG-*b*-PCL (5k-*b*-5k) in near-critical trifluoromethane. The concentrations of polymer and drug in solution are 1 and 0.1 wt%, respectively. (Reproduced from Tyrrell *et al.*[20] with permission from the American Chemical Society.)

CP (and hence the capacity) dependence on the amount of dimethyl ether present. Also, as depicted in Figure 13.13, the size of the micellar region scales nicely with solvent composition.

These results serve to show that using mixed solvents allows not only reduction of the operating pressure, but also gives better control over the solvent selectivity and capacity. While these results illustrate an improved NCM control, the next development hurdle was to demonstrate that NCM can be used to make drug-loaded micelles. If so, can NCM allow for improved drug loading?

13.5   Loading PEG-*b*-PCL with a Cancer Drug

Towards this end, Tyrrell *et al.*[20] investigated PEG-*b*-PCL (5k-*b*-5k) with tamoxifen in trifluoromethane. An example of their results is shown in Figure 13.14. The CP of the drug is lower than that of the copolymer, which

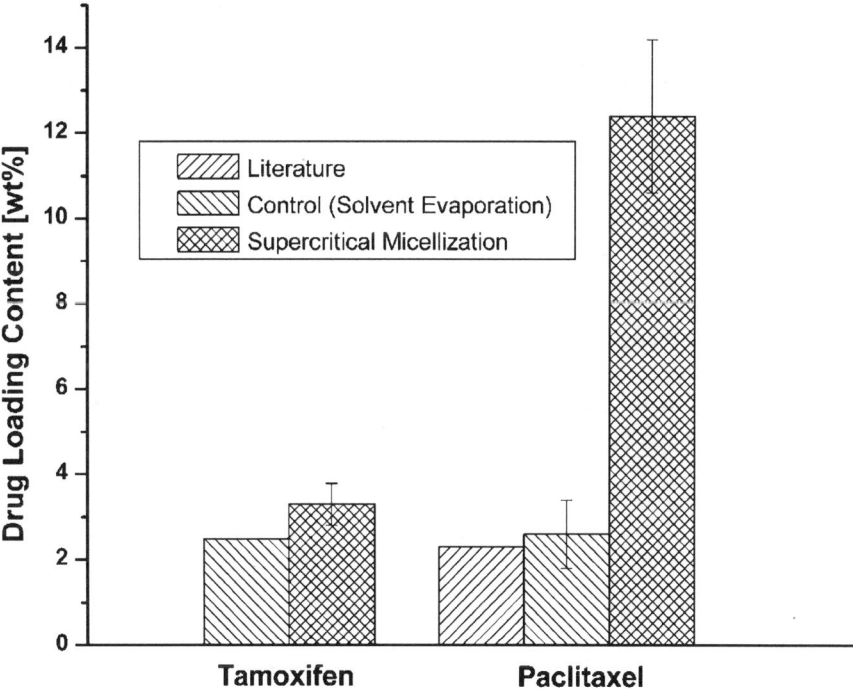

Figure 13.15   Drug loading content of tamoxifen and paclitaxel in PEG-*b*-PCL (5k-*b*-5k). Error bars represent the standard deviation of $n = 3$ for tamoxifen and $n = 5$ for paclitaxel. Literature values for paclitaxel[22] were obtained by a solvent evaporation method using THF. A literature reference was unavailable for tamoxifen in PEG-*b*-PCL. (Reproduced from Tyrrell *et al.*[20] with permission from the American Chemical Society.)

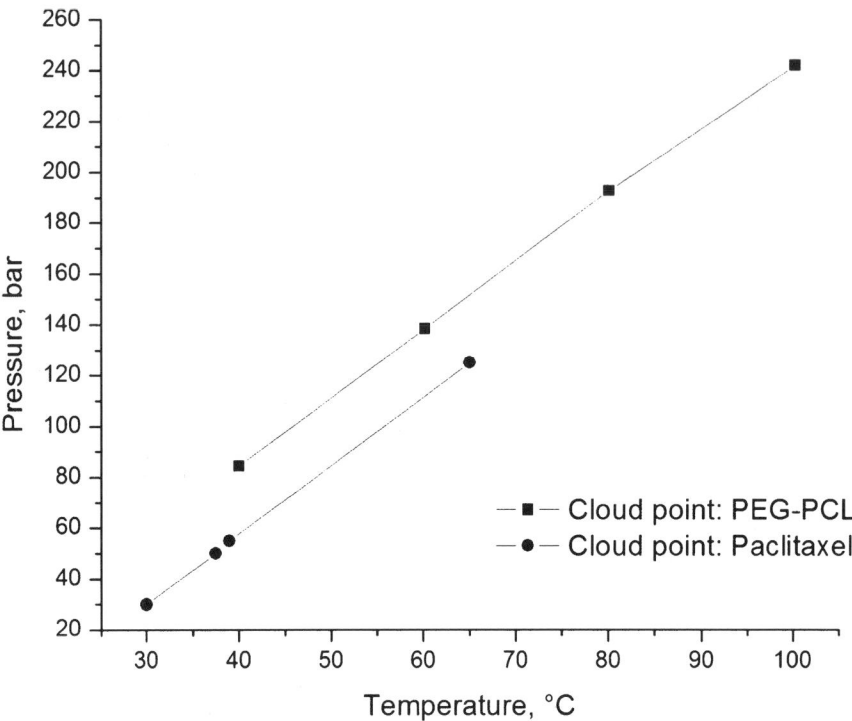

Figure 13.16 Cloud pressures of paclitaxel and PEG-*b*-PCL (5k-*b*-5k) in near-critical dimethyl ether. The concentrations of polymer and drug in solution are 1 and 0.3 wt%, respectively.

implies that the drug is still likely to be in solution when the polymer is precipitated. This leads to a low driving force for the drug to enter into the micelle core. As expected from this low driving force, the corresponding drug loading is about 3.3%, as shown in Figure 13.15 on the left, which is slightly higher than that obtained from the conventional solvent evaporation method.

However, what would happen if the drug CP fell between the MP and the CP of the polymer? As it turns out, paclitaxel satisfies this requirement with PEG-*b*-PCL (5k-*b*-5k). Since paclitaxel does not dissolve in trifluoromethane but easily dissolves in dimethyl ether, as shown in Figure 13.16, their mixture (70% dimethyl ether) allows us to bring the drug CP to the desired pressure range, between the copolymer CP and MP, as shown in Figure 13.17.

As a result, the drug precipitates from solution while micelles are present, which should drive it to the micelle core. This higher driving force results in a much higher drug-loading content, about 12.4%, as is seen from Figure 13.15 on the right. This is a substantial improvement over conventional solvent evaporation, which yields about 2.6% drug-loading content. However, the drug release profile, as seen in Figure 13.18, still exhibits burst release

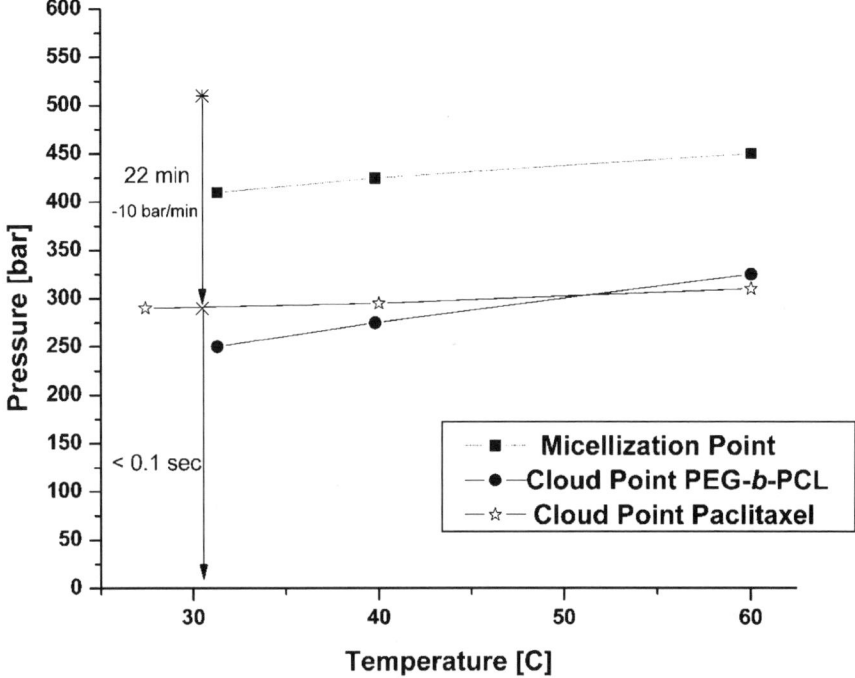

Figure 13.17 Cloud pressures of paclitaxel, PEG-*b*-PCL (5k-*b*-5k), and mixture and micellization pressure (MP) of PEG-*b*-PCL (5k-*b*-5k) in near-critical dimethyl ether/trifluoromethane (70/30) mixture. The concentrations of polymer and drug in solution are 1.5 and 0.15 wt%, respectively. The *arrows* indicate the pressure/time profile of the micellization/depressurization process used to recover drug-loaded micelles. (Reproduced from Tyrrell *et al.*[20] with permission from the American Chemical Society.)

characteristic of conventional approaches, which posed a new challenge for NCM.

13.6 NCM: A Remedy for Burst Release?

Tyrrell *et al.*[21] attempted to address this burst release challenge using a triblock copolymer. A simplistic idea underpinning the conventional methods is that, when diblock micelles are formed in liquid solution, all the drug will be encapsulated in the core of the micelle due to the core-forming block's higher affinity for the drug. As suggested by Tyrrell *et al.*,[21] for all such conventional methods, much of the drug can be deposited on the outside of the core, which can lead to burst release.

By contrast, a triblock micelle, if made synchronously with drug precipitation, forms a middle layer that can coat and hence protect the

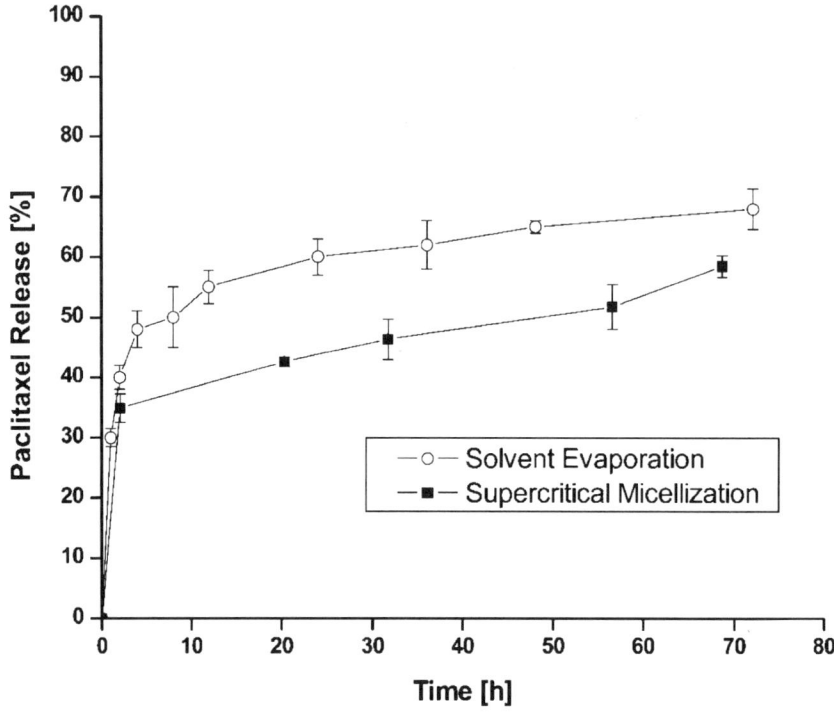

Figure 13.18 Paclitaxel release from PEG-*b*-PCL micelles prepared by solvent evaporation (*open circles*) and supercritical micellization (*squares*) into distilled water at pH = 7.4 and *T* = 37 °C. *Error bars* represent the standard deviation of *n* = 4 experiments. (Reproduced from Tyrrell *et al.*[20] with permission from the American Chemical Society.)

drug that failed to penetrate the core. This would require a sequential collapse of blocks, which in turn requires a control of the phase behavior that is simply unavailable by conventional methods, but attainable by the NCM method.

Tyrrell *et al.*[21] chose to prove this concept using FDA-approvable blocks, such as corona-forming PEG, core-forming PCL, and middle-layer-forming poly(L-lactic acid) (PLLA) or poly(D,L-lactic acid) (PDLLA) or a random copolymer of PLLA and PDLLA. Figure 13.19 suggests that trifluoromethane has a much lower capacity for PCL than for any of the other homopolymers. In the temperature range of interest, 20–60 °C, PLLA has a CP above that of PEG, and PDLLA has a CP below PEG. So, if the triblock were to consist of a PEG block (to form the corona in water), a PCL block (to form the core), and a middle block of PLLA, a sequential collapse may be viable.

Such a sequential collapse is illustrated in Figure 13.20. The top plot for PEG-*b*-(PCL-*co*-PDLLA) displays only two transitions: one from homogenous to

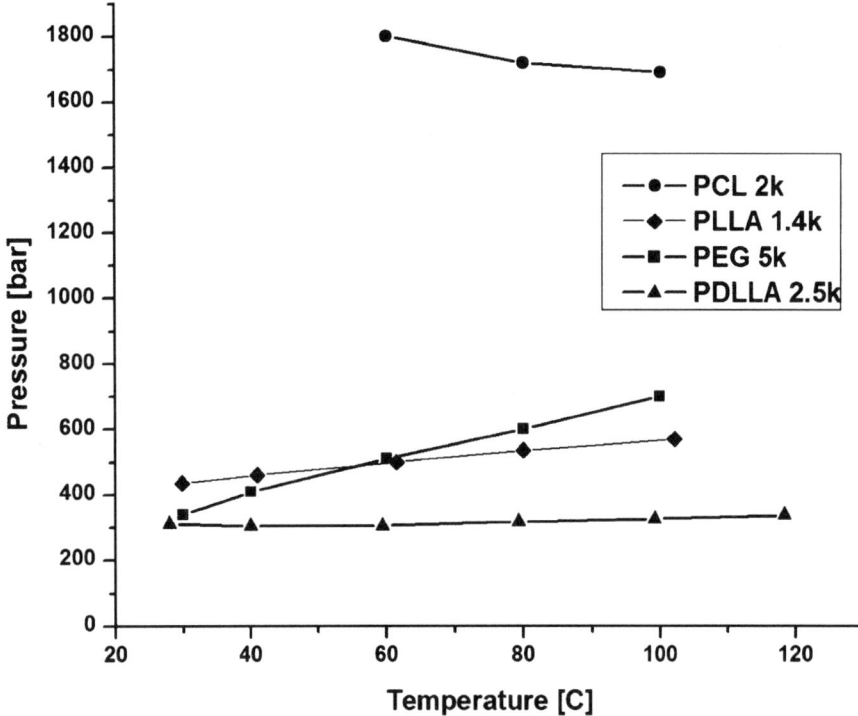

Figure 13.19 Cloud pressures of PCL, PLLA, PEG, and PDLLA in trifluoro-methane. The concentrations in solution are 1 wt%. (Reproduced from Tyrrell *et al.*[21] with permission from the American Chemical Society.)

micellar solution, and one from micellar to bulk precipitation, as expected for a diblock. The bottom plot for the PEG-*b*-PLLA-*b*-PCL triblock shows three distinct transitions. The first transition is from homogenous to micellar state, due to PCL aggregation. The second transition is from the micellar to another micellar state, due to PLLA collapse onto the PCL core. The final one is from micellar to bulk precipitation, due to micelle aggregation. These results for all temperatures are summarized in Figure 13.21. The solvent composition was chosen such that the drug CP fell between the micellization and middle block collapse pressures. Figure 13.22 shows that such a sequential block collapse indeed produces particles that exhibit very little, if any, burst release. This confirms the hypothesis that a precise control of block collapse and drug aggregation, which is virtually impossible with conventional preparation methods but attainable with the NSM method, can indeed produce particles made of FDA-approvable blocks that exhibit not only high drug-loading content and efficiency but also burst-free release.

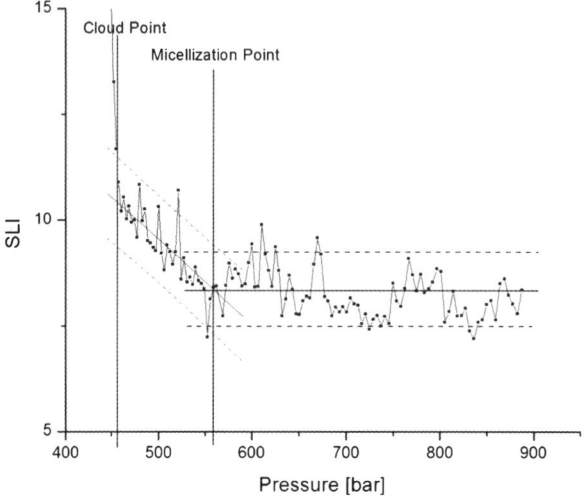

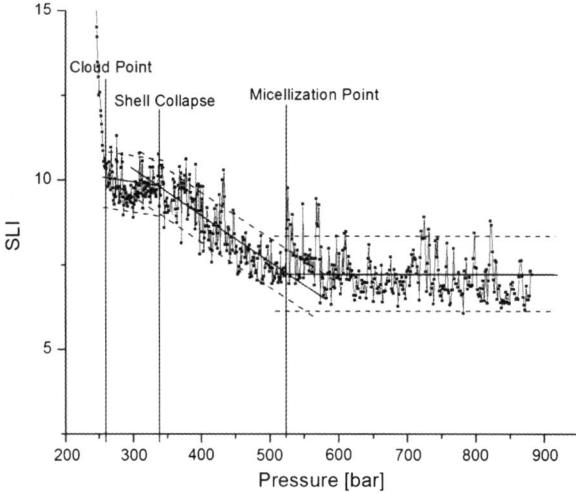

Figure 13.20   Scattered light intensity as a function of pressure for PEG-*b*-(PDLLA-
co-PCL) [5k-*b*-(2k-*co*-3k)] (*top plot*) and PEG-*b*-PLLA-*b*-PCL (5k-*b*-
1.5k-*b*-1.5k) (*bottom plot*). Both are in a 1:1 mix of trifluoromethane/
dimethyl ether solution at 40 °C and polymer concentrations of 2 wt%.
(Reproduced from Tyrrell *et al.*[21] with permission from the American
Chemical Society.)

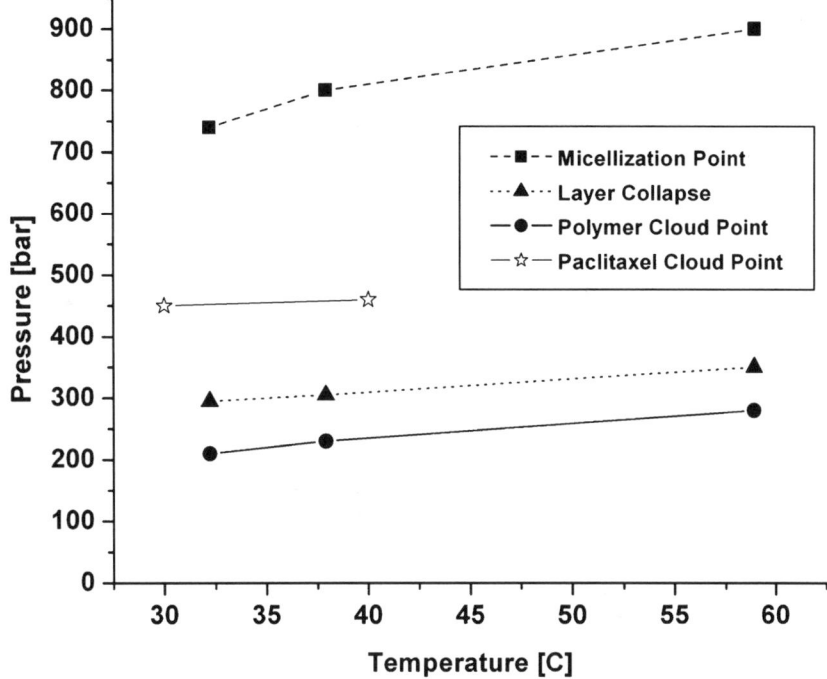

Figure 13.21 Cloud pressure and micellization pressure of PEG-*b*-PLLA-*b*-PCL (5k-*b*-1.5k-*b*-1.5k) and CP of paclitaxel in a 1:1 mix of trifluoromethane/dimethyl ether. Polymer and drug concentrations were 1.5 wt% and 0.15 wt%, respectively. (Reproduced from Tyrrell *et al.*[21] with permission from the American Chemical Society.)

13.7 Conclusion and Future Research Questions

Near-critical micellization (NCM) can produce solvent-free polymeric micelles useful for drug delivery systems. The NCM-made PEG-*b*-PCL micelles were found not only to be dispersible in water, but also to have drug loading content up to six times higher than that attainable from conventional methods. Even more promising, NCM has been demonstrated to produce sequentially collapsed triblock copolymer micelles, which is impossible to accomplish using conventional methods. Such triblock micelles, made of PEG-*b*-PLLA-*b*-PCL loaded with paclitaxel encapsulated by the middle PLLA layer, were found to exhibit virtually no burst release that is typical of diblock copolymer micelle formulations, while retaining the enhanced drug loading content and drug loading efficiency (Table 13.1).

These early NCM results are promising, but there are many intriguing research questions still to be answered. For example, how to further increase the drug loading by optimizing the process conditions and multi-block structure? How to probe the micelle structure *in situ* at high pressures? How to

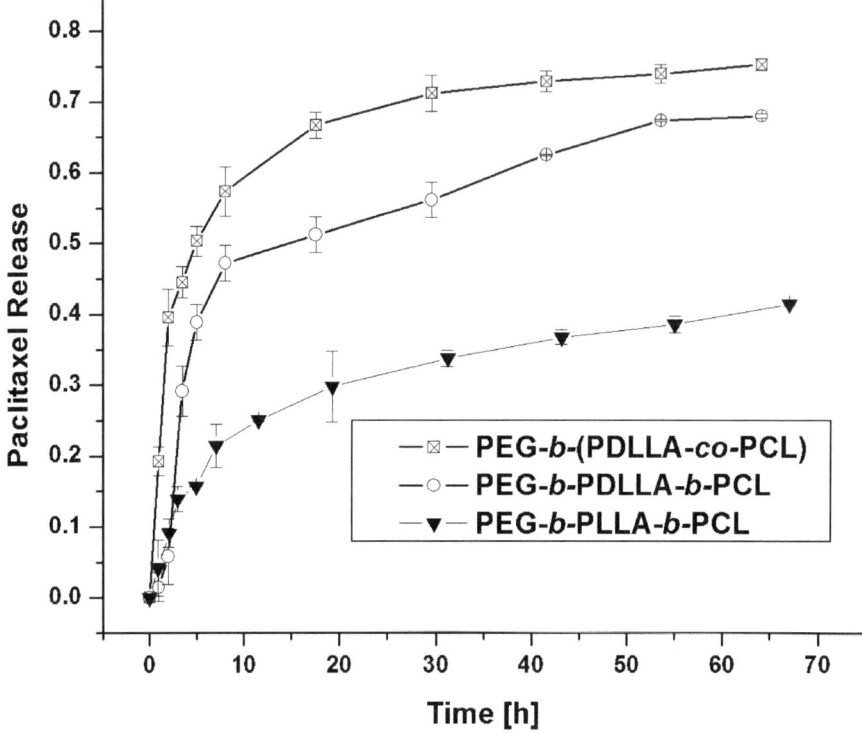

Figure 13.22   Cumulative paclitaxel release from a diblock, PEG-*b*-(PDLLA-*co*-PCL) [5k-*b*-(2k-*co*-3k)], and two triblocks, PEG-*b*-PDLLA-*b*-PCL (5k-*b*-1.5k-*b*-1.5k) and PEG-*b*-PLLA-*b*-PCL (5k-*b*-1.5k-*b*-1.5k), in distilled water at pH = 7.4 and T = 37 °C. (Reproduced from Tyrrell *et al.*[21] with permission from the American Chemical Society.)

extend this method to other polymer–drug formulations? How to scale-up NCM safely to kilogram or more product quantities? How to achieve a consistently uniform product quality necessary for FDA approvals?

Table 13.1   Drug loading and efficiency of diblock and triblock copolymers.[a]

Polymer	MW	Loading content	Encapsulation efficiency	Overall loading efficiency
PEG-*b*-PCL	5k-5k	12.4%	82.1%	87.0%
PEG-*b*-(PDLLA-*co*-PCL)	5k-(2k-*co*-3k)	10.5%	100.0%	80.2%
PEG-*b*-PDLLA-*b*-PCL	5k-1.5k-1.5k	10.9%	106.0%	72.3%
PEG-*b*-PLLA-*b*-PCL	5k-1.5k-1.5k	10.4%	98.6%	77.1%

[a]Data shown taken from work by Tyrrell *et al.*[20,21]

Acknowledgements

The authors would like to acknowledge financial support from the U.S. National Science Foundation (CBET-0828472 and CBET-1034530) and the National Science Foundation of China Distinguished Young Scholar Award (50888001).

References

1. A. V. Kabanov and V. Y. Alakhov, *Amphiphilic Block Copolymers: Self-Assembly and Applications*, Elsevier, Amsterdam, 1997.
2. K. Kataoka, G. Kwon, M. Yokoyama, T. Okano and Y. Sakurai, *J. Controlled Release*, 1992, **24**, 119.
3. T. Koga, S. Zhou, B. Chu, J. L. Fulton, S. Yang, C. K. Ober and B. Erman, *Rev. Sci. Instrum.*, 2001, **72**, 2679.
4. G. Kwon, M. Naito, M. Yokoyama, T. Okano, Y. Sakurai and K. Kataoka, *J. Controlled Release*, 1997, **48**, 195.
5. S. B. La, T. Okano and K. Kataoka, *J. Pharm. Sci.*, 1996, **85**, 85.
6. G. Kwon, *Crit. Rev. Ther. Drug Carrier Syst.*, 2003, **20**, 357–301.
7. V. Torchilin, *J. Controlled Release*, 2001, **73**, 137–172.
8. Z. Tyrrell, Y. Shen and M. Radosz, *Prog. Polym. Sci.*, 2010, **35**, 1128–1143.
9. G. Gaucher, M.-H. Dufresne, V. Sant, N. Kang, D. Maysinger and J.-C. Leroux, *J. Controlled Release*, 2005, **109**, 169–188.
10. R. Gref, Y. Minamitake, M. Peracchia, V. Trubetskoy, V. Torchilin and R. Langer, *Science*, 1994, **263**, 1600–1603.
11. P. Vangeyte, S. Gautier and R. Jérôme, *Colloids Surf., A*, 2004, **242**, 203–211.
12. J. L. Kendall, D. A. Canelas, J. L. Young and J. M. DeSimone, *Chem. Rev.*, 1999, **99**, 543–563.
13. W. Winoto, H. Adidharma, Y. Shen and M. Radosz, *Macromolecules*, 2006, **39**, 8140–8144.
14. A. K. C. Chan, P. Russo and M. Radosz, *Fluid Phase Equilib.*, 2000, **173**, 149.
15. M. Łuszczyk and M. Radosz, *J. Chem. Eng. Data*, 2003, **48**, 226–330.
16. S. Tan, W. Winoto and M. Radosz, *J. Phys. Chem. C*, 2007, **111**, 15752–15758.
17. C. Allen, Y. Yu, D. Maysinger and A. Eisenburg, *Bioconjugate Chem.*, 1998, **9**, 564–572.
18. Z. Tyrrell, W. Winoto, Y. Shen and M. Radosz, *Ind. Eng. Chem. Res.*, 2009, **48**, 1928–1932.
19. J. Green, Z. Tyrrell and M. Radosz, *J. Phys. Chem. C*, 2010, **114**, 16082–16086.
20. Z. Tyrrell, Y. Shen and M. Radosz, *J. Phys. Chem. C*, 2011, **115**, 11951–11956.
21. Z. Tyrrell, Y. Shen and M. Radosz, *Macromolecules*, 2012, **45**, 4809–4817.
22. X. Shuai, T. Merdan, A. Schaper, F. Xi and T. Kissel, *Bioconjugate Chem.*, 2004, **15**, 441–448.

Subject Index

References to tables and charts are in **bold** type

1,1'-carbonyldiimidazole, 78
1,1,4,7,10,10-hexamethyltriethylene-
tetramine (TFA), **240**
1,2-dioleoyl-*sn*-glycero-3-phosphoe-
thanolamine (DOPE), 44, 49
1,2-distearoyl-*sn*-glycero-3-phosphoe-
thanolamine (DSPE), 45, 93
1-(2-aminoethyl)piperazine tripropionylhydrazine, 128
1-ethyl-3-(3-dimethylaminopropyl)-
carbodiimide (EDC), 70, 232,
239, 241
2-(*N*-1-tritylimidazol-4-yl)-*N*-(6-gly-
cidyloxyhexyl)acetamide, 76
2-iodobenzoic acid, 91
2-methacryloyloxyethyl phosphorylcholine (MPC), **229**, 233–7, 239
2R2S
  defined, 34
  dextran, 46
  doxorubicin (DOX), 36, 41, 43, 45
  folic acid (FA), 50
  glutathione (GSH), 38, 41, 44
  liposomes, 35, 40, 44–6, 48–9
  paclitaxel (PTX), 46
  plasmid DNA (pDNA), 49
  poly(ε-caprolactone) (PCL), 36–7,
    39–41, 44, 46, 49
  poly(ethylene glycol) (PEG), 36–8,
    40–1, 44–5, 50
  reticuloendothelial system (RES),
    45–6

  spleen, 45–6
  transactivator of transcription
    (TAT), 50
3'-dithiobispropanoic acid (DTPA), 70
3-aminopropyl methacrylamide
  (APMA), 104
4',6-diamidino-2phenylindole
  (DAPI), **182, 273**
4-acetylbenzoic acid (4AB), 253
5-ethyl-5-(hydroxymethyl)-1,3-
  dioxan-2-one (EHDO), 69–70
6-lauroxyhexyl lysinate (LHLN), 93

absorption, biodistribution, metabolism, and elimination
  (ADME), 51, 245, 248
accelerated blood clearance (ABC),
  46, 149, 227
acetone, 281
acid ceramidase (AC), 167
aconitic acid (Aco), 224
actin, 269
acylhydrazone bonds, 128
adenosine, 95
adenovirus (Ad), 108–9, 265, 270–1,
  272
adipose tissue, 267
adjuvant therapy, 14
adypocytes, 268
affinity therapy, 14
age-related macular degeneration
  (AMD), 159

albumin, 26, 40, 46, 227, 239
alveolar epithelial cells, 79
amidation reaction, 79, 82, 84
aminolaevulinic acid (ALA), 83
ammonia, 82, 177
amphiphilic brush polymers, 40–1
anaerobic respiration, 127
angiogenic factors, 4, 27, 86, 159
angiopep, 85
antennapedia (Antp), 90–1
antibiotic selection, 275
antibody staining, 11, **197**
antigen presenting cells (APCs), 192, 196, 204
antisense oligonucleotides, **173,** 174, 179, 190, 192
aptamers, 50, 249
arginine, 88, 91, 134, 240, *also see* polyarginine
arginine-glycine-aspartic acid (RGD)
  active targeting, 26–7
  drug delivery, 11
  gene vectors, 73–4, 79, 87–8, 111
  hyperbranched polymers (HBP), 134
  siRNA, 165–6, 180
  targeting, 153
arginine-grafted bioreducible polymer (ABP), 108–9
Argonaute 2, 158
arm first method
  see one-pot method
arthritis, 201
artificial viruses, 110
asialofetuin (ASF), 222
asialoglycoprotein receptor (ASGPR), 25
aspraragine (Asn), 88
atom-transfer radical polymerization (ATRP), 233, 237
avidin, **74,** 75, 93, 172

β-lapachone (β-lap), 22
β-thioester bonds, 251–2

backbone modifications, 123, 125, 196
bacteriorhodopsin, 152
Bevasiranib, 159
BEXXAR®, 14
bile duct, 51, 223
binding site barrier, 25, 48
biodetection, 125
biofilm, 11
bioluminescent, 122, 135
biotin, 84, 93
bisacrylamides, 70, 95
blood-brain barrier (BBB), 84–5, 93, 153
bone, 92, 106, 108, 267
bone marrow, 22, 89, 95, 108, 162, 267–8, 270
bone marrow stromal cells (BMSCs), 95–6, 108
boronic acid, 76, 134
  also see phenylboronic acid
bovine serum albumine (BSA), 46, 239
brain stroke, 88
branched PEI (BPEI), 66, 100
breast
  current paradigm, 4
  drug delivery, 10, 14
  gene therapy, 268
  gene vectors, 92, 103
  hyperbranched polymers (HBP), 133
  polymersomes, 147
  siRNA, 167, 179
  targeting, 20, 25, 27
brush configuration, 40–1, 45, 80
bulk precipitation, 297
butane-1,4-diol diglycidyl ether (BDE), 76, 126
butane-2,3-dione, 128
butyl methacrylate (BMA), 171, 233–4

Caenorhabditis elegans, 158
CALB enzyme, 152

calcium, 90, 94, 145–6, 164–5, 265
calcium phosphate, 90, 164–5
camptothecin (CPT), 36–8, 52, 236–7,
    246–53
cancer statistics, 20–1
Candida albicans, 199
carboxybetaine (CB), 228, 232
carboxymethyldextran, 249
cardiovascular disease, 63
carrier dependent factors, 2, 4–5
cartilage, 75, 108, 267
castor oil, 4
cathepsin, 153
cationic shell-crosslinked knedel-like
    (cSCKs), 103–4, 210–1
caveolae-mediated pathway, 87, 95
CBMA-based polymers, 237, 239
cell-penetrating peptides (CPP),
    89–90
chain conformation, 231
chain growth, 122–3
charge reversal technique, 44, 49–50
chemiluminometer, **266**
chemokines, 265, 269–70, 272
Chinese herbal medicine, 191
chlorodifluoromethane, 289
chlorotoxin (CTX), 85, 101
cholesterol, 92–3, 159
cholic acid, 90, 102, 213
chondrocyte, **74**, 75, 268
cisplatin, 86, 127, 137, 253
citraconic acid (Cit), 224
click chemistry, 38, 73, 83–4, 137,
    235–6
cloud pressure (CP), 282–4, 286, **288**,
    289–94, 296, **299**
CMVL cationic lipid, 93–4
colitis, 203
colon, 133, 203, 268
complex micelles with mixed shells
    (CMMs), 217
complexation interaction, 124
confocal laser scanning microscope
    (CLSM), 70, 134, **146, 182**
connective tissue, 268

copper, 145
coprecipitation, 36
core-shell corona (CSC), 208–9, 211,
    213–4, 217–8
coumarin, 106
coxsackievirus, 109
Cremephor EL®, 4
critical aggregation concentration
    (CAC), 145
critical micelle concentration (CMC),
    39–40, 163, 167, 172, 210, 213,
    233
CT-2103 prodrug, 249, **250**
curcumin (Curc), 37, 252–4, **255**
cyclen, 93
cystamine, 70, 105, 149, 237
cysteines, 26, 80, 90
cystic fibrosis, 63
cytokines
    gene therapy, 264–5, 270–1, **272,**
        274–5
    gene vectors, 93, 111
    mesenchymal stem cells (MSC),
        261, **263**
    oligonucleotide complexes, 198, **202**
    siRNA, 162–3

dectin-1, 192, 196, **197, 199, 202,** 204
degree of branching (DB), 126, 133
dendrimer entrapped nanoparticles
    (DENPs), 99
dendritic units, 122
deoxythymidine (dTs), 159
desmin, 76
dextran
    2R2S, 46
    CBMA-based polymers, 239
    gene vectors, 68–9, 95
    ODN complex applications, 199
    polymersomes, 146, 151
    prodrug, 248–9
    targeting, 23
dextran sodium sulfate (DSS), 199,
    203
dichloroethane, 209

DiCPT prodrug, 37–8, 251
didodecyldimethylammonium brom-
    ide (DDAB), 109
diisopropylaminoethyl methacrylate
    (DPA), 237
dimethyl 3,3'-dithiobispropionimi-
    date (DTBP), 177
dimethylaminoethyl methacrylate
    (DMAEMA), 171–2, 233
dioleoylphosphatidylethanolamine
    (DOPE), 94
dioleylspermine, 104
diosgenin, 93
dipropylenetriamine (DPT), 170–1,
    223
disulfide crosslinked micelles
    (SSCLM), 212
dithiothreitol (DTT), 38, 239
DNA cell specific complexes
    kidney, 190
    liposomes, 190
    macrophages, 192
docetaxel (DTX), 211–2, **213**
DOTAP, 109
Doxil®
    see doxorubicin (DOX)
doxorubicin (DOX)
    2R2S, 36, 41, 43, 45
    drug delivery, 13
    gene vectors, 96, 101, 111
    hyperbranched polymers (HBP),
        128–9, 134, 137
    micelles design, 207, 214–6
    polymersomes, 146–8, 151, 153–4
    prodrug, 246–9, 253
    siRNA, 163, 166, 179–80
    targeting, 22, 25
drug delivery
    arginine-glycine-aspartic acid
        (RGD), 11
    breast, 4, 10, 14
    doxorubicin (DOX), 13
    enhanced permeability and reten-
        tion (EPR) effect, 3–5, 7, 9–11,
        12

folate, 4
kidney, 3
liposomes, 2, 5, 7, 12
macrophages, 7
paclitaxel (PTX), 4, 13
poly(ethylene glycol) (PEG), 7
reticuloendothelial system (RES), 3
spleen, 3
transactivator of transcription
    (TAT), 4
drug-diffusion resistance, 35

efflux drugs, 42
efflux pumps, 13, 21
EGFP sequence, 93
Ehrlich, Paul, 2, 6, 28
electrostatic adsorption, 49, 65
electrostatic attraction, 77, 79, 96,
    104, 124, 221
electrostatic bandpass, 48
encapsulation efficiency (EE), 179
endocytic pathway, 64
endometrial, 133
enhanced permeability and retention
    (EPR) effect
    current paradigm, 3–5
    drug delivery, 7, 9, 11, **12**
    gene vectors, 85
    hyperbranched polymers (HBP),
        128, 132–3, 137
    micelles design, 207, 212
    near-critical micellization (NCM),
        281
    prodrug, 248–9
    siRNA, 163, 169, 176, 183
    translational nanomedicine, 32–3
enzyme linked immunosorbent assay
    (ELISA), 197, **200, 202**
epidermal growth factor (EGF), 24,
    84, 236, 269
epidermal growth factor receptors
    (EGFR), 68, 134, 153, 236,
    268–9
epigenetic instability, 13
epithelial / endothelial barrier, 67

ERBB2, 25
Erbitux®, 24
Erlotinib®, 20
erythropoietin, 151
ethanolamine (EA), 101
ethylenediamine (ED), 101–2
excellular matrix (ECM), 4, 6, 11, 46–8
exonuclease, 161
experimental autoimmune encephalomyelitis (EAE), 201
EZN-2208 prodrug, 249, **250**

factor VIII antibody, 175
Fas gene, 89
fetal bovine serum (FBS), 97, 239
fiber-mutant, 265
fibrin gel, 108
fibroblasts, 25, 76, 80, 148, 172, 235, 270
fibronectin attachment protein of mycobacterium bovis (FAP-B), 79
Fick's first law, 145
film thickness, 231
floating piston, 282
flow cytometry, 135, 166, 197
fluorenylmethyl chloroformate (Fmoc), 237
fluorescein, 39, 149, 182
fluorescein isothiocyanate (FITC), 196, **197, 202**
fluorescence, 7, 40, 100, 166–7, 178, 180, 196–7
    also see (GFP)
fluorescence-resonance energy-transfer (FRET), 40
fluorouracil, 137
folate
    drug delivery, 4
    gene therapy, **269**
    gene vectors, **74,** 75, 79, 102, 109, 111
    hyperbranched polymers (HBP), 134

micelles design, 207, 214–5
    polymersomes, 147
    siRNA, 164, 180
    targeting, 25–7
folate receptor (FR), 4, 25, 75, 102, 214, 269
folic acid (FA)
    2R2S, 50
    CBMA-based polymers, 237
    gene vectors, 68, 73, 79, 104
    hyperbranched polymers (HBP), 122, 124, 133–4
    micelles design, 214–5
    polymersomes, 153
    prodrug, 249
    siRNA, 180
    targeting, 24
Food and Drug Administration (FDA), 8, 36, 51, 160
fusogenic peptides, 44, 49, 90, 176

galactosamine, 25, 249
galactose, 73, 79, 152, 222
gastrointestinal tract, 128, 220
Gaussian curves, **8, 193,** 194
gel retardation assay, 179
gel-permeation chromatography, 192
gelatin, 108, 267
gene therapy
    breast, 268
    cytokines, 264–5, 270–1, **272,** 274–5
    folate, **269**
    immunogenicity, 265, 268
    Lipofectamine, 265–6
    lung, 271–3
    transactivator of transcription (TAT), 265–6, 274
gene vectors
    arginine-glycine-aspartic acid (RGD), 73–4, 79, 87–8, 111
    breast, 92, 103
    cytokines, 93, 111
    dextran, 68–9, 95
    doxorubicin (DOX), 96, 101, 111

enhanced permeability and retention (EPR) effect, 85
folate, **74,** 75, 79, 102, 109, 111
folic acid (FA), 68, 73, 79, 104
glutathione (GSH), 70, 79
immunogenicity, 68, 105, 108
kidney, 70, 74, 88
Lipofectamine, 68, 70, 76, 80, 83, 93, 97, 103–4, 107
liposomes, 64, 92–3, 99, 109
low molecular weight (LMW), 43, 49, 86, 99–100, 103, 106–11, 111
lung, 67, 77, 92–3
macrophages, 67
messenger RNA (mRNA), 67
plasmid DNA (pDNA), 67, 76–7, 79, 92, 99, 108, 110
plasmid pDNA, 99, 108, 110
poly(ε-caprolactone) (PCL), 72, 96–7, 111
poly(ethylene glycol) (PEG), 66–8, 72, 75, 84–5, 88, 92–4, 96, 102, 105, 111
spleen, 89
transactivator of transcription (TAT), 68, 87, 90, 111
genomic instability, 13
glial cell, 85
glioblastomas, 76, 101, 103
gliomas, 75, 85, 101, 106, 147, 153, 268–72
globulins, 40, 46
glomerular filtration, 45
glucose, 4, 66, 68, 191
glucuronic acid, 169
glutathione (GSH)
    2R2S, 38, 41, 44
    gene vectors, 70, 79
    hyperbranched polymers (HBP), 129
    micelles design, 212–3
    polymersomes, 148
    prodrug, 248, 252
    siRNA, 169, 173–4, 177
    targeting, 26

glyceraldehyde 3-phosphate dehydrogenase (GAPDG), 68, 172, **197**
glycidyl methacrylate (GMA), 101, 128, 237
glycine (Gly), 25, 88, 134, 240
glycolysis, 127
glycosaminoglycan (GAG), 169
glycylphenylalanylleucylglycine (GFLG), 43, 248
glycyrrhetinic acid, 102
gold nanoparticles (GNP), 2, 26–7, 46, 48, 99, 109
good manufacturing practice (GMP), 34, 51
GOX enzyme, 152
GPEI magnetic nanoparticles, 101, 264
graft from method, 124
graft to method, 124
grafting, 66–7, 73, 76–7, 83, 99, 102–3, 111, 179
green fluorescent protein (GFP), 67, 93, 148, 167, 175, 177, 179
griseofulvine (GRF), 234–5

heart, 181
HeLa cells, 74, 76, 78, 83–4, 88, 96–7, 100–2, 235–6
hematoxylin and eosin (H&E), **201, 203, 273**
hemocompatibility, 228
hemoglobin, 146
heparin, 45, 76, 84, 169
hepatitis, 199, **201,** 236
hepatitis B surface antigen (HBsAG), 236
hepatoma cells, 77, 84, 88, 171, 236
hexafluoroethane, 290–1
high-pressure dynamic light scattering, 284
hight molecular weight, 66, 73, 106
histidine
    see polyhistidine
HIV, 68, 90

HMGB1 nuclear protein, 97
homopolymers, 130, 179, 228, 286, **287,** 292, 296
horse spleen ferritin (HSF), 149
host-guest interactions, 124, 135–6, 139
HRP enzyme, 152
human adipose-derived stem cells (hADSCs), 106–7
human embryonic kidney (HEK) cells, 79, 94, 98, 162, 167
human epidermal growth factor receptor 2 (HER2), 4, 8–9, 14, 26–7, 90
human fibroblast cells (HFW), 235
human prostate carcinoma cells, 174, 176
human umbilical vein endothelial cells (HUVEC), 88, 240
hyaluronan, 152, 154
hyaluronic acid (HA), 74–5, 111, 169
hyaluronic acid-spermine conjugates (HHSCs), 169
hybrid modifications, 123–5
hydrogen bonding interaction, 41, 124, 127, 136
hydroxylation camouflage, 69
hyperbranced polymers (HBP)
  enhanced permeability and retention (EPR) effect, 128, 132–3
hyperbranched homopolymer, 130, 179.228, 286, **287,** 292, 296
hyperbranched poly(amid-amine), 123
hyperbranched poly(ester-amine), 123
hyperbranched poly(sulfone-amine), 123
hyperbranched poly[3-methyl-3-(hydroxymethyl)oxetane] (HPMHO), 127
hyperbranched polyacylhydrazone (HPAH), 128
hyperbranched polyglycerols (HPGs), 133, 139

hyperbranched polymers (HBP)
  arginine-glycine-aspartic acid (RGD), 134
  breast, 133
  doxorubicin (DOX), 128–9, 134, 137
  enhanced permeability and retention (EPR) effect, 137
  folate, 134
  folic acid (FA), 124, 133–4
  glutathione (GSH), 129
  immunogenicity, 134, 137
  kidney, 128, 132–3
  liposomes, 139
  lung, 133
  paclitaxel (PTX), 137
  poly(ethylene glycol) (PEG), 134
  reticuloendothelial system (RES), 128
hyperbranched polymers HBP)
  folic acid (FA), 122
hyperbranched polyphosphate (HPHDP), 130–1
hypercholesterolemia, 159
hypersensitivity reactions (HSRs), 52

ibuprofen, 209, **210,** 219, **220**
IgG, 46, **202, 224**
IgM, 45
Imatinib®, 20
imidazolium, 93
imidazolyl, 76
imidazoquinolines, 163
imine reaction, 74
iminothiolane, 177
immmunoglobulins, 46
immobilized porcine pancreas lipase (IPPL), 99
immune evasion, 10, 21, 183
immunoediting, 10
immunogenicity
  gene therapy, 265, 268
  gene vectors, 68, 105, 108
  hyperbranched polymers (HBP), 134, 137

mesenchymal stem cells (MSC), **263**
oligonucleotide complexes, 201
siRNA, 160, 165
immunoliposomes, 27
immunomodulation, 14
immunostaining, 2
importin, 91
Independent Data Monitoring Committee, 159
inducible nitric oxide synthase (iNOS), 211
inflammatory bowel disease (IBD), 199, 201
inhibitory concentration (IC$_{50}$), 171, 236
integrins, 73–4, 79, 88, 111, 134, 165, 180, 270
interleukin (IL), 167, 181, **272**
interstitial fluid pressure (IFP), **6,** 11, 22, 33, 46
intestines, 128, 220
intra-cytosol release mechanism, 43–4
intra-lysosome release mechanism, 42–3
investigational new drug (IND), 159
Invitrogen, 265
iron oxide, 2, 85, 101
ischemia/reperfusion (I/R), 88

Japan, 191–2

KALA peptide, 68, 87, 91, 97
kidney
   current paradigm, 3
   DNA cell specific complexes, 190
   gene vectors, 70, 74, 88
   hyperbranched polymers (HBP), 128, 132–3
   nanocarrier production process, 51
   prodrug, 250
   siRNA, 160–2, 181
kinase, 88, 179, 269
kinetic stability, 39

knockdown, 68–9, 90, 101, 160, 167–8, 170–2, 178, 223
Kuhn, Thomas, 14
Kupffer cells, 162

L-carnosine, 70
lactoferrin, 85, 153
lactose, 77, 222
lamina propria (LP), **202**
lasers, 70, 134, 182, 283–4
lectin, 133, 152, 192
lentinan, 191
lentiviruses, 265
leucine, 85, *also see* polyleucine
leukemia, 5, 20
levulinic acid (LEV), 253
ligand-exchange reactions, 100
light units (LU), **266**
linear units, 122
Lipofectamine
   gene therapy, 265–6
   gene vectors, 68, 70, 76, 80, 83, 93, 97, 103–4, 107
   siRNA, 168–9
lipopolysaccharide (LPS), 199, **200–2**
lipoprotein receptor-related protein-1 (LRP1), 85
liposomes
   2R2S, 35, 40, 44–6, 48–9
   DNA cell specific complexes, 190
   drug delivery, 2, 5, 7, 12
   gene vectors, 64, 92–3, 99, 109
   hyperbranched polymers (HBP), 139
   mesenchymal stem cells (MSC), 264–5
   polymersomes, 144–6, 154
   prodrug, 251
   siRNA, 160, 163, 168
   targeting, 21, 25–7
   translational nanomedicine, 32
Lipotap, 109
liver, 3, 132, 162, 181
   also see hepatoma cells
locked nucleic acid (LNA), 162

low molecular weight (LMW)
   gene vectors, 43, 49, 86, 99–100,
      103, 106–11, 111
low-density lipoproteins (LDLs), 159
lower critical solution temperature
      (LCST), 77, 128, 149, 209, 218
luciferase, 93, 98, 165, 171, 178, 222–
      3, **266**
lumen, 146
lung
   gene therapy, 271–3
   gene vectors, 67, 77, 92–3
   hyperbranched polymers (HBP),
      133
   mesenchymal stem cells (MSC),
      264
   micelles design, 211
   prodrug, 249
   targeting, 20
luteinizing hormone-releasing hor-
      mone (LHRH), 86, 176
lymphatic drainage, 21, 132
lymphocytes, 68
lymphokine, 201

macrophages
   DNA cell specific complexes, 192
   drug delivery, 7
   gene vectors, 67
   mesenchymal stem cells (MSC),
      263
   oligonucleotide complexes, 196–9,
      201, 203
   siRNA, 161–2
   targeting, 22
   Zwitterionic polymers, 227
magic bullet, 2, 5–6, 8–9, 14, 24, 28
magic shotgun, 13
magnetic nanoparticles (GPEI), 101
magnetic resonance imaging (MRI),
      27, 100–1
magnetic-fluorescence nanocompo-
      sites, 100
magnetofection, 125
maleimide, 26, 173

mannose, 73, 79
marrow-isolated adult multilineage
      inducible (MIAMI), 106
Matrigel invasion assay, 268
matrix metalloproteinases (MMPs),
      4, 49
maximum tolerated dose (MTD),
      247–8
membrane penetration mechanism, 4
mesenchymal stem cells (MSC)
   cytokines, 261, **263**
   immunogenicity, **263**
   liposomes, 264–5
   lung, 264
   macrophages, **263**
   reticuloendothelial system (RES),
      264
mesenteric lymph node (MLN), **202**
messenger RNA (mRNA)
   gene vectors, 67
   micelle design, 211, 216
   oligomer chains of ethylene glycol
      (OEG), 197
   siRNA, 158–60, 160–2, 165, 172,
      175–6
metastases, 11, 14, 49, 134, 253, 268,
      271–3
metastasize, 13
metastatic potential, 12
methotrexate, 137
methylthiirane, 70
micellar nanoparticles (MNPs),
      166–7
micelle complex design
   doxorubicin (DOX), 207
   enhanced permeability and reten-
      tion (EPR) effect, 207
   folate, 207
   paclitaxel (PTX), 207
   poly(ethylene glycol) (PEG), 207
micelle core-shell-corona
   doxorubicin (DOX), 214–6
   enhanced permeability and reten-
      tion (EPR) effect, 212
   folate, 214–5

folic acid (FA), 214–5
glutathione (GSH), 212–3
lung, 211
plasmid DNA (pDNA), 211, 214
poly(ε-caprolactone) (PCL), 213–5
poly(ethylene glycol) (PEG), 208,
    211, 213–5
transactivator of transcription
    (TAT), 212
micelle design
    messenger RNA (mRNA), 211,
        216
    plasmid (pDNA), 211, 214, 221–2
    plasmid DNA (pDNA), 211, 214
    poly(ε-caprolactone) (PCL), 220–1
    poly(ethylene glycol) (PEG), 217–
        9, 221–3
micelle dissociation kinetics, 39
micelle dissociation rate, 39
micellization pressure (MP), 282–4,
        286, 288–90, **291**, 294, **295**
Michael addition, 70, 72, 82, 135
Michael polyaddition, 96
miconazole (MCZ), 234
microbubbles, **262**, 264
microemulstions, 2, 7
micrometastases, 268
migration inhibitory factor (MIF),
        199, 201, **202**, 203
minicircle DNA (mcDNA), 75
monocyte chemo-attractants, 270
monomer effective interaction
        energy, 39
mononuclear phagocyte system
        (MPS), 3, 7, 22, 45–6, 160–1,
        190, 227, **262**
Monte Carlo simulation, 39
mucin, 79
multi-angle light scattering analysis,
        129, 192, **237**, 284
multidrug-resistant (MDR) cancer
        cells, 21, 111, 165–6, 180
multifunctional envelope-type nano
        device (MEND), 49

murine, 9–10, 77, 163, 167, 181, 268–
    9
muscle, 92, 267
mushroom configuration, 45
Mycobacterium bovis, 79
Mylotarg®, 4, 8
myocardium, 268
myoglobin, 146

N,N,N-trimethylated-chitosan
        (TMC), 80
N,N-dimethyldipropylenetriamine
        (DP), 165
N,N-dimethylformamide (DMF), **240**
N-(2-hydroxyethyl methacrylamide)-
        oligolactates (HEMAmLac),
        40
N-(2-hydroxypropyl)methacrylamide
        (HPMA), 25, 43–5, 51, 104,
        247–50, 256
N-acetyl-D-glucosamine, 77, 169
N-acetylglucosamine (GlcNAc), **74,**
        76
N-hydroxysuccinimide (NHS), 70,
        232, 239, 241
N-nitrophenylcarbonyloxyethyl
        methacrylate (NPMA), 236
nano-emulsions, 262
nanocarrier production process
    kidney, 51
    poly(ε-caprolactone) (PCL), 51
    poly(ethylene glycol) (PEG), 52
nanocavities, 81, 138
nanodelivery vehicles (NDV), 234,
        236–7
nanogel, 128, 239–40
nanomedicine defined, 21
nanopumps, 209
nanorods, 23
nanospheres, 23
nanotubes, 45–6, 99
natural killer (NK) cells, 263
near critical micell micellization
        (NCM)

enhanced permeability and reten-
　　tion (EPR) effect, 281
paclitaxel (PTX), 293–5, 296, 299–
　　300
poly(ε-caprolactone) (PCL), 284–5,
　　287, 290–3, 293–4, **295,** 296–7,
　　298, 299, 300
poly(ethylene glycol) (PEG), **287,**
　　292, 294–5, 297, 300
near-infrared (NIR), 146
neuroepithelial tissues, 11
neuronal cells, 74, 84
neurotoxicity, 4, **263,** 268
nicotinic acetylcholine receptor
　　(nAchR), 84
niosomes, 265
NK012 prodrug, 253, **254**
NKTR-105 prodrug, 249
Nobel Prize, 159
non-small-cell lung cancer (NSCLC),
　　20, 22
nonfouling, 146, 228–9, 239–41
NOVAFECT chitosans, 80
nuclear localization signal (NLS), 68,
　　87, 91, 97
nuclear membrane barrier
　　91
nuclear translocation, 91

oligoalkylaminosiloxane, 76
oligoarginine, 104
oligomer chains of ethylene glycol
　　(OEG), 37–8, 197, 231, 251–3,
　　255
oligonucleotide complexes
　　cytokines, 198, **202**
　　immunogenicity, 201
　　macrophages, 196–9, 201, 203
oligonucleotides (ODN), 174, 192,
　　196, 199–201
oncogenes, 13, 159
one-pot method, 105, 121, 123, **237**
opsonization, 3, 7, 45, 50, 249
optical fibers, 283–4
oral drug delivery, 218, 220, 235

oscillating magnetic field, 128–30
osmotic factors, 11, 65, 170–1
osteoarthritis, 75
osteoblasts, 268, 270
ovarian cancer, 10, 79, 176, **251**

P-glycoprotein (P-gp), **42,** 111, 165–6,
　　180
paclitaxel (PTX)
　　2R2S, 46
　　drug delivery, 4, 13
　　hyperbranched polymers (HBP),
　　　137
　　micelles design, 207
　　near critical micellization (NCM),
　　　296, 299–300
　　PEG-*b*-PCL loading, 293–5
　　phosphorylcholine-based poly-
　　　mers, 234, 237
　　polymersomes, 146
　　prodrug, 246
　　siRNA, 179, 181–2
pancreas, 22–3, 92, 98, **272**
parent drug, 245–7, 249
Parkinson's disease, 63
Pearce, Homer, 10
Pegamotecan, 249
pegylated immuno-lipopolyplex
　　(PILP), 93
PEGylation, 7, 22, 66–8
pentaethylenehexamine (PEHA), 84
phagocytosis
　　see mononuclear phagocytic sys-
　　　tem (MPS)
phase-guided assembly, 151, **152**
phenotypic expansion, 14
phenylboronic acid, **74,** 76
　　also see boronic acid
phosphate buffered saline (PBS), 172,
　　181, **199–201, 203,** 239
phosphodiester (PO), 162, 194–6
phosphorothioate (PS), 104, 162, 192,
　　194, 196, 211
phosphorylcholine-based polymers
　　paclitaxel (PTX), 234, 237

poly(ε-caprolactone) (PCL), 235
photoluminescence, 135
physical targeting
see targeting
pinocytosis, 132
PK1 (PHPMA-DOX), 247, 249
PK2 (PK1-galactosamine), 25, 249
placenta, 267
plasmacytoid dendritic cells (pDCs),
    163
plasmid DNA (pDNA)
    2R2S, 49
    gene vectors, 67, 76–7, 79, 99, 108,
        110
    micelle design, 211, 214
plasmid pDNA
    2R2S, 49
    gene vectors, 67, 76–7, 79, 92, 99,
        108, 110
    micelle design, 211, 214, 221–2
platelets, 228, 270
platinum, 254–5, **256**
pluronics, 38, 40, 86, 92, **93**
PNU166945, 247
polo-like kinae 1 (Plk1), 179, 181
poly(2,4,6-trimethoxybenzylidene-
        pentaerythritol carbonate
        (PTMBPEC), 148
poly(2-aminoethyl ethylene phos-
        phate (PPEEA), 166, 181, **182**
poly(2-methacryloyloxyethyl phos-
        phorylcholine (pMPC), 228,
        229, 233–5
poly(2-methyl-2-oxazoline
        (PMOXA), 103
poly(2-vinylpyridine) (P2VP), 147,
        208
poly(3-amino-2-hydroxypropyl
        methacrylate) (PAHPMA),
        104
poly(4-vinylpyridine) (P4VP), 209–
        10, 217
poly(5-methyl-5allyloxycarbonyl-tri-
        methylene carbonate
        (PMAC), 98–9

poly(α-benzyl carboxylate ε-capro-
        lactone) (PBCL), 41
poly(α-carboxyl-ε-caprolactone)
        (PCCL), 41
poly(α-propargyl carboxylate-ε-
        caprolactone) (PPCL), 38
poly(acrylamidoethylamine) (PAEA),
        211
poly(acryloyl carbonate) (PAC), 38
poly(amido amine) (PAMAM), 42,
        49, 82–5, 99, 135, 139, 172–3,
        211
poly(aminoethyl methacrylate)
        (PAEMA), 104
poly(β-amino esters) (PBAEs), 103
poly(β-benzyl-L-aspartate) (PBLA),
        170
poly(carboxybetaine acrylamide)
        (pCBAA), **229, 231**
poly(carboxybetaine methacrylate
        (pCBMA), 228, **229, 231**
poly(carboxybetaine) (PCB), 232–3,
        239, **240**
poly(D,L-lactic acid) (PLA), 22, 88,
        106, 134, 218–9, 235, 253
poly(D,L-lactide) (PDLLA), 38–40,
        153, 296, **297, 300**
poly(ε-caprolactone) (PCL)
    2R2S, 36–7, 39–41, 44, 46, 49
    gene vectors, 72, 96–7, 111
    micelles design, 213–5, 220–1
    nanocarrier production process, 51
    near critical micellization (NCM),
        284–5, 287, **288,** 289, 290–3,
        293–4, **295,** 296–7, **298,** 299, **300**
    phosphorylcholine-based poly-
        mers, 235
    polymersomes, 146, 151, 152
    siRNA, 165–7, 179–81
poly(ethylene glycol) (PEG)
    2R2S, 36–8, 40–1, 44–5, 50
    drug delivery, 7
    gene vectors, 66–8, 72, 75, 84–5, 88,
        92–4, 96, 102, 105, 111

hyperbranched polymers (HBP), 134

micelles design, 207, 208, 211, 213–5, 221–3

micelles designs, 217–9

nanocarrier production process, 52

near critical micellization (NCM), **287**, 292, 294–5, 297, 300

polymersomes, 146, 148, 150

prodrug, 247, 249, 253–4

siRNA, 159, 164, 168, 170–3, 175–7, 180

targeting, 22, 26

Zwitterionic polymers, 227, 229, 241

poly(ethylene oxide) (PEO), 111, 145–7, 149, 151, 153, 165, 180, 208

poly(ethylenimine sulfide) (PEIS), 65–8, 70–2, 180

poly(γ-benzyl L-glutamate) (PBLG), 73–4

poly(glutamic acid) (PGA), 80, 96–7, 147–8, 180, 215, 248–9, 253

poly(hexafluorobutyl methacrylate) (PHFMA), 105

poly(L-amino acid) (PLAA), 36–7

poly(L-aspartic acid) [P(Asp)], 36, 41, 96, 247

poly(L-lysine) (PLL), 49, 86, 103, 111, 139, 167, 177–8, 223

poly(L-phenylalanine) (PPhe), 211–2

poly(L-succinimide) (PSI), 73

poly(lactic-co-glycolic acid) (PLGA), 41, 107–10, 219–20, 233, 239–40, 247

poly(mercaptoethyl ethylene phosphate) (PPE), 41, 213–4

poly(methacryloxyethyl trimethylammonium chloride (PMOTAC), 105

poly(methacryloyl sulfadimethoxine (PSD), 50

poly(methacryloylglycinamide) (PMAG), 247

poly(*N*-hexyl stearate L-aspartamide (PHSA), 40

poly(*N*-isopropylacrylamide) (PNIPAM), 41, 128, 149, 208–10, 217–21

poly(propylacrylic acid) (PPAA), 43–4

poly(propylene oxide) (PPO), 102, 219–20

poly(propylenesulfide) (PPS), 148

poly(sulfobetainemethacrylate) (pSBMA), 228, **229**

poly(*t*-butylacrylate) (P*t*BA), 217

poly[(2,4-dinitrophenyl)thioethyl ethylene phosphate (PPE), 41

poly[2,4,6-trimethoxybenzylidene-1,1,1-tris(hydroxymethyl)ethanemethacrylate (PTTMA), 150–1

poly[2-(2-aminoethoxy)ethoxy]phosphazene (PAEP), 102

poly[2-(2-aminoethylamino)ethyl methacrylate] (PAEAEMA), 104

poly[2-(2-methoxyethoxy)ethyl methacrylate] (PMEO2), 77

poly[2-(diisopropylamino)ethyl methacrylate (PDPA), 145, 148, 151, 237

poly[2-(dimethylamino)ethyl methacrylate] (PAMA), 221

poly[2-(methacryloyloxy)ethyl-phosphorylcholine] (PMPC), 145, 148, 235, 237

poly[2-(*N*,*N*-diethylamino)ethyl methacrylate] (PDEA), 36, 148–50

poly[*N*,*N*-bis(acryloyl)cystamine] (PBAC), 105

poly[*N*-(2-hydroxypropyl)methacrylamide] (PHPMA), 247–50, 256

poly[*N*-(3-aminopropyl)methacrylamide] (PAPMA), 41

polyarginine, 152

polycondensation, 122, 128, 254
polycurcumin (PCurc), 254
polydispersity (PD), 7–8, 51, 77, 126
polydispersity index (PDI), 8
polyhistadine (PHis), 43, 50
polyion complex (PIC), 41, 171, 177–8, 208, 221–5
polyion complex micelles (PICMs), 172
polyisoprene, 149, 282, 286
polylactide (PLLA), 44, 296–300
polyleucine, 152
polymeric hydrogels, 99
polymeric micelles for siRNA
  enhanced permeability and retention (EPR) effect, 163, 169, 176
polymersomes
  arginine-glycine-aspartic acid (RGD), 153
  breast, 147
  dextran, 146, 151
  doxorubicin (DOX), 146–8, 151, 153–4
  folate, 147
  folic acid (FA), 153
  glutathione (GSH), 148
  liposomes, 144–6, 154
  paclitaxel (PTX), 146
  poly(ε-caprolactone) (PCL), 146, 151, 152
  poly(ethylene glycol) (PEG), 146, 148, 150
  spleen, 149
polystyrene, 46, 105, 208, 211, 282, 286
polyurethane (PU), 76, 172
polyvinylpyrrolidone (PVP), 22, 51
potassium, 128, 222, **231**
potassium poly(vinyl sulfate) (PVSK), 222
pressure-temperature phase diagram, **286–7, 291**
primary cardiomyocyte (PCM), 89
prodrug
  dextran, 248–9
  doxorubicin (DOX), 246–9, 253

  enhanced permeability and retention (EPR) effect, 248–9
EZN-2208, 249, **250**
folic acid (FA), 249
glutathione (GSH), 248, 252
kidney, 250
liposomes, 251
lung, 249
NK012, 253, **254**
NKTR-105, 249
paclitaxel (PTX), 246
poly(ethylene glycol) (PEG), 247, 249, 253–4
reticuloendothelial system (RES), 248–9
SN38, 246, 249, 253
Pronectin, 267
propylacrylic acid (PAA), 95, 145–6, 149–51, 164, 171, 209–10, 224
prostate, 153, 174, 176, 271–2
protamine sulfate, 96
protein transduction domains (PTDs), 90, 152
proton sponge mechanism, 43, 65, 82, 169
proton-transfer polymerization (PTP), 123, 126, 128
pullulan, 265, **266,** 267

quantum dots (QD), 11, 99–100, 122, 125, 135, 146, 160, 214

rabies virus, **74,** 75, 84
rabies virus glycoprotein (RVG), 75, 84–5
radical coupling method, 80
radiofrequency magnetic hyperthermia, 147
recapitulation, 14
receptor-mediated endocytosis, 24, 50, 79, 111, 169, 171, 176
receptor-mediated processes, 26, 87, 249
reducible poly(β-amino ester) (RPAE), 43

refractive index (RI), 192 4
refractometer, 192
renal threshold, 45, 249–50
reticuloendothelial system (RES)
    2R2S, 45–6
    current paradigm, 3
    hyperbranched polymers (HBP),
        128
    mesenchymal stem cells (MSC),
        264
    prodrug, 248–9
    siRNA, 161–3, 183
    targeting, 22, 24, 26
    translational nanomedicine, 33
    Zwitterionic polymers, 227
retroviruses, 265
reversible addition-fragmentation
        chain transfer (RAFT), 80,
        104, 135, 171
rhodamine
    see TAMRA
ribonucleic acid interference (RNAi),
        101, 158–9, 162, 171, 181
ribonucleotide reductase (RRM2),
        159
ring-opening multibranching poly-
        merization (ROMBP), 123
ring-opening reaction mechanism
        (ROP), 123, 136
Ringsdorf model, 3
Ringsdorf, Helmut, 3
RNA interference (RNAi), 101, 158–
        9, 162, 171, 181
RNA-inducing silencing complex
        (RISC), 158

Saccharomyces cervisiae, 199
scale-up ability, 24, 34, 51–3, 137,
        252, 256, 300
scattered-light intensity (SLI), 283–4,
        285, 286
Schizophyllum commune (SPG), 191–
        2, **193,** 194, 196–9, **200–2,**
        203–4
scintigraphy, 9

SEC chromatograms, 192, **193**
selectins, 270
self-assembling prodrugs, 37, 250,
        252, 256
self-condensing ring-opening poly-
        merization (SCROP), 123,
        130, 133, 139
self-condensing vinyl polymerization
        (SCVP), 123, 139
shell-crosslinked knedle-like (SCKs),
        210–2
short hairpin RNA (shRNA), 158
silica, 99, 102
silica nanotubes (SNTs), 99
silicic acid, 102
single-wall carbon nanotubes
        (SWNT), 45–6
siRNA
    arginine-glycine-aspartic acid
        (RGD), 165–6, 180
    breast, 167, 179
    cytokines, 162–3
    doxorubicin (DOX), 163, 166, 179–
        80
    enhanced permeability and reten-
        tion (EPR) effect, 183
    folate, 164, 180
    folic acid (FA), 180
    glutathione (GSH), 169, 173–4, 177
    immunogenicity, 160, 165
    kidney, 160–2, 181
    Lipofectamine, 168–9
    liposomes, 160, 163, 168
    macrophages, 161–2
    messenger RNA (mRNA), 158–60,
        160–2, 165, 172, 175–6
    paclitaxel (PTX), 179, 181–2
    poly(ε-caprolactone) (PCL), 165–7,
        179–81
    poly(ethylene glycol) (PEG), 159,
        164, 168, 170–3, 175–7, 180
    reticuloendothelial system (RES),
        161–3, 183
    spleen, 162

transactivator of transcription
(TAT), 165–6, 180
Sirna-027, 159
size-exclusion chromatography
(SEC), 39, 192
small interfering RNA
see siRNA
SN38 prodrug, 246, 249, 253
solvent-evaporation method, 36, **37**,
235, 281, **293, 294, 296**
SOX trio, 107
spermine, 104, 111, 165, 169, 265–7
spleen
2R2S, 45–6
current paradigm, 3
gene vectors, 89
polymersomes, 149
siRNA, 162
targeting, 22
translational nanomedicine, 32–3,
52
spotted polymersomes, 145, **146**
stable core-surface crosslinked
micelles (SCNs), 40
stable nucleic acid lipid particle
(SNALP), 159
statistical associated fluid theory
(SAFT1), 286
stearic acid (SA), 40, 79, 179
stearyl methacrylate (SMA), 237
step growth, 122–3
steric hindrance, 45, 49
streptavidin antibody, 93, 172
stromal cell-derived factor-1 (SDF-
1), 271
stromal cells, 11, 95, 267, 270–1
succinic anhydride (SA), 127, 172,
179–80, 230–1
sulfobetaine (SB), 231
supepararmagnetic iron oxide
(SPIO), 85, 101
supercritical micellization, **296**
superparamagnetic polymeric micelle
(SPPM), 26–7
supramolecular interaction, 122

surface packing, 231
synchrotron X-ray irradiation, 68

T cells, 11, 192, 261, 263
tamoxifen, 37, 237, **292,** 293
targeting
arginine-glycine-aspartic acid
(RGD), 26–7
breast, 20, 25, 27
dextran, 23
doxorubicin (DOX), 22, 25
folate, 25–7
folic acid (FA), 24
glutathione (GSH), 26
liposomes, 21, 25–7
lung, 20
macrophages, 22
poly(ethylene glycol) (PEG), 22, 26
reticuloendothelial system (RES),
22, 24, 26
spleen, 22
targeting drug delivery systems
(TDDS), 264, 273
Taxol®, 4
tendon, 267
ter-ephthaldicarbaldehyde (TDA), 41
terminal groups replacing (TGR),
124
terminal modifications, 123, 124, 127
terminal units, 122
tetanus toxin, **74,** 75
tetraethylene glycol (TEG), 73
tetraethylenepentamine (TP), 165
tetrafluoroethane, 290–1
tetramethylrhodamine (TAMRA),
182, 196, 197
tetrandrine, 153
theranostics, 147, 232, 239, 241
therapeutic index, 1, 15, 245, 273
thermoresponsive backbone, 126
thiirane, 70
thiolation reaction, 70, 79
three dimensional (3D), 99, 107–8,
266–7

three-layer onion-structured nanoparticles (3LNPs), 36, **50**
toll-like receptors (TLR), 162–3, 196,
198–9
trafficking, 21, 95, 152, 160
transactivator of transcription (TAT)
2R2S, 50
drug delivery, 4
gene therapy, 265–6, 274
gene vectors, 68, 87, 90, 111
micelles design, 212
siRNA, 165–6, 180
transduction, 90, 108, 152, 162, 271,
274
transferrin (TF), 25, 27, 50, **74**, 75–6,
85, 159
transferrin/avidin/biotinylated PEI
(TABP), 75
transforming growth factor (TGF),
108, 111, 162, 270
translational nanomedicine
enhanced permeability and retention (EPR) effect, 32–3
liposomes, 32
reticuloendothelial system (RES), 33
spleen, 32–3, 52
transmitted-light intensity (TLI),
283–4, 286
Trastuzumab®, 14, 20
triethanolamine, 128
Trifectin®, 166
trifluoroacetic acid, 240
trifluoromethane, 36, **37, 284–5, 287–
8,** 289–96, **297–9**
triglycerides, 159
trimethylolpropane, 128
triphosphate (TPP), 168
tumor dependent factors, 5
tumor necrosis factor (TNF), 153,
167, 181, 271

tumor necrosis factor-related apoptosis-inducing ligand
(TRAIL), 111, 271, **272**
turmeric, 252

ultaviolet light (UV), 41, 149–51,
192–4, 240
ultrasound, 126, 264
umbilical, 88, 240
UV chromatograms, 192–3

valine, 168
vascular endothelial growth factor
(VEGF), 159, 168–9, 174–6,
179–80
vascular permeability, 11–12, 23
very low-density lipoproteins
(VLDLs), 159
vidarabine, 234
vimentin, 76
vinylpyridine, 147, 208, 217
viruses artificial, 110

window chambers, 5, 23

X-ray, 68
xenotransplantation, 268

zeta potential, 63, 167, **173,** 179
Zevalin®, 14
zinc, 149
Zwitterionic polymers
defined, 232
macrophages, 227
poly(ethylene glycol) (PEG), 227,
229, 241
reticuloendothelial system (RES),
227